Ecosystem-Based Management in Practice

Ecosystem-Based Management in Practice

Editor: Leanna Perry

Callisto Reference
www.callistoreference.com

Callisto Reference,
118-35 Queens Blvd., Suite 400,
Forest Hills, NY 11375, USA

Visit us on the World Wide Web at:
www.callistoreference.com

ISBN: 978-1-64116-784-0 (Hardback)

Cataloging-in-publication Data

Ecosystem-based management in practice / edited by Leanna Perry.
p. cm.
Includes bibliographical references and index.
ISBN 978-1-64116-784-0
1. Ecosystem management. 2. Biodiversity conservation. I. Perry, Leanna.

QH75 .E26 2023
333.951 6--dc23

Table of Contents

Preface

The purpose of the book is to provide a glimpse into the dynamics and to present opinions and studies of some of the scientists engaged in the development of new ideas in the field from very different standpoints. This book will prove useful to students and researchers owing to its high content quality.

Ecosystem-based management (EBM) refers to an integrated and science-based approach which emphasizes on the management of natural resources. It focuses on maintaining the resilience, health and diversity of ecosystems that enables humans to utilize the products and services they offer in a sustainable manner. EBM is an adaptive method to manage human activities that seeks to ensure the coexistence of healthy, fully functioning ecosystems and human societies. It can be used to study both terrestrial and marine ecosystems along with the challenges they pose. Marine EBM aims to manage marine resources in a way that safeguards ecosystem health in order to provide the ecosystem services required by people. This book aims to shed light on the applications of ecosystem-based management. It will also provide interesting topics for research which interested readers can take up. The book will serve as a valuable source of reference for graduate and post-graduate students.

At the end, I would like to appreciate all the efforts made by the authors in completing their chapters professionally. I express my deepest gratitude to all of them for contributing to this book by sharing their valuable works. A special thanks to my family and friends for their constant support in this journey.

Editor

1

The Final Ecosystem Goods & Services (FEGS) Approach: A Beneficiary-Centric Method to Support Ecosystem-Based Management

Theodore H. DeWitt, Walter J. Berry, Timothy J. Canfield, Richard S. Fulford, Matthew C. Harwell, Joel C. Hoffman, John M. Johnston, Tammy A. Newcomer-Johnson, Paul L. Ringold, Marc J. Russell, Leah A. Sharpe, and Susan H. Yee

T. H. DeWitt (✉)
US Environmental Protection Agency, Pacific Ecological Systems Division, Newport, OR, USA
e-mail: dewitt.ted@epa.gov

W. J. Berry
US Environmental Protection Agency, Atlantic Coastal Environmental Sciences Division, Narragansett, RI, USA

T. J. Canfield
US Environmental Protection Agency, Groundwater Characterization and Remediation Division, Ada, OK, USA

R. S. Fulford · M. C. Harwell · L. A. Sharpe · S. H. Yee
US Environmental Protection Agency, Gulf Ecosystem Measurement and Modeling Division, Gulf Breeze, FL, USA

J. C. Hoffman
US Environmental Protection Agency, Great Lakes Toxicology and Ecology Division, Duluth, MN, USA

J. M. Johnston
US Environmental Protection Agency Office Research and Development, Environmental Processes Division, Athens, GA, USA

T. A. Newcomer-Johnson
US Environmental Protection Agency, Watershed and Ecosystem Characterization Division, Cincinnati, OH, USA

P. L. Ringold
US Environmental Protection Agency, Pacific Ecological Systems Division, Corvallis, OR, USA

M. J. Russell
US Environmental Protection Agency, Center for Computational Toxicology and Exposure, Gulf Breeze, FL, USA

Abstract Ecosystem-Based Management (EBM) and other social-ecological environmental management frameworks recognize that most environmental problems are ultimately social problems, requiring the reconciliation of human needs with the limits of ecosystem productivity and resilience. Using a social-ecological perspective in management practice can be greatly facilitated by identifying the attributes of ecosystems that are directly used, enjoyed, or appreciated by people connected to the environmental issue at hand. These are the *final ecosystem goods and services* (FEGS), which are specific to ecosystem types and how people use or appreciate ecological attributes of those ecosystems. This article: (1) reviews the conceptual basis of a FEGS approach for linking people's well-being to ecosystems; (2) describes how FEGS are identified, organized, and measured using classification systems, and metrics and indicators; and (3) presents examples of how the FEGS approach can be integrated into EBM and other decision-making frameworks.

Lessons Learned

- FEGS are the subset of ecosystem services that are directly used, enjoyed, or appreciated by people. Individual FEGS are identified as the biophysical attributes found within a given ecosystem that are used, enjoyed, or appreciated for a specific purpose.
- FEGS facilitate identifying, quantifying, and assigning value to biophysical attributes of ecosystems that are of greatest relevance to people who care about or depend on those ecosystems.
- FEGS are useful for communicating with stakeholders and policy-makers about how people obtain specific benefits from specific biophysical attributes of an ecosystem.
- Tools have been developed to identify FEGS within all types of ecosystems found on earth, for working with stakeholders to prioritize which FEGS are of greatest concern within a given decision context, and to identify mathematical models useful for estimating FEGS production.

Needs to Advance EBM

- Greater awareness within the EBM community of practice, including developing case-study applications, of the usefulness of FEGS and the availability of tools useful for identifying, prioritizing, and quantifying them.
- A standardized list of metrics or indicators for each FEGS, based on the attributes of ecosystem types that each beneficiary class uses, enjoys, or appreciates. Site-specific metrics or indicators could then be developed from those generic attributes.
- Integration of the FEGS tools (e.g., NESCS Plus, FEGS Scoping Tool, Rapid Benefits Indicators, EcoService Models Library) to facilitate identification of priority FEGS, relevant metrics and indicators for FEGS endpoints and benefits, and models for estimating responses of those FEGS to environmental change or stressors.

1 Introduction

Humankind depends wholly on nature for its well-being. An array of ecosystem goods and services, often available at no apparent cost, sustains our health, economy, and society. Those aspects of nature that people benefit from include productive soil for farmers, clean and safe water for swimmers, and inspirational landscapes for artists. The key feature of the examples in this list are that they link a good or service that is provided by nature to a specific type of beneficiary, or user group. Those ecosystem products and processes that are directly used, enjoyed, or appreciated by people are identified as *Final Ecosystem Goods and Services* (FEGS) (Boyd and Banzhaf 2007; Boyd et al. 2016). Those FEGS are a subset of all ecosystem goods and services (e.g., MEA 2005; Haines-Young and Potschin 2018) distinguished as the *final* "endpoints" in nature's production networks that people directly use (Fig. 1). The production of FEGS is dependent on "supporting" and "regulating" ecological functions; these intermediate processes (Potschin-Young et al. 2017) are critically important to human well-being, for without them, FEGS would not exist. This is the essence of the FEGS approach: making explicit the biophysical attributes of ecosystems from which specific *beneficiaries* obtain a specific benefit. Beneficiaries are "the interests of an individual (i.e., person, group, and/or firm) that drive active or passive consumption and/or appreciation of ecosystem services resulting in an impact (positive or negative) on their welfare" (Nahlik et al. 2012). For example, beneficiaries are recreational anglers who fish for wild fish for food or pleasure, industrial processors who use water for cooling or product manufacturing, or artists who use attributes of nature for inspiration to produce art (Landers and Nahlik 2013). Beginning from this beneficiary perspective, we identify and quantify the biophysical attributes of ecosystems that people use or appreciate in order to achieve a wide

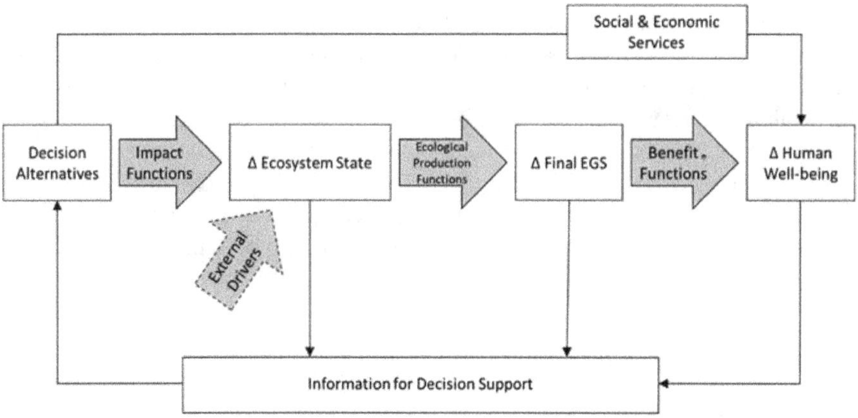

Fig. 1 Conceptual model for how changes to the state of an ecosystem (e.g., its biophysical attributes) and consequent changes in the production of final ecosystem goods and services (FEGS) influence environmental decision-making via impacts to human well-being. For more on this conceptual model, see Harwell and Molleda (2018)

range of benefits. In this chapter, we describe steps being taken to incorporate FEGS into decision-making and policies using: a classification system that identifies the beneficiaries and ecosystem attributes for each FEGS; methods to measure or estimate (through modeling) FEGS stocks, production and value; and decision-support tools to facilitate integrating FEGS into decision making.

Identifying, quantifying, and forecasting the stocks, production, and value of FEGS are important for demonstrating the relevance of natural systems to the public and policy makers, and they are fundamental to understanding the capacity of an area, or an ecosystem, to produce natural goods and services used by or useful to people. Characterization of FEGS is thus a valuable communication tool that provides support and added justification for ecosystem protection in decision making, including tradeoff analyses. At the same time, the identification of FEGS facilitates investigation into the social-ecological interactions between human actions, ecosystem condition, human well-being, a renewed reason to better understand how ecosystems function, and a basis for prioritizing which ecological attributes and processes should be incorporated into environmental management and policy.

The FEGS approach requires a subtle but important shift in perspectives in ecosystem services science, from identifying ecological goods and services that are important to human well-being (MEA 2005), to identifying what attributes of ecosystems people use, enjoy, or appreciate to fulfill a specific interest, and recognizing that the attributes that people use in pursuit of those interests differ across ecosystem types (Ringold et al. 2013). In other words, the FEGS approach is predicated on the fact that people use or depend on ecosystems in different ways, contingent on what they are doing or needing at a given moment and where those people are located (e.g., within what type of ecosystem). An important aspect of this beneficiary perspective is the inclusion of a wide range of human interests, including those related to culture and spirituality, as well as to health, economic, and overall well-being. By recognizing the importance of the human relationship to ecosystems, the beneficiary perspective addresses the concern that traditional resource management emphasizes goods and services (i.e., using metrics such as fisheries yield or recreational days) without explicitly considering the social context of a problem and the depth of the relationship between beneficiaries and the environment (Grumbine 1994).

In this chapter, we: (1) review the concept of the FEGS approach to human-nature interactions; (2) describe how FEGS are identified, organized, and measured using classification systems, and metrics and indicators; and (3) present examples of how the FEGS approach can be integrated into decision making, specifically to support Ecosystem-Based Management (EBM).

2 The FEGS Approach

The FEGS approach (Fig. 2) is motivated by the idea that identifying the biophysical attributes of ecosystems that are relevant to people will facilitate holistic benefit and economic assessment (Boyd et al. 2015). A key challenge for scientists is to identify

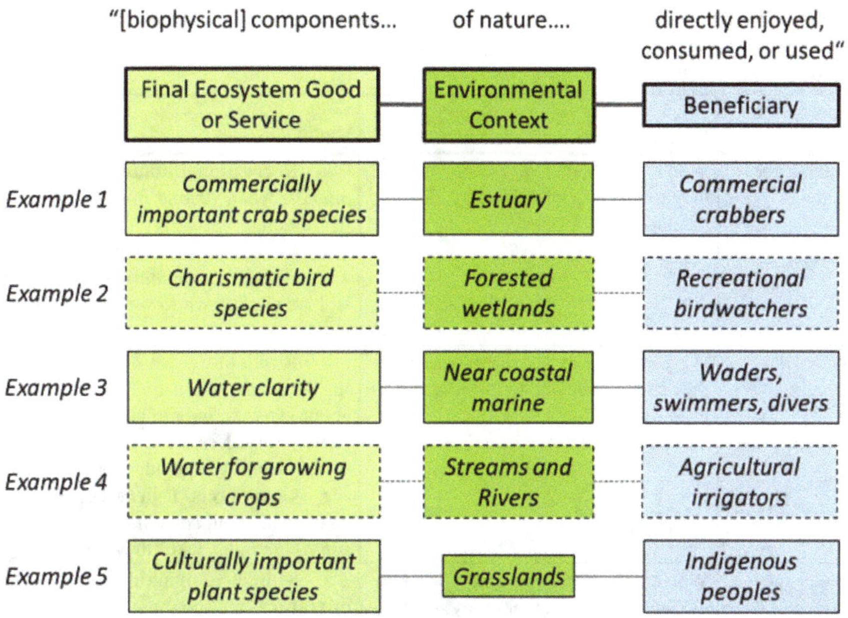

Fig. 2 Illustration and examples of the three elements needed to define FEGS. Note that many, if not most, beneficiaries depend directly on multiple biophysical attributes of multiple ecosystems rather than the single ones shown here

and measure biophysical attributes that are most relevant to human well-being (Boyd and Banzhaf 2007; Landers and Nahlik 2013; Boyd et al. 2015). In the FEGS approach, specification of the beneficiary must come before identification of which biophysical attributes should be measured. The benefits that groups of people obtains from nature are determined by how they use, enjoy, or otherwise depend on nature. To facilitate the identification of FEGS endpoints, people are grouped into beneficiary classes that describe their interests which drive the consumption, use, or appreciation of ecosystem goods or services (Landers and Nahlik 2013). Then, the ways that a given beneficiary class interacts with a given ecosystem to obtain those good and services determines the biophysical component of a specific FEGS. Note that FEGS do not include the ecological components or processes that are necessary to produce it, which are intermediate processes (Haines-Young and Potschin 2018). Additionally, it is important to acknowledge that FEGS for one beneficiary (e.g., water temperature for an aquaculturalist) may be an intermediate process for another (e.g., a recreational angler).

The FEGS approach can bring clarity to environmental management by translating intermediate ecological processes into FEGS by asking the following questions (Table 1):

- "*Who* are the beneficiaries?"
- "How and where (i.e., in what ecosystem type) do they use, enjoy or appreciate nature?"

Table 1 Examples of how starting with the beneficiaries can help bridge intermediate ecosystem processes with their beneficiaries (**bold**), the FEGS attributes they use or enjoy (*italics*), and the relevant environmental context (<u>underline</u>)

Intermediate processes	Beneficiary-oriented questions	FEGS approach
Habitat for fauna	Who are the beneficiaries and what do they use, appreciate, or enjoy about habitat?	• **Recreational hunters** hunt *game animals* when visiting <u>forested areas</u> in the region • **Recreational birders** want to see a *specific species of charismatic bird* and are thus drawn to <u>wetlands</u> in the region
Water quality regulation	Who are the beneficiaries and what do they use, appreciate, or enjoy about water quality?	• **Residents, dependent on drinking water aquifers**, are concerned about *water salinity* in <u>groundwater</u> • **Snorkelers** are concerned about *water turbidity* in popular <u>coastal waters</u> • **Commercial fishermen** are concerned about *contaminants in edible fish tissue* in the <u>lakes</u> they frequent
Water quantity regulation	Who are the beneficiaries and what do they use, appreciate, or enjoy about water quantity?	• **Municipal drinking-water plant operators** care about the *reliable availability of fresh water* from <u>streams</u> • <u>Coastal</u> **home owners** directly understand the value of shoreline protection through dunes and vegetation in *reducing the probability of property damage due to erosion by waves*

- “What ecological end-products (EEPs) do they use, enjoy or appreciate”
- “Where is the EEP located (i.e., ecosystem type) that is used, enjoyed, or appreciated?”

In a given lake, for example, while recreational boaters might be concerned about water clarity, odor, or having sufficient water depth in which to operate a boat, an industrial processor will be primarily concerned with water corrosiveness, presence of biofouling organisms, and the reliability of water quantity. Taking a FEGS approach helps ensure that the full range of benefits is considered by identifying meaningful biophysical indicators or metrics to be monitored, valued, and reported. A FEGS approach also facilitates communication about what matters to people by ensuring that key issues or stakeholders are not overlooked, and by allowing management decisions to include things people care about and understand (Yee et al. 2017).

2.1 FEGS Classification System

Several frameworks have been proposed for classifying ecosystem services (e.g., MEA 2005; Haines-Young and Potschin 2018; Landers and Nahlik 2013; US EPA 2015). Finisdore et al. (2018) describes a suite of classification systems and outlines the range of benefits of using classification systems. The goals of these classification systems are to develop a common, shared language in an interdisciplinary field and provide a consistent framework for identifying, organizing, and accounting for ecosystem services. Two classification systems for FEGS have been developed by the US EPA to identify the types of uses, needs, or desires that a beneficiary seeks to obtain from a given ecosystem type from specific biophysical attributes present there. A FEGS approach facilitates the creation of information useful for valuation and helps to minimize double counting and valuation problems regarding intermediate processes that are not clearly distinguished from FEGS (Ojea et al. 2012; Nahlik et al. 2012). The FEGS Classification System (FEGS-CS) was developed to help "determine those specific ecosystem attribute(s) associated with the specific FEGS that the beneficiary values" such that "these can directly lead to identifying appropriate metrics and indicators for FEGS" (Landers and Nahlik 2013). The FEGS-CS was created primarily to aid in organizing ecological metrics and indicators that would provide meaningful input to environmental benefit assessment and policy decisions. The National Ecosystem Services Classification System (NESCS) was developed by environmental economists (US EPA 2015). The NESCS comprehensively and uniquely identifies distinct categories of FEGS to support analysis of how changes in ecosystems affect human well-being by applying, adapting, and combining the principles underlying existing economic accounting systems for market goods and services, primarily for use in environmental accounting (Russell et al. 2020). The NESCS defines the ecological attributes of an environment that flow as inputs to human uses (both market and non-market) to both emphasize human reliance on these flows and to illustrate how changes in policy could affect those flows and the well-being they provide to human users of those environments. The NESCS also links to standard accounting systems (such as the North American Industry Classification System, NAICS; https://www.census.gov/eos/www/naics/) and more directly to existing economic valuation practices than does the FEGS-CS (US EPA 2015). The US EPA is nearing completion of a merged classification system (NESCS Plus) to leverage the best of both systems. The NESCS Plus will reduce confusion caused by parallel classification systems and be consistent with prior systems so that it remains relevant for audiences of FEGS-CS and NESCS.

2.2 FEGS Metrics and Indicators

The metrics and indicators associated with FEGS are important for incorporating definable benefits that people receive from nature into elements of decision-making

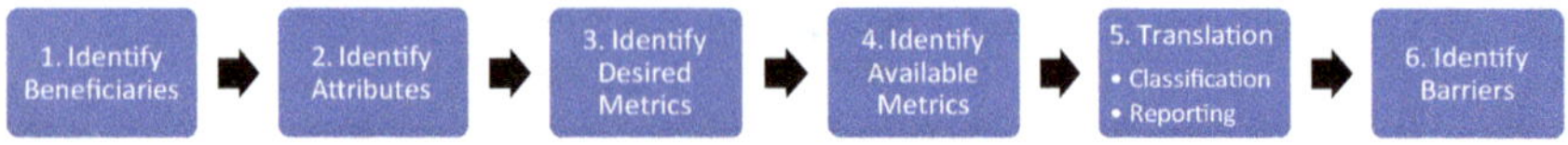

Fig. 3 Developing FEGS metrics and indicators for a given environment type is a 6-step process starting with beneficiaries and the attributes of nature that they value

processes (e.g., issue identification, options analysis, communication), as well as overall decision-making processes such as EBM and Structured Decision-Making (SDM).). In particular, the ecosystem-based framework underpinning FEGS-based classification systems allows analysts to view comprehensively the biophysical attributes (e.g., wild food, drinkable water, specific organisms) and resulting human benefits (e.g., nutrition, recreation, improved health, spiritual enrichment) provided by an ecosystem (Landers and Nahlik 2013). Further, FEGS-based classification systems identify classes of beneficiaries or users (e.g., commercial harvesters, recreational anglers, boaters) that potentially benefit from FEGS when they interact with nature. A simultaneous comparison of the different FEGS that might benefit each beneficiary or user group affected by a management option provides a complete summary and representation of the ways in which people may be impacted by management decisions. Synergies and trade-offs among groups of beneficiaries are expected. That is, an increase in a good or service of value to one group could either support a related increase to another beneficiary group or result in a reduction of a good or service to another group. For example, an increase in clean water for swimmers might also benefit recreational anglers, but an increase in complex reef structure that provides superior habitat to produce fish may be desirable to scuba divers (i.e., tourism), but undesirable for commercial and recreational fishing (e.g., due to gear entanglement).

Developing metrics and indicators of FEGS for a given ecosystem type begins with identifying which beneficiary groups directly interact with that ecosystem, then considers what specific components of nature are directly used, enjoyed, or appreciated by each beneficiary group, and moves to considering how those components could be measured directly (Fig. 3; Ringold et al. 2013). Typically, each beneficiary group will directly use, appreciate, or enjoy multiple biophysical attributes within a given ecosystem, with each attribute represented by one or more metrics. Those attributes may be identified by considering what individual people directly perceive or interact with from the environment. For example, an individual partaking in recreational angling within an estuary may care directly about the taxa and size of the fish she might catch, whether the fish she catches are safe to eat, whether the conditions at the site are safe, and/or whether the aesthetics of the site are appealing (Table 2). That is because an individual can use different attributes of an ecosystem depending which one of many "roles" she is playing (e.g., catch and release fisherman, subsistence fisherman, experiencer of natural place, etc.) while doing an activity as complex as recreational angling. Direct measurement of some attributes may be difficult or expensive (e.g., angler success in fishing), necessitating the use of surrogate metrics that approximate the suite of valued attributes (e.g., percent

Table 2 Developing FEGS metrics and indicators for recreational anglers in estuaries based on the attributes of nature that anglers care about. For illustrative purposes of a full suite of "what matters" topics, two non FEGS rows are also included (marked with **)

Ecosystem	Beneficiary	What matters directly to this beneficiary?	Attribute category	Attribute	Sub-attribute	Ideal metrics	Currently available metrics & indicators
Estuaries	Recreational anglers	Is it safe to go out?**	Water	Water movement	Wave intensity	Wave height, speed & direction	Wave height, current speed & current direction
			Weather	Wind strength/ speed	Wind intensity	Wind speed & direction	Wind speed & wind direction
		Will I catch fish I want to eat?	Fauna	Edible fauna	Taxa & presence	Game fish species presence or abundance, size, distribution, diversity	Game fish presence/ absence, catch per unit effort, harvest (biomass)
		Is there a chance I will become sick from eating caught fish?	Fauna	Edible fauna	Concentration of pathogens/ toxins/contaminants/parasites	Biotoxins, pathogens, parasites, chemical contaminants	Alerts from the FDA, fish consumption advisories
		Will my hook or net get fouled? **	Soil & substrate	Substrate quality	Structure	Marine hazards, shoreline/ benthos complexity	Marine hazards, local reports
		Is this area aesthetically enjoyable?	Environmental aesthetics	Scents, viewscapes, soundscapes	Water quality	Water color, algae taxa, water clarity & smell, presence of hypoxia or anoxia, sound environment	Monitoring data for water quality, sound, or odors, local reports
					Viewscape	Natural shoreline landscapes, unobstructed waterscapes	Percent natural land cover on shoreline, percent of water surface area as built structures (from aerial or satellite imagery)

of natural land cover along shorelines) while acknowledging the limitations of using surrogates. An important distinction here is ideal metrics versus available, or surrogate, metrics. While we need to specify the ideal metrics, we must then quantify them within budget, time, and available data constraints.

Identifying those metrics that closely represent how beneficiaries use attributes of nature facilitates the translation of data obtained with those metrics into information that analysts, stakeholders, and the public can use in their decision processes (Wainger and Mazzotta 2011). Balanced against that ideal is the pragmatic desire to use both existing metrics (i.e., to minimize methodology development) and existing data. Wainger and Mazzotta (2011) suggest that ideal FEGS metrics are:

- Easily understood by non-experts
- As close to the FEGS as possible (e.g., the taxa, size, condition, and abundance of game fish, as opposed to the total number of all fish at a location)
- Readily available (e.g., existing methodology and data sets)
- Available for large areas
- Available at a user-appropriate scale as defined by the management or user question(s) and by the beneficiary perceptions (e.g., a recreational angler might want data on the scale of an individual fishing spot, while a fisheries regulator might need information at a statewide or regional scale).

The number of FEGS metrics that need to be measured depends on the problem requirements. The list of all potential FEGS metrics for any ecosystem type can be quite extensive if the interests of all potential beneficiaries are considered. From an operationalizing perspective, however, analysts, stakeholders, or managers may determine that the classification of beneficiaries or ecosystem types (e.g., for each of the FEGS classification systems) may be too general for the problem at hand. That could drive the need for a narrower specification of beneficiaries, ecosystem sub-types, or ecosystem attributes, and consequently, a refined set of relevant metrics. The FEGS Scoping Tool (Sharpe et al. 2020) was developed to help stakeholders or managers collaboratively and transparently identify common interests and prioritize the attributes of nature that are most valued across beneficiary groups. Analysts can draw from existing sets of FEGS metrics or build from those to identify metrics relevant to quantifying and sustaining the stakeholders' most valued benefits of nature.

3 Operationalizing the FEGS Approach

3.1 Role of FEGS in Ecosystem-Based Management (EBM)

The FEGS approach can provide several important tools for EBM. First, the categorization of beneficiaries of FEGS helps to specify who the stakeholders are whose wants, needs, desires, and perceptions need to be addressed (Ringold et al. 2013; Landers et al. 2016). Those beneficiaries may live within or adjacent to an ecosystem (e.g., coastal communities) or distantly, and may come from diverse socio-economic groups. The potential for heterogeneity in their values needs to be considered when assigning weights to biophysical outcomes. Second, a FEGS-based classification system informs the selection of metrics or indicators useful for identifying, communicating, and quantifying the ecological attributes for application in assessing or monitoring FEGS, their production, or their use or appreciation by people. Third, identification of beneficiaries and associated FEGS informs the selection and development of models to forecast production of FEGS (i.e., ecological production functions; Wainger and Mazzotta 2011; Bruins et al. 2017), the delivery of benefits to people (i.e., benefit functions; Wainger and Mazzotta 2011; Villa et al. 2014; Bousquin and Mazzotta 2020), and integrated ecological-social well-being frameworks (Schlueter et al. 2012).

Identification of the beneficiaries of FEGS within an area of interest, and the ecological attributes that the beneficiaries use, helps decision makers understand the types and magnitudes of the tradeoffs involved in policy options, how beneficiaries will be affected by changing conditions within the ecosystem, the biophysical features of the ecosystem that are important to those beneficiaries, where and how beneficiaries experience those biophysical features, and where and how those biophysical features are produced. For example, the US EPA's Remediation-to-Restoration-to-Revitalization (R2R2R) program for contaminated-site clean-up actions around the U.S. Great Lakes inherently uses a beneficiary-centric, FEGS approach by working towards each community's vision for their desired human-nature outcomes (i.e., revitalization), and incorporates those outcomes into the goals of each phase of the clean-up and restoration (Williams and Hoffman 2020). In combination, this information can inform prioritization and local management relative to ecosystem goals and help to align decision-making with local values.

3.2 Integrating FEGS Into a Structured Decision Making (SDM) Framework, and Relevance to EBM

Structured Decision Making (SDM) provides an organizing framework to formally integrate FEGS, or any other approaches and tools into EBM (Gregory et al. 2012; also see Sharpe et al. 2020). The use of SDM places a strong emphasis on problem structuring by clarifying the problem, identifying objectives (i.e., separating those objectives to accomplish at the end from those objectives that are important ways to reach end objectives), and developing meaningful measures (Marcot et al. 2012; Maseyk et al. 2017). The use of SDM is an alternative to technical assessments or cost-benefit analysis, which may be done along with SDM, but are not required. Without clarifying "what really matters" upfront, resources can be wasted collecting the wrong information for the wrong problem, leading to irrelevant or misleading assessments (Carriger et al. 2013). A focus on what stakeholders' value, in contrast, can lead to more creative and effective outcomes. The FEGS approach can facilitate management decisions and actions that have a higher likelihood of acceptance across a variety of stakeholders because they are based on the stakeholders' priorities.

An SDM process includes a series of steps similar to other decision processes (Table 3) and can be used to identify where FEGS concepts can be integrated into an EBM decision process even while using other decision frameworks, such as AQUACROSS (Piet et al. 2017, 2020), an EBM policymaking framework (Cormier et al. 2017), or integrated ecosystem assessment (Foley et al. 2013). Each decision framework has a unique set of steps, but FEGS concepts can still be integrated at many places within each process (Table 3).

The FEGS approach advances the ability to identify, articulate, measure, and assess the potential role of relevant ecosystem goods and services in a given decision context. Using FEGS metrics or indicators, for example, may provide more relevant assessment endpoints for EBM and enable communication of benefits to humans better than endpoints that are difficult to link to human use (e.g., total primary production, pH, species diversity, etc.). The use of FEGS metrics should also be useful for regulatory purposes, such as risk assessment endpoints (Munns et al. 2017). Additionally, having a consistent approach for defining EBM terminology (Arkema et al. 2006) and clear articulation of EBM principles (Delaclamara et al. 2020) will help practitioners identify how to incorporate ecosystem services and decision strategies for a given EBM context. Further, the FEGS approach and tools presented here help EBM practitioners identify users or beneficiaries who will be affected by environmental changes at a site due to changes in the specific ecological attributes that those groups derive benefit from, for a given decision context (Fig. 4).

Within a larger SDM framework, the FEGS approach can also assist EBM practitioners identify and prioritize stakeholders to bring into the decision process (i.e., using the FEGS Scoping tool; Sharpe et al. 2020). Because FEGS are the link between biophysical condition and socio-economic benefits to people, this approach is compatible with socio-ecological systems frameworks (Elliott and O'Higgins 2020; Piet et al. 2020). By connecting ecosystem services directly to human health and well-being endpoints within a structured framework, the FEGS approach lends itself to systems analysis (including linkage frameworks, network analysis, and Health Impact Assessments (HIA)); Robinson and Culhane 2020; Williams and Hoffman 2020) for identifying and evaluating key stressors in the system or vulnerabilities to ecosystem-services supply that may need to be prioritized for management. Additionally, the FEGS approach helps advance efforts to develop classification systems useful for a range of EBM practitioners (Culhane et al. 2020). These classification systems can be leveraged to identify ecosystem-services focused metrics and indicators that can inform the evaluation of management alternatives.

Decision analysis tools, such as means-ends networks, direct ranking, swing weighting, or consequence tables (Gregory et al. 2012) can be used to identify management actions that might improve ecosystem services production or where ecosystem services may be a means to achieving other social or economic objectives. Because FEGS are closely linked to human beneficiaries, the FEGS approach fosters the identification and application of relevant tools across the SDM cycle (Fig. 4; Yee et al. 2017), such as for prioritizing and measuring economic benefits of ecosystem

Table 3 How a FEGS perspective can be integrated into other EBM decision frameworks. Acronyms: SDM (structured decision making); AQUACROSS (AQUAtic biodiversity & ecosystem services aCROSS European Union policies); EBM (ecosystem-based management)

SDM framework[a] *FEGS concepts*	AQUACROSS framework[b]	EBM policymaking process[c]	Integrated ecosystem assessment[d]
1. **Clarify decision context** *What FEGS might be impacted?* *What FEGS beneficiaries should be included as stakeholders?*	Phase 1: Societal goals • Key threats; policies; synergies and conflicts	1. Strategic goal-setting • Prioritize threats to ecosystems • Identify opportunities to improve socioeconomic and ecosystem status	1. Scoping • Identify goals and threats
2. **Define objectives and performance measures** *Which objectives are FEGS or their benefits?* *How will they be measured?*	Phase 2: Description of the Socio-ecological System • Ecological & social system criteria	2. Tactical objectives • Evaluations of ecological, cultural, social, and economic impacts • Indicators and performance measures	2. Indicators and reference levels • Identify measures for attributes of management interest
3. **Develop alternatives** *Are any FEGS means to achieve objectives?*	Phase 3: Planning an EBM response • Management strategy – Management measures – Policy instrument	• Development of management strategies	3. Risk analysis • Link indicators and threats • Identify overlaps between activities and impacts
4. **Estimate consequences** *What ecosystem services models or data are needed to estimate consequences?*	Phase 4: Implementation, monitoring and evaluation • Assessment of current state using functional relationships, time-series, spatial comparisons • Forecasting and scenarios to predict consequences	3. Management measures • Assess implementation feasibility • Evaluate ability of proposed management actions to achieve objectives	4. Management strategy evaluation • Assess options
5. **Evaluate trade-offs** *How much loss or gain in FEGS benefits is acceptable?* *Are FEGS beneficiaries differentially impacted?*	• Evaluation of management plans – Compare alternative strategies to "business as usual"	• Comparison of alternative management actions against weighted objective priorities	• Choose approach
6. **Implement, monitor, review** *Did the decision lead to measurable change in FEGS?* *Were there unforeseen impacts on FEGS to consider next time?*	• Evaluate effectiveness, efficiency, and equity of outcomes	4. Adaptive Management • Effective monitoring plan • Evaluate management effectiveness and alternative options • Recommendations to improve scientific input	5. Monitoring and evaluation • Track indicators and assess change

Table modified from Yee et al. (2017). (a) Yee et al. (2017); (b) Piet et al. (2017); (c) Cormier et al. (2017); (d) Foley et al. (2013)

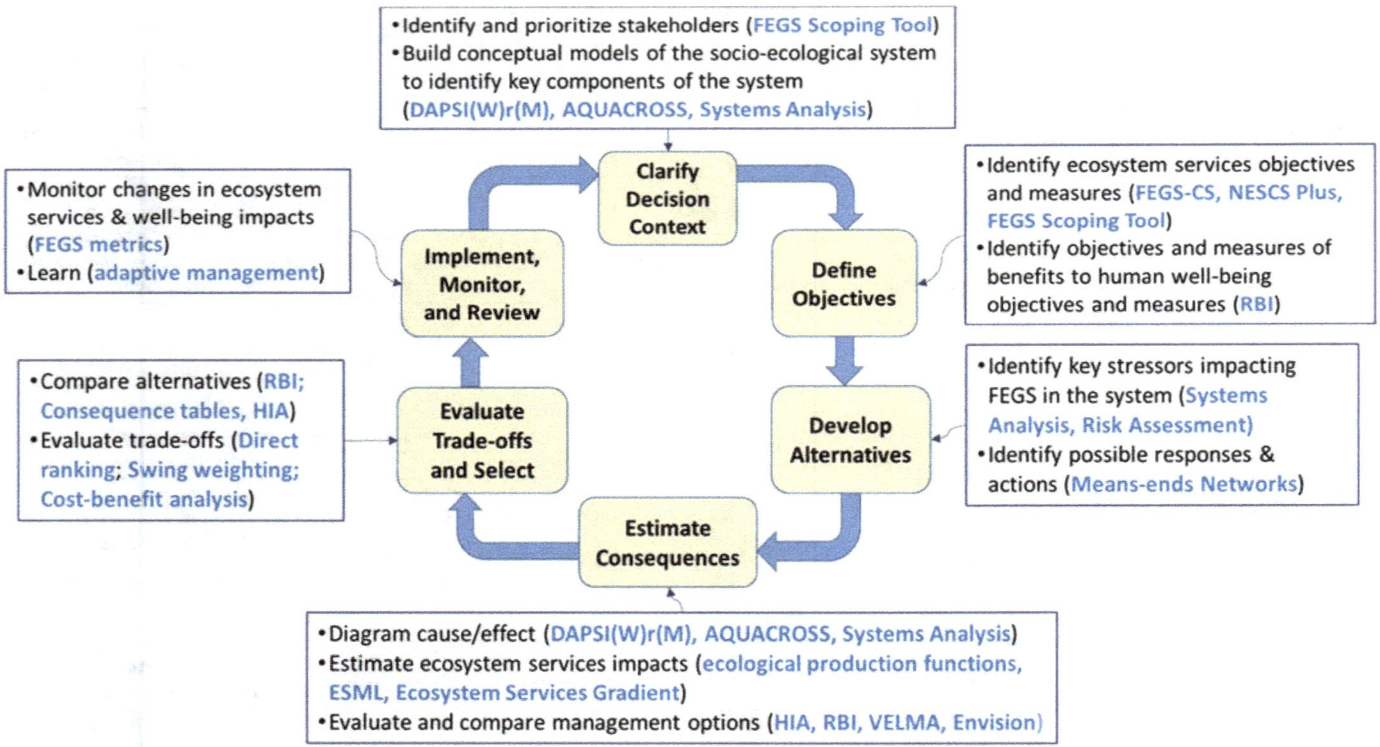

Fig. 4 Examples of tools (in bold blue; references in the text) that support integrating FEGS (Final Ecosystem Goods & Services) into different phases of Structured Decision Making (SDM). Tool acronyms modified from Yee et al. (2017). Acronyms: AQUACROSS (AQUAtic biodiversity & ecosystem services aCROSS European Union policies); DAPSI(W)r(M) (Drivers, Activities, Pressures, State, Impacts (on Welfare), Responses (as Measures); ESML (Eco Service Models Library); FEGS-CS (FEGS Classification System); HIA (Health Impact Assessment); NESCS (National Ecosystem Services Classification System); RBI (Rapid Benefits Indicators); VELMA (Visualizing Ecological Land anagement Assessments); see text for explanation

services using the related Rapid Benefits Indicators (RBI) approach (Bousquin and Mazzotta 2020).

Ecological production functions (e.g., models useful for estimating ecosystem services) can be utilized in an EBM context to quantitatively predict changes in ecosystem services in response to changes in a system, management alternatives for a decision context (Fulford et al. 2020; Lewis et al. 2020). In some cases, expert knowledge and synthesis of available data may be sufficient, or used in combination with quantitative modelling, to communicate how usage of natural resources changes in response to changes in ecological conditions (e.g., ecosystem services gradients; Yee et al. 2020). Different models may be integrated and applied within decision support systems (e.g., Envision, VELMA; McKane et al. 2020) to evaluate and compare the impacts of alternative management options on ecosystem services or other relevant objectives.

4 Summary

This chapter outlined the suite of concepts, methods, and tools that comprise the FEGS approach and demonstrates steps taken to incorporate the approach into decision-making for EBM. The FEGS approach focuses on advancing both the field of ecosystem services science and the utilization of decision-making frameworks to the field of EBM (Delacámara et al. 2020). One important application of the FEGS approach is to aid in guiding the development of ecological endpoints and metrics useful for assessment of stocks or site/system conditions for EBM. The foundational research across elements of the FEGS approach has been made operational across a suite of case studies and decision contexts, including examining alternatives in coastal forest management in the U.S. Pacific Northwest (McKane et al. 2020), cleanup of contaminated sites in the Great Lakes (Williams and Hoffman 2020), resiliency planning following natural disasters (Myer and Johnston 2020), restoring large ecosystems such as the Everglades (Gibble et al. 2020), and examining ecosystem management practices among different future climate scenarios in the Lower Mekong Basin (Johnston et al. 2020).

The FEGS approach can also be useful to EBM practitioners through the development of strategic communication messages (Harwell et al. 2020) for stakeholders and the public regarding the benefits associated with pending or implemented decisions. In particular, the FEGS approach aids in communicating which people within a system (i.e., beneficiary groups) will be affected by changes to the condition of the environment owing to how those changes will affect the benefits (i.e., health, economic, or social) people obtain from nature (Fig. 1). For example, recent work has advanced the field of biological condition gradients used to characterize and communicate the status of an ecosystem in relation to thresholds of meaningful change and potential management actions to incorporate ecosystem services (Yee et al. 2020). Overall, the approach of starting with human beneficiaries and asking what they use, appreciate, or enjoy about ecosystems holds promise for advancing ecosystem goods and services science to support environmental decision making, particularly ecosystem-based management.

Disclaimer This chapter has been subjected to Agency review and has been approved for publication. The views expressed in this paper are those of the author(s) and do not necessarily reflect the views or policies of the U.S. Environmental Protection Agency.

References

Arkema, K. K., Abramson, S. C., & Dewsbury, B. M. (2006). Marine ecosystem-based management: From characterization to implementation. *Frontiers in Ecology and the Environment, 4* (10), 525–532.

Bousquin, J., & Mazzotta, M. (2020). Rapid benefit indicator tools. In T. O'Higgins, M. Lago, & T. H. DeWitt (Eds.), *Ecosystem-based management, ecosystem services and aquatic biodiversity: Theory, tools and applications* (pp. 309–332). Amsterdam: Springer.

Boyd, J., & Banzhaf, S. (2007). What are ecosystem services? The need for standardized environmental accounting units. *Ecological Economics, 63*, 616–626.

Boyd, J. W., Ringold, P. L., Krupnick, A. J., Johnston, R. J., Weber, M., & Hall, K. (2015). *Ecosystem services indicators: Improving the linkage between biophysical and economic analyses.* RFF DP 15-40, Resources for the Future, Washington, DC.

Boyd, J., Ringold, P., Krupnick, A., Johnston, R. J., Weber, M. A., & Hall, K. (2016). Ecosystem services indicators: Improving the linkage between biophysical and economic analyses. *International Review of Environmental and Resource Economics, 8*(3–4), 359–443. https://doi.org/10.1561/101.00000073.

Bruins, R. J. F., Canfield, T. J., Duke, C., Kapustka, L., Nahlik, A. M., & Schäfer, R. B. (2017). Using ecological production functions to link ecological processes to ecosystem services. *Integrated Environmental Assessment, 13*(1), 52–61.

Carriger, J. F., Fisher, W. S., Stockton, T. B., & Sturm, P. E. (2013). Advancing the Guánica Bay (Puerto Rico) watershed management plan. *Coastal Management, 41*(1), 19–38.

Cormier, R., Kelble, C. R., Anderson, M. R., Allen, J. I., Grehan, A., & Gregersen, O. (2017). Moving from ecosystem-based policy objectives to operational implementation of ecosystem-based management measures. *ICES Journal of Marine Science, 74*, 406–413.

Culhane, F. E., Robinson, L. A., & Lillebø, A. I. (2020). Approaches for estimating the supply of ecosystem services: Concepts for ecosystem-based management in coastal and marine environments. In T. O'Higgins, M. Lago, & T. H. DeWitt (Eds.), *Ecosystem-based management, ecosystem services and aquatic biodiversity: Theory, tools and applications* (pp. 105–126). Amsterdam: Springer.

Delacámara, G., O'Higgins, T., Lago, M., & Langhans, S. (2020). Ecosystem-based management: moving from concept to practice. In T. O'Higgins, M. Lago, & T. H. DeWitt (Eds.), *Ecosystem-based management, ecosystem services and aquatic biodiversity: Theory, tools and applications* (pp. 39–60). Amsterdam: Springer.

Elliott, M., & O'Higgins, T. G. (2020). From the DPSIR, the D(A)PSI(W)R(M) emerges... a butterfly-'protecting the natural stuff and delivering the human stuff'. In T. O'Higgins, M. Lago, & T. H. DeWitt (Eds.), *Ecosystem-based management, ecosystem services and aquatic biodiversity: Theory, tools and applications* (pp. 61–86). Amsterdam: Springer.

Finisdore, J., Rhodes, C., Haines-Young, R., Maynard, S., Wielgus, J., Dvarkas, A., Houdet, J., Quétier, F., Ding, H., Soulard, F., Van Houtven, G., & Rowcroft, P. (2018). *Expanding the field of ecosystem services practitioners—18 benefits from using classification systems*. Sustainable flows working paper. December 2018.

Foley, M. M., Armsby, M. H., Prahler, E. E., Caldwell, M. E., Erickson, A. L., Kittinger, J. N., Crowder, L. B., & Levin, P. S. (2013). Improving ocean management through the use of ecological principles and integrated ecosystem assessments. *BioScience, 63*(8), 619–631.

Fulford, R. S., Heymans, S. J. J., & Wu, W. (2020). Mathematical modelling for ecosystem-based management (EBM) and ecosystem goods and services (EGS) assessment. In T. O'Higgins, M. Lago, & T. H. DeWitt (Eds.), *Ecosystem-based management, ecosystem services and aquatic biodiversity: Theory, tools and applications* (pp. 275–290). Amsterdam: Springer.

Gibble, R., Miller, L., & Harwell, M. C. (2020). Using stakeholder engagement, translational science and decision support tools for ecosystem-based management in the Florida Everglades. In T. O'Higgins, M. Lago, & T. H. DeWitt (Eds.), *Ecosystem-based management, ecosystem services and aquatic biodiversity: Theory, tools and applications* (pp. 517–542). Amsterdam: Springer.

Gregory, R., Failing, L., Harstone, M., Long, G., McDaniels, T., & Ohlson, D. (2012). *Structured decision-making: A-practical guide to environmental management choices*. West Sussex, UK: Wiley-Blackwell.

Grumbine, R. E. (1994). What is ecosystem management? *Conservation Biology, 8*, 27–38.

Haines-Young, R., & Potschin, M. (2018). *Common international classification of ecosystem services (CICES) V5.1 guidance on the application of the revised structure*. Fabis Consulting, UK. Retrieved October 28, 2019, from https://cices.eu/content/uploads/sites/8/2018/01/Guidance-V51-01012018.pdf.

Harwell, M. C., & Molleda, J. L. (2018). *FY 16 output SHC 2.61.1 ecosystem goods and services production and benefits case studies report*. EPA/600/R-18/189. Gulf Breeze, FL: U.S. Environmental Protection Agency.

Harwell, M.C., Molleda, J. L., Jackson, C. A., & Sharpe, L. (2020). Establishing a common framework for strategic communications in the natural sciences. In T. O'Higgins, M. Lago, & T. DeWitt (Eds.), *Ecosystem-based management, ecosystem services and aquatic biodiversity: Theory, tools, and applications* (pp. 165–188). Amsterdam: Springer.

Johnston, J. M., Zomer, R., & Mingcheng, W. (2020). Predicting future vegetated landscapes under climate change: Application of the environmental stratification methodology to protected areas in the Lower Mekong Basin. In T. O'Higgins, M. Lago, & T. DeWitt (Eds.), *Ecosystem-based management, ecosystem services and aquatic biodiversity: Theory, tools, and applications* (pp. 561–580). Amsterdam: Springer.

Landers, D., & Nahlik, A. (2013). *Final ecosystem goods and services classification system (FEGS-CS)*. EPA/600/R-13/ORD-004914. Washington, DC: U.S. Environmental Protection Agency.

Landers, D., Nahlik, A., & Rhodes, C. R. (2016). The beneficiary perspective—benefits and beyond. In M. Potchin, R. Haines-Young RFish, & K. Turner (Eds.), *Routledge handbook of ecosystem services* (pp. 74–87). New York, NY: Routledge.

Lewis, N. S., Marois, D. E., Littles, C. J., & Fulford, R. S. (2020). Projecting changes to coastal and estuarine ecosystem goods and services—Models and tools. In T. O'Higgins, M. Lago, & T. DeWitt (Eds.), *Ecosystem-based management, ecosystem services and aquatic biodiversity: Theory, tools, and applications* (pp. 235–254). Amsterdam: Springer.

Marcot, B. G., Thompson, M. P., Runge, M. C., Thompson, F. R., McNulty, S., Cleaves, D., Tomosy, M., Fisher, L. A., & Bliss, A. (2012). Recent advance in applying decision science to managing national forests. *Forest Ecology and Management, 285*, 123–132.

Maseyk, F. J. F., Mackay, A. D., Possingham, H. P., Dominati, E. J., & Buckley, Y. M. (2017). Managing natural capital stocks for the provision of ecosystem services. *Conservation Letters, 10*, 211–220.

McKane, R. B., Brookes, A., Djang, K., Halama, J., Barnhart, B., Russell, M., Vache, K., & Bolte, J. (2020). A community-based decision support tool for flexible, interactive assessments that quantify tradeoffs in ecosystem goods and services for alternative decision scenarios. In T. O'Higgins, M. Lago, & T. DeWitt (Eds.), *Ecosystem-based management, ecosystem services and aquatic biodiversity: Theory, tools, and applications* (pp. 255–274). Amsterdam: Springer.

MEA (Millennium Ecosystem Assessment). (2005). *Ecosystems and human well-being: Synthesis*. Washington, DC: Island Press.

Munns, W. R., Jr., Poulsen, V., Gala, W. R., Marshall, S. J., Rea, A. W., Sorensen, M. T., & von Stackelberg, K. (2017). Ecosystem services in risk assessment and management. *Integrated Environmental Assessment and Management, 13*, 62–73.

Myer, M., & Johnston, J. M. (2020). Models and mapping tools to inform resilience planning after disasters: A case study of hurricane Sandy and Long Island (NY) ecosystem services. In T. O'Higgins, M. Lago, & T. DeWitt (Eds.), *Ecosystem-based management, ecosystem services and aquatic biodiversity: Theory, tools, and applications* (pp. 417–430). Amsterdam: Springer.

Nahlik, A. M., Kentula, M. E., Fennessy, M. S., & Landers, D. H. (2012). Where is the consensus? A proposed foundation for moving ecosystem service concepts into practice. *Ecological Economics, 77*, 27–35.

Ojea, E., Martin-Ortega, J., & Chiabai, A. (2012). Defining and classifying ecosystem services for economic valuation: The case of forest water services. *Environment: Science and Policy, 19*, 1–15.

Piet, G., Delacámara, G., Gómez, C. M., Lago, M., Martin, R., & van Duinen, R. (2017). *Making ecosystem-based management operational: Deliverable 8.1 executive summary*. Report as part of the Horizon 2020 project AQUACROSS (Knowledge, Assessment, and Management for AQUAtic Biodiversity and Ecosystem Services aCROSS EU policies). Retrieved October 28, 2019, from https://aquacross.eu/sites/default/files/AQUACROSS%20Executive%20Summary%20D8.1_v2_18062018.pdf.

Piet, G., Delacamara, G., Kraan, M., Röckmann, G. C., & Lago, M. (2020). Advancing aquatic ecosystem-based management with full consideration of the social-ecological system. In T. O'Higgins, M. Lago, & T. H. DeWitt (Eds.), *Ecosystem-based management, ecosystem services and aquatic biodiversity: Theory, tools and applications* (pp. 17–38). Amsterdam: Springer.

Potschin-Young, M., Czucz, B., Liquete, C., Maes, J., Rusch, G. M., & Haines-Young, R. (2017). Intermediate ecosystem services: An empty concept? *Ecosystem Services, 27*, 124–126.

Ringold, P., Boyd, J., Landers, D., & Weber, M. (2013). What data should we collect? A framework for identifying indicators of ecosystem contributions to human well-being. *Frontiers in Ecology and the Environment, 11*, 98–105.

Robinson, L.A. & Culhane, F.E. (2020). Linkage frameworks: An exploration tool for complex systems in ecosystem-based management. In T. O'Higgins, M. Lago, & T. DeWitt (Eds.), *Ecosystem-based management, ecosystem services and aquatic biodiversity: Theory, tools, and applications* (pp. 213–234). Amsterdam: Springer.

Russell, M. J., Rhodes, C., Sinha, R. K., Van Houtven, G., Warnell, G., & Harwell, M. C. (2020). Ecosystem-based management and natural capital accounting. In T. O'Higgins, M. Lago, & T. H. DeWitt (Eds.), *Ecosystem-based management, ecosystem services and aquatic biodiversity: Theory, tools and applications* (pp. 149–164). Amsterdam: Springer.

Schlueter, M., McAllister, R. R. J., Arlinghaus, R., Bunnefeld, N., Eisenack, K., Hoelker, F., Milner-Gulland, E. J., Müller, B., Nicholson, E., Quaas, M., & Stöven, M. (2012). New horizons for managing the environment: A review of coupled social-ecological systems modeling. *Natural Resource Modeling, 25*(1), 219–272.

Sharpe, L., Hernandez, C., & Jackson, C. (2020). Prioritizing stakeholders, beneficiaries and environmental attributes: A tool for ecosystem-based management. In T. O'Higgins, M. Lago, & T. H. DeWitt (Eds.), *Ecosystem-based management, ecosystem services and aquatic biodiversity: Theory, tools and applications* (pp. 189–212). Amsterdam: Springer.

US EPA (United States Environmental Protection Agency). (2015). *National ecosystem services classification system (NESCS): Framework design and policy application*. EPA-800-R-15-002. Washington, DC: United States Environmental Protection Agency.

Villa, F., Bagstad, K. J., Voigt, B., Johnson, G. W., Portela, R., Honzak, M., & Batker, D. (2014). A methodology for adaptable and robust ecosystem services assessment. *PLoS ONE, 9*(3), e91001.

Wainger, L., & Mazzotta, M. (2011). Realizing the potential of ecosystem services: A framework for relating ecological changes to economic benefits. *Environmental Management, 48*, 710–733.

Williams, K. C., & Hoffman, J. C. (2020). Remediation to restoration to revitalisation: Ecosystem-based management to support community engagement at clean-up sites in the Laurentian Great Lakes. In T. O'Higgins, M. Lago, & T. H. DeWitt (Eds.), *Ecosystem-based management, ecosystem services and aquatic biodiversity: Theory, tools and applications* (pp. 543–560). Amsterdam: Springer.

Yee, S., Bousquin, J., Bruins, R., Canfield, T. J., DeWitt, T. H., de Jesús-Crespo, R., Dyson, B., Fulford, R., Harwell, M. C., Hoffman, J., Littles, C. J., Johnston, J. M., McKane, R. B., Green, L., Russell, M., Sharpe, L., Seeteram, N., Tashie, A., & Williams, K. (2017). *Practical strategies for integrating final ecosystem goods and services into community decision-making*. EPA/600/R-17/266. Gulf Breeze, FL: U.S. Environmental Protection Agency.

Yee, S., Cicchetti, G., DeWitt, T. H., Harwell, M. C., Jackson, S. K., Pryor, M., Rocha, K., Santavy, D. L., Sharpe, L., & Shumchenia, E. (2020). The ecosystem services gradient: A descriptive model for identifying thresholds of meaningful change. In T. O'Higgins, M. Lago, & T. DeWitt (Eds.), *Ecosystem-based management, ecosystem services and aquatic biodiversity: Theory, tools and applications* (pp. 291–308). Amsterdam: Springer.

2

Projecting Changes to Coastal and Estuarine Ecosystem Goods and Services—Models and Tools

Nathaniel S. Lewis, Darryl E. Marois, Chanda J. Littles, and Richard S. Fulford

Abstract Coasts and estuaries provide an abundance of ecosystem goods and services (EGS) to humans worldwide. Models that track the supply, demand, and change in EGS within these ecosystems provide valuable insights that have applications in the context of land-use planning, decision-making, and coastal community engagement. However, developing models for use in coastal and estuarine ecosystems is challenging given the multitude and variability of potential input variables, largely due to their dynamic nature and extensive use. Models that can incorporate scenarios of environmental change to forecast changes in EGS endpoints are highly valuable to decision-makers, but only a minor proportion of available EGS models offer this utility. In this chapter, we describe the domain of models most useful to coastal decision-makers, present models at multiple scales that can predict EGS changes, and examine specific examples that epitomize this utility. We also highlight common difficulties in modeling coastal and estuarine EGS and propose suggestions for integrating EGS models into the coastal management decision-making process during times of increasing environmental change.

Lessons Learned

- Identifying the most suitable model(s) given the scale(s) of a particular question or goal is paramount in the modeling process
- Uncertainty is an inherent component of modeling that should be well-communicated by users to avoid misinterpretation of results

N. S. Lewis (✉) · D. E. Marois
Pacific Ecological Systems Division, U.S. Environmental Protection Agency, Newport, OR, USA
e-mail: nate.lewis@oregonstate.edu

C. J. Littles
Environmental Resources Branch, U.S. Army Corps of Engineers, Portland, OR, USA

R. S. Fulford
US Environmental Protection Agency, Gulf Ecosystem Measurement and Modeling Division, Gulf Breeze, FL, USA

- The complexity, as well as the time, cost, and data requirements, of many EGS models are barriers to widespread implementation

Needs to advance EBM

- To improve model relevancy and usability, resource managers and other stakeholders must be part of the model development process to identify important decision metrics
- Tiered or coupled models can allow for the identification and inclusion of multiple drivers as well as cumulative impacts on coastal and estuarine EGS, which is a critical need
- Development of a standardized framework for model implementation would increase the adoption of EGS models into coastal and marine planning processes

1 Modeling Changes in Coastal and Estuarine EGS

Coastal and estuarine ecosystems host some of the most dynamic and productive habitats in the world (Costanza et al. 1997; Costanza et al. 2014), including: seagrass beds, mangroves, coral reefs, salt marshes, sandy beaches, and dunes (Barbier et al. 2011). Each of these habitats provides a unique combination of benefits across the broad ecosystem goods and services (EGS) categories (i.e., provisioning, regulating, cultural, and supporting services) set forth by the Millennium Ecosystem Assessment (2005) (e.g., Marois and Mitsch 2015; Kassakian et al. 2017). As environmental changes propagate through these ecosystems, the provision of important EGS may also be affected. Changes can result from natural or anthropogenic impacts and come in a myriad of shapes and forms, including: sea-level rise, pollution, invasive species, storm events, and detrimental fishing practices.

Ecosystem alterations can be the product of a multitude of factors, occur at different spatiotemporal scales, and affect EGS of all types. Integrating these dynamic variables poses a challenge when modeling EGS endpoints. However, the quantity of models and tools that explicitly predict changes to EGS based on alterations to input variables make up a minor proportion of all models applied in coastal and estuarine ecosystems. Some of this disparity may be due to the relatively recent development of the EGS concept, as well as the inherent complexity and variability of these models. In this chapter, we present a suite of models that exemplify the approach of predicting EGS changes at different scales, outline the domain of models that may offer the most utility to coastal decision-makers, present examples epitomizing this utility, and highlight common difficulties across coastal and estuarine EGS models. We conclude with suggestions for integrating EGS models into the coastal management decision-making process during times of increasing environmental change.

Developing models for application in coastal and estuarine ecosystems can be challenging given their dynamic nature and the multitude of potential inputs and outputs; however, models provide valuable tools that can be used for many purposes, such as visualizing species distribution, assessing chemical fluxes, explaining habitat-species relationships, or identifying spatial patterns. The utility of models increases substantially for resource managers when environmental changes can be simulated within these tools to predict changes to EGS.

Although there are many EGS models (Bagstad et al. 2013b; Turner et al. 2016; Gret-Regamey et al. 2017; Little et al. 2017), this chapter focuses on models that provide predictions of how EGS may be affected as coastal ecosystems undergo change. We further narrow this focus to models that demonstrate a high degree of utility to decision-makers based on relevant spatial scales and endpoints. Within these criteria, we present example models that are well-established and demonstrate these traits with different degrees of complexity and quantities of modeled services to show the utility of this approach across a diverse spectrum of models.

Coastal and estuarine EGS models can range from simulating services provided by a specific site or habitat (Mendoza et al. 2017; Harris et al. 2018), to estimating global EGS values (Costanza et al. 1997; Boumans et al. 2015). Here, we limit our focus to models with spatial scales ranging from local (e.g., estuary or bay) to regional (e.g., U.S. Pacific West Coast), as they are the most likely to provide useful information when making resource management decisions (Turner et al. 2016). These models also vary greatly in their final output or endpoint. Some estimate relative changes in the degree of EGS provision (Hanson et al. 2012; Harris et al. 2018), while others estimate monetary values of EGS (Carr et al. 2018). The example models we discuss in this chapter all provide outputs that are quantitative and informative (but not necessarily monetary), as these provide the most benefit to decision-makers at local and regional scales (Ruckelshaus et al. 2015).

Within these established bounds of endpoint and spatial scale there still exists a wide range of model complexity. The spatial and/or temporal resolution, system dynamics considered, and type of EGS simulated each affect model complexity. In the following sections, we explore five example models representing different combinations of spatial scale, quantity of EGS, and complexity, each occupying different niches within the domain of EGS models that can predict change and inform decisions. The first model, HexSim, is a mechanistic model that describes living populations by tracking individuals; the second, XBeach, is a mechanistic model for estimating shore protection; the third, Atlantis, is a whole system model used in fishery management; the fourth, InVEST, contains many sub-models that can predict delivery of a suite of EGS; and the fifth, ARIES, uses machine learning to trace ecosystem service flows to beneficiaries.

2 HexSim Model

Ecosystem services under management are frequently tied to changes in both the behavior and demographics of living resources. Predicting these changes is an important part of decision making, and object-oriented and Individual-Based Models (IBMs) play an important role, both in making these predictions and in operationalizing the answers as a tool for policy. IBMs are mechanistic models that have typically been used to evaluate the movement, growth, and mortality of living populations, by tracking individuals rather than the population as a whole (DeAngelis and Rose 1992). IBMs are complex by nature because they describe an individual's state, and therefore the spatial resolution of any input data must match individual behavior (e.g., meter), while ensuring the spatial scale of the model remains relevant to decision making (e.g., whole estuary). IBMs are inherently data intensive because the mechanistic descriptions of movement, feeding, growth, and mortality are all necessary to properly track a group of individuals. Yet, given sufficient input data, IBMs have proven useful for predicting population-level change, particularly in cases where average population conditions are not sensitive to management.

IBMs track the state and response of individual agents, then combine the outcomes into a population level distribution for the chosen response variable (e.g., total biomass, individual size). Agents are defined by the context of the question but are generally individual organisms of a population or demographic group (e.g., individual anglers). In spatially explicit cases, the agent can be a spatial grid cell as is the case in the HexSim IBM (Rustigian et al. 2003; Fulford et al. 2011). Such spatial models are optimized for the study of population distribution in response to heterogeneous landscapes. Input data must include initial agent state for key variables, as well as function parameters describing how agents respond to environmental variability and management-based change. IBMs are by nature an iterative suite of interconnected forcing functions. Model output is temporally and spatially explicit as the model tracks individual deviations from an initial state, which can be summarized at any time during a model simulation.

The HexSim model (Schumaker et al. 2004; Schumaker and Brookes 2018) was used to predict the impact of seagrass management on fishery resources in Tampa Bay (Fulford et al. 2016). Seagrass restoration was combined with a suite of other habitat components, including water temperature and hypoxia, to predict how individual habitat selection and subsequent growth were impacted. Management decisions were evaluated in the context of multiple environmental factors based on their cumulative impact on growth and production of aquatic resources that provide ecosystem services. The IBM also informed the delivery of ecosystem services to stakeholders in that individual angler behavior was modeled in response to changes in fish distribution and apparent availability of habitat as fishing grounds. This latter bioeconomic component of the model was based on angler preference data (Fulford et al. 2016).

HexSim in Tampa Bay was a coupled movement and bioenergetics model for fish and a coupled movement and fishing success model for anglers. Input data included fish distribution prior to habitat change, habitat-specific fish growth and movement functions, angler distribution, and an angler choice function based on distance from access points and daily fishery return. Fish moved first, and their growth and mortality were predicted based on habitat choice. Anglers chose fishing locations based on distance from access points and knowledge of optimal fishing habitat and their catch rate was predicted based on a probability model and predicted overlap between anglers and catchable fish (Fulford et al. 2016).

IBMs are most useful in cases where the mean output value is insensitive to manageable change in the system (e.g., habitat restoration). A good example is the mean fish growth rate in the Tampa recreational fishery in response to seagrass habitat restoration (Fulford et al. 2016). Compensatory behavior and the multi-faceted nature of habitat selection greatly ameliorated the mean fish growth response, even with a large change in seagrass coverage. As a result of this insensitivity, the impact of successful seagrass restoration on fishery harvest was highly difficult to identify. As an alternative, an IBM approach was used to track individual behavioral response to habitat distribution, then bioenergetics sub-models were used to translate fish distribution into population production.

An IBM approach, such as the HexSim model described here, is both data-intensive and complicated enough that managers rarely attempt its application without assistance. That limits the utility of this approach to environmental decision making in data rich contexts (Rose et al. 2010; Rose et al. 2015). However, the availability of data and expertise for IBM use is higher than it has ever been and growing. These models require a high level of expertise to apply, yet the subtle nature of management related change and the confounding influence of other environmental variables make the use of IBMs much more informative and highly desirable for decision making.

3 XBeach Model

Models that can simulate the link between an ecosystem's structure and its ability to deliver ecosystem services are key to bridging the gap between ecology and economics. These types of models are often highly mechanistic and can occasionally be adapted for integration into tools for estimating coastal EGS (Bruins et al. 2017). XBeach is one such model that has been used to predict the ability of coral reefs to deliver the service of coastal protection under various conditions (Quataert et al. 2015; Pearson et al. 2017; Harris et al. 2018). It is a one-dimensional, morphodynamic model that simulates how ocean waves travel across complex, near-coast topographies and propagate up to the shore (Roelvink et al. 2009). XBeach was originally designed for wave propagation on beaches, but with some modification it has been expanded to simulate accurate reef hydrodynamics (Pomeroy et al. 2012; Van Dongeren et al. 2013). These capabilities allow it to estimate the extent to which coral reefs reduce wave run-up (how high a wave rises above mean sea level as it reaches the coastline) at local and regional scales. XBeach can also predict changes in the provision of reef shoreline protection under scenarios of environmental change such as sea level rise (Harris et al. 2018).

If a coral reef ecosystem is significantly diminished in size or health, its valuable coastal protection service could be completely lost (Sheppard et al. 2005). These protective services can also be reduced if coral reefs are not able to grow vertically at a rate comparable to rates of sea level rise (Harris et al. 2018). Within XBeach, reef degradation scenarios can be simulated by altering coastal bathymetry and surface roughness inputs, while sea level rise can be simulated by increasing offshore wave-height drivers. Resulting changes in the model output of wave run-up can then be interpreted from a coastal hazard perspective, with higher run-up values leading to a greater risk of damage to coastlines, particularly during storm events that further increase wave heights (Quataert et al. 2015).

Harris et al. (2018) demonstrated the utility of XBeach in predicting coastal protection services, by applying the model in reef sites in French Polynesia. Model parameters were calibrated using measured data from pressure loggers in cross-reef transects. They varied model inputs of sea level, reef vertical accretion or erosion, and reef surface roughness (corresponding to reef structural complexity). Monte Carlo simulations, in which the model was run multiple times with inputs varying randomly across a range of values, were used to address variability that was not captured by XBeach. Overall, results showed that the combined effects of sea level rise, reef erosion, and reduction in reef structural complexity led to wave heights that were 2.4 times greater than those under present conditions (Harris et al. 2018). These findings not only quantified the amount of protection offered by reefs under current conditions, but also provided coastal land managers with predictions of how waves may impact coastal areas in the future should the extent of coral reefs decline.

XBeach model output primarily consists of an estimated reduction in wave height, which by itself may not be informative enough for coastal decision-makers to draw worthwhile conclusions. However, when mechanistic models like XBeach are incorporated into larger modeling frameworks, its output can be used as an input for another model predicting levels of ecosystem service provision (Bagstad et al. 2013b). For example, the reduction in wave height could be converted into a reduced frequency of flooding, which may be more informative. Further, this reduction in flooding frequency could be converted into monetary value using methods similar to those described by Barbier (2016). This use of well-established and validated mechanistic models to deliver predictions of coastal EGS can be effective for incorporating EGS considerations into coastal management decisions and policy.

4 Atlantis Model

In whole-system (i.e., end-to-end) models (Travers et al. 2007), biophysical, socio-economic, and industrial components and processes are considered, along with associated feedbacks and interactions between components (Fulton et al. 2011; Kaplan et al. 2012; Weijerman et al. 2015; Marshall et al. 2017). One such model is Atlantis, developed for use in Management Strategy Evaluation, it supports Ecosystem-Based Management (EBM) of multiple fisheries on a regional scale (Link et al. 2010; Fulton et al. 2011). Atlantis is a complex, spatially explicit, hierarchical model containing interconnected submodels to evaluate potential management actions, policy changes, and ecosystem tradeoffs under various scenarios (Kaplan et al. 2012; Fulton et al. 2014; Marshall et al. 2017).

A biophysical submodel is the primary submodel within Atlantis, which follows nutrient flows (primarily nitrogen and silica) and distribution through the food-web to large marine mammals via ecological processes (e.g., production, predation, recruitment) and physical features (e.g., hydrodynamics, seabed types, water properties) (Link et al. 2010; Fulton et al. 2011). Output data from the biophysical model feed into an exploitation submodel that considers human uses and impacts, such as fishing, pollution, development, and other environmental changes (Link et al. 2010; Fulton et al. 2011). A monitoring and assessment submodel then utilizes outputs from the biophysical and exploitation models to simulate scenarios, from which adaptive management options and the associated uncertainty can be assessed (Link et al. 2010; Fulton et al. 2011). Finally, the monitoring and assessment output is fed into a management submodel, which consists of potential management rules defined by the user (e.g., restrictions, quotas, limits) that respond to the inputs from the assessment model (Link et al. 2010; Fulton et al. 2011).

Endpoints of Atlantis can include metrics such as biomass, concentration, catch, effort, or revenue (Link et al. 2010; Fulton et al. 2011; Kaplan et al. 2012; Marshall et al. 2017). Marshall et al. (2017) built upon previous applications (Kaplan et al. 2012; Weijerman et al. 2015) and utilized Atlantis to project the impacts of ocean acidification scenarios on fisheries EGS in the California Current ecosystem. To achieve this, future pH projections and biological pH sensitivities were plugged into the biophysical submodel; pH projections were derived from a separate model, whereas pH sensitivities were obtained from a meta-analysis of experimental results (Busch and McElhany 2016). Results demonstrated that a projected 0.2-unit decrease in pH negatively affected biomass, catch, and resulting revenue of most fishery management units considered. Atlantis predicted that state-managed functional groups would experience the greatest pH effects on revenue, largely due to the strong indirect effects on the valuable Dungeness crab fishery.

Although the full capability of Atlantis is impressive, utilization of the entire model without identifying key drivers can lead to over-parameterization (Fulton et al. 2011). There is also an inherent tradeoff between generality and precision with this broad ecosystem-based approach (Plaganyi 2007; Link et al. 2010). These limitations provide some context as to why Atlantis should not be used to determine specific management rules or actions, such as setting quotas (Fulton et al. 2011; Fulton et al. 2014). Atlantis researchers and developers have continued to improve and enhance the model however, to address some of model's weaknesses and create a more cohesive user community (Weijerman et al. 2016).

Properly implemented, end-to-end models like Atlantis can provide tremendous benefits to users, including the comparison of direct, indirect, and cumulative ecosystem effects across multiple fisheries under a variety of future scenarios (Fulton et al. 2011; Kaplan et al. 2012; Weijerman et al. 2015). Atlantis has been successfully implemented in regional-scale studies around the world (Weijerman et al. 2016), at scales of up to 1.475 million km^2 (Marshall 2017). These studies have demonstrated that Atlantis is a valuable tool to forecast changes in fisheries-related

EGS resulting from drivers of ecosystem change, including pollution, fishing, and climate change (Weijerman et al. 2015; Marshall et al. 2017). The scale and scope at which this spatially explicit model can be applied is consistent with the concepts of EBM, which is relevant for ecosystems not restricted by political boundaries, and thus important to the management of coastal and estuarine EGS.

5 InVEST Model Suite

Coastal decision-makers must often consider different types of EGS, thus models capable of analyzing tradeoffs between different types of EGS under different scenarios are valuable in the context of coastal planning. Recently, some 'decision support tools' have been developed to aid in this process by connecting a suite of EGS models within ecological and socioeconomic frameworks (Bagstad et al. 2013b; Gret-Regamey et al. 2017). In a recent review by Ochoa and Urbina-Cardona (2017), InVEST (Integrated Valuation of Ecosystem Services and Tradeoffs), an open-source tool developed by the Natural Capital Project (NCP, www.naturalcapitalproject.org), was one of the most widely cited resources for spatially modeling EGS. InVEST was developed with the explicit purpose of integrating natural capital into decision making (Daily et al. 2009). InVEST started as a general EGS model for terrestrial and freshwater systems, but in recent years, several marine and coastal models have been incorporated into the toolset (Tallis and Polasky 2009; Guerry et al. 2012). Similar EGS decision support tools include MIMES, LUCI, and ARIES (Bagstad et al. 2013b), the latter of which is discussed in the following section.

InVEST models can be applied from the local to regional scale and provide a variety of endpoints, making them highly useful for land-use planners and other decision-makers. Most input data must be spatially explicit and can include maps of biophysical information (e.g., elevation, habitat, species distributions, etc.), as well as socio-economic indicators (e.g., population density, property values, industry costs, etc.) (Tallis and Polasky 2009; Guerry et al. 2012). Submodels use this input to predict changes in EGS such as coastal protection, fisheries, and recreation. Submodels employ several different approaches for estimating EGS, sometimes with an option to select the 'tier' of submodel (i.e., higher tiers being more complex) depending on the desired endpoints, data requirements, and effort investment (Guerry et al. 2012). Two submodels for habitat risk and water quality provide interconnections between EGS submodels and allow users to consider tradeoffs when prioritizing management across different EGS. The final submodel predicts the quantity of EGS delivered or applies valuation methods to provide monetary outputs, such as the value of sequestered carbon (Guerry et al. 2012). Individual EGS submodels can also be used independently if desired.

InVEST has been applied in many coastal planning contexts (Kim et al. 2012; Guannel et al. 2015; Oleson et al. 2017), though fewer coastal studies have taken advantage of the tool's submodel interconnections to analyze tradeoffs. The government of Belize partnered with the NCP in one such case study and used InVEST models to facilitate the development of an Integrated Coastal Zone Management (ICZM) plan (Arkema et al. 2015; Ruckelshaus et al. 2015; Verutes et al. 2017). With input from stakeholders and government agencies, the scientific modeling team identified three key ecosystem services (tourism, coastal protection, and the spiny lobster fishery) to analyze tradeoffs under three development scenarios. Simulating these scenarios in InVEST ultimately informed the development of an ICZM plan that was approved by the Belizean government in 2016 (Verutes et al. 2017).

The NCP continues to develop InVEST models in an on-going effort to improve connectivity with other complex models (e.g., Atlantis), enhance utility of submodels, and add new EGS models. Still, InVEST is most ideal for users with at least some spatially explicit data available from the area of interest. Gathering the spatial data required to run some of InVEST's coastal models can take considerable time and effort (e.g., see Bayani and Barthelemy 2016) and may not be feasible for all local decision makers. However, the ability to run core functions with less data and disable the more advanced model components helps to minimize this issue (Bayani and Barthelemy 2016).

6 ARIES Model Suite

A fundamental need to account for the spatial connectivity of ecosystem service supply and demand, with an explicit link to beneficiaries, led to development of the Artificial Intelligence for Ecosystem Services (ARIES) platform (Bagstad et al. 2013a; Villa et al. 2014a). ARIES has been refined through case studies, machine-learning, and technological advancements that now facilitates automated model customization to meet the needs of various decision makers (Villa et al. 2014a; Martínez-López et al. 2018).

As with InVEST, ARIES relies on spatially explicit input data and generates output that accounts for the location of beneficiaries, thereby closing the loop between ecological production functions that track resources (i.e., supply), to EGS of direct benefit to human users (i.e., demand) (Bagstad et al. 2013a; Villa et al. 2014a). ARIES accommodates both deterministic models and Bayesian networks to facilitate mapping of ecosystem service provision (Bagstad et al. 2014). Service Path Attribution Networks (SPANs) contain the ontologies for how services accrue in the system and are used to evaluate relationships between sources, sinks, use, and *'carriers'*. Carriers are the direct link between ecosystems and people that can facilitate provisioning benefits like drinking water, or preventive benefits like floodwater (Johnson et al. 2012; Bagstad et al. 2013a; Villa et al. 2014a, b). While not explicitly an IBM like HexSim, the flow of benefits is tracked using an agent-based approach via the discretized amount of a carrier from a source, to a use region or possible sink (Villa et al. 2014a, b). Carriers can be either physical (e.g., water) or informational (e.g., aesthetic views) and underlying SPAN models that vary based on the type of benefit, contain the rules dictating how a carrier moves in the system, including its absorption or ultimate delivery to an end user (Johnson et al. 2012; Villa et al. 2014a). Tangible benefits result from the accumulation of carriers by beneficiaries (Bagstad et al. 2013a).

The ARIES model suite, accessible through the open source software k.LAB (Knowledge Laboratory Integrated Development Environment), includes five baseline "Tier 1" ecosystem service models (Martínez-López et al. 2018). These base models (crop pollination, flood regulation, outdoor recreation, carbon storage, and sediment regulation) can be applied anywhere in the world without any new input due to existing datasets already built into the framework. In a basic model run, the user selects the spatiotemporal context, model resolution, ecosystem service of interest, and potential scenario conditions (optional). While users with data of greater spatial or temporal resolution can get further refined output, it is not necessary (Bagstad et al. 2014; Martínez-López et al. 2018). The supply component of the models quantifies the potential provision of a benefit, but it does not necessarily capture people's decisions to utilize ecosystem services (Villa et al. 2014a). Additionally, the models for crop pollination, outdoor recreation, and flood regulation estimate demand relative to other locations, thus output reflects ranked indicators, not biophysical values (Martínez-López et al. 2018).

ARIES models have been used to evaluate multiple ecosystem services including water provision, water quality, air quality, flood regulation, climate regulation, and aesthetic views (Bagstad et al. 2014; Ochoa and Urbina-Cardona 2017). For example, Bagstad et al. (2014) used ARIES to assess how aesthetic views translated to homeowner property values in Puget Sound, Washington (USA). Source features included mountains or waterbodies that added value, whereas view obstructions or blight areas were modeled as sinks (i.e., detracting from view quality). The flow models, which included distance decay functions, computed visibility along lines of sight between use locations and sources or sinks. Final results were a ratio between the values accrued by homeowners, relative to those of the entire landscape (Bagstad et al. 2014). A web-based ARIES 'explorer' is slated for release in 2020 and should facilitate greater accessibility and subsequent application in multiple settings (Martínez-López et al. 2018). The flexibility of the k.LAB software package and integration of more data-driven models (e.g., see Willcock et al. 2018) means that ARIES is easily customizable, and a growing user community may facilitate output more tailored to coastal beneficiaries and decision-makers.

Suites of models like ARIES and InVEST that have options to model multiple EGS over space and time, have indeed proven beneficial for coastal planning (Bagstad et al. 2014; Verutes et al. 2017). The true strength of these tools is their ability to take a comprehensive look at EGS and evaluate tradeoffs between potentially competing coastal EGS (Bagstad et al. 2013b). Integrating ecosystem service considerations into the coastal planning process can provide sustained benefits to coastal communities, and tools like ARIES and InVEST enable EGS to be more readily considered (Arkema et al. 2015). The ability to visualize EGS tradeoffs in a spatial context is beneficial to communicating the importance of ecosystems to stakeholders and this may help to optimize coastal planning by aligning scientific research with community priorities (Bayani and Barthelemy 2016; Verutes et al. 2017).

7 Common Difficulties, Emerging Issues, and Future Directions

Environmental changes are occurring at an unprecedented rate within coastal and estuarine ecosystems. Models capable of predicting alterations to coastal and estuarine EGS based on changes within these dynamic systems have much to offer scientists, resource managers, and decision-makers. Though these models vary in spatial scale, quantity of EGS modeled, and complexity, they share a number of common challenges.

7.1 Common Difficulties

Scale is important to understanding the context and applicability of model output. The importance and relevance of coastal and estuarine EGS span across the spectrum of spatial, temporal, and governance scales (Costanza et al. 2017). Identifying the best-suited model(s) given the scale(s) of a particular question or goal is paramount in the modeling process (Carpenter et al. 2009; Turner et al. 2016), as is utilizing data that corresponds to the same scale(s). Scale-dependence is tied to the accuracy and transferability of a model—some models can be applied across multiple spatial and/or temporal scales (Boumans et al. 2015; Francesconi et al. 2016; Lewis et al. 2019), whereas others are constrained to a single place and/or time. Difficulties can arise in interpreting outputs when existing models are applied at new scales or locations, an issue that is particularly relevant to coastal ecosystems, which can span a range of spatial and temporal scales (Swaney et al. 2012). Similar difficulties occur when gathering and applying model input data from a variety of sources and scales. Upscaling and downscaling are therefore common data manipulations in EGS modeling. For instance, results from small-scale experiments or processes will often be scaled-up for use in larger-scale models because of resource constraints that limit the scope of experiments and measurements (Craft et al. 2009; Peck et al. 2016). Downscaling, on the other hand, allows coarse output from global climate or systems models to be utilized in smaller-scale models (Peck et al. 2016). In both cases, scaling the original data introduces uncertainties (Andrew et al. 2015; Cheung et al. 2016) that impact model accuracy.

Uncertainty is an inherent component of modeling that should be acknowledged by users to avoid misinterpretation of results (Bagstad et al. 2013b; Peck et al. 2016). Communication of uncertainty to community partners and stakeholders is therefore vital (Guerry et al. 2012; Ruckelshaus et al. 2015); see Fulford et al. (2020) for a general discussion on communicating this uncertainty as an estimation of risk. Unfortunately, this uncertainty has not often been assessed systematically in coastal and estuarine ecosystem-scale models (Weijerman et al. 2015; Cheung et al. 2016). As the scale and complexity of a model increases, so typically does the uncertainty, which can be difficult to quantify for multiple interconnected submodels. Models that project changes in biological resources based on environmental alterations are also subject to uncertainties from model parameters, model structure, internal variability, and multiple scenarios (Arkema et al. 2015; Cheung et al. 2016; Marshall et al. 2017). Monte Carlo simulations, as demonstrated by Harris et al. (2018), are one approach that can provide deterministic models with an assessment of uncertainty due to errors in input data or parameter calibration. Tools based on probabilistic frameworks that facilitate a transparent characterization of uncertainty are also gaining traction (Bryant et al. 2018; Willcock et al. 2018). Although qualitative or bounded assessments of uncertainty provide some value to users, improved methods of assessment will need to be developed to progress toward deeper understanding of the uncertainty in applications of these complex models within coastal and estuarine ecosystems (Fulton et al. 2011; Cheung et al. 2016).

The complexity, as well as the time, cost, and data requirements, of many EGS models are barriers to widespread implementation (Chan and Ruckelshaus 2010; Plaganyi et al. 2011; Bagstad et al. 2013b; Willcock et al. 2018). A tradeoff exists between the complexity required for prediction accuracy and resource (i.e., time and cost) investment (Chan and Ruckelshaus 2010), the optimal balance varies among situations (see Fulford et al. (2020) for a general discussion on the tradeoff between parsimony and realism in regard to necessary complexity). The applicability of models requiring an extensive amount of input is limited for data-poor locations (Link et al. 2010; Bayani and Barthelemy 2016; Turner et al. 2016). A model's application can also be hindered in locations with ample data if a model's input data requirements are overly stringent (Bagstad et al. 2013b). As EGS models for coastal and estuarine systems become functionally more comprehensive and complex, the expertise needed to correctly apply them can become an obstacle to implementation. Improving a model's user interface may only superficially improve its usability, which may not increase the likelihood that it is ultimately applied correctly. The inclusion of adequate model documentation will aid in the application of the model by bringing transparency to its assumptions and limitations, while also providing adequate instructions and validation exercises (Bagstad et al. 2013b).

7.2 Emerging Issues and Future Directions

EBM emerged as an important and effective strategy for managing natural resources and EGS, in large part because of the explicit inclusion of humans within the ecosystem (Rosenberg and McLeod 2005; Guerry et al. 2012), which has increased the complexity and difficulty in ecosystems modeling. Although many EGS models have progressed, the consideration of human benefits that are difficult to monetize (e.g., cultural, existence, and subsistence values) is an area that is lacking (Chan and Ruckelshaus 2010; Plaganyi et al. 2011; Guerry et al. 2012; Liquete et al. 2013; Turner et al. 2016). The breadth of user groups and social behaviors considered needs to be increased to represent socio-ecological systems more realistically (Plaganyi et al. 2011), which will include potential interactions, feedbacks, and tradeoffs between groups and uses (e.g., see Fulford et al. 2016).

There is no single model that will fit the needs of every user, nor should there be; the best model will be the one most appropriate for the situation, data, and question at hand (Plaganyi et al. 2011; Peck et al. 2016; Turner et al. 2016). To improve model relevancy and usability, resource managers and other stakeholders must be part of the model development process to identify important decision metrics (Bagstad et al. 2013b; Costanza et al. 2017) and define the most practical management alternatives to test (Fulton et al. 2014; Turner et al. 2016). Decision-makers should have models available for different uses, scales of complexity, spatial resolution, and number of services, preferably models that can be coupled (Peck et al. 2016). Searchable inventories of EGS models like the EcoService Models Library (found at https://www.epa.gov/eco-research/ecoservice-models-library) may aid managers in the initial process of selecting the appropriate one. Perhaps the most optimal solution is a modeling framework that allows users to choose which components, groups, and interactions to include, similar to the modular designs of InVEST and ARIES (Guerry et al. 2012; Bagstad et al. 2014). With some modification, these tiered or coupled models can allow for the identification and inclusion of multiple drivers and cumulative impacts on coastal and estuarine EGS, which is a critical need (Carpenter et al. 2009; Chan and Ruckelshaus 2010; Little et al. 2017).

The concept of coupling models highlights another issue—error or uncertainty from input datasets can propagate through the model and significantly affect results. Validation provides users with a measure of confidence in model output, increasing the likelihood that a model's results will be accepted and used to inform the decision process (Bagstad et al. 2013b; Andrew et al. 2015). However, the availability of suitable data to validate and inform understanding is generally lacking (Mach et al. 2015; Plaganyi et al. 2011). For instance, ecological properties or processes within EGS models have often been parameterized with coarse data (Andrew et al. 2015), whereas many satellite-based land cover datasets have not been validated at all (Song 2018). Validation of outputs from the simulation of potential future scenarios is particularly difficult as the conditions simulated may not currently exist. The EGS model comparison study by Sharps et al. (2017), demonstrates the extensive validation (utilizing measured data for flow, water quality, soil carbon, and above-ground biomass) that is ideal when applying complex EGS models such as InVEST or ARIES.

While progress continues to be made on these issues, the lack of a standardized framework for implementation may continue to limit widespread adoption of EGS models into coastal and marine planning processes (Daily et al. 2009). In recent years, scientists working with decision-makers have been developing methods to formalize the inclusion of EGS in the planning process (Arkema et al. 2015; Ruckelshaus et al. 2015). Incorporating the estimation of impacts on EGS into the existing framework of Environmental Impact Assessments may be a promising route (Karjalainen et al. 2013). The DPSIR (Driver, Pressure, State, Impact, Response) is another, more conceptual, framework that has shown great progress incorporating EGS in decision making (Cranford et al. 2012; Kelble et al. 2013; Elliot and O'Higgins 2020). Starting in 2015, the European interdisciplinary research project AQUACROSS developed an assessment framework for the EBM of Europe's aquatic ecosystems and applied it in eight case studies (www.aquacross.eu). As more studies demonstrate the utility of EGS modeling within ecosystem-based management situations, knowledge and experience from these cases can be used as reference to further develop these frameworks (Forst 2009). The tools provided by the growing field of EGS modeling will undoubtedly aid scientists and decision-makers as they establish, validate, and apply innovative approaches to planning for estuarine and coastal change.

This chapter has been subjected to Agency review and has been approved for publication. The views expressed in this paper are those of the author(s) and do not necessarily reflect the views or policies of the U.S. Environmental Protection Agency.

Disclaimer This chapter has been subjected to Agency review and has been approved for publication. The views expressed in this paper are those of the author(s) and do not necessarily reflect the views or policies of the U.S. Environmental Protection Agency.

References

Andrew, M. E., Wulder, M. A., Nelson, T. A., & Coops, N. C. (2015). Spatial data, analysis approaches, and information needs for spatial ecosystem service assessments: A review. *GIScience & Remote Sensing, 52*(3), 344–373.

Arkema, K. K., Verutes, G. M., Wood, S. A., Clarke-Samuels, C., Rosado, S., Canto, M., Rosenthal, A., et al. (2015). Embedding ecosystem services in coastal planning leads to better outcomes for people and nature. *Proceedings of the National Academy of Sciences, 112*(24), 7390–7395. https://doi.org/10.1073/pnas.1406483112.

Bagstad, K. J., Johnson, G. W., Voigt, B., & Villa, F. (2013a). Spatial dynamics of ecosystem service flows: A comprehensive approach to quantifying actual services. *Ecosystem Services, 4*, 117–125. https://doi.org/10.1016/j.ecoser.2012.07.012.

Bagstad, K. J., Semmens, D. J., Waage, S., & Winthrop, R. (2013b). A comparative assessment of decision-support tools for ecosystem services quantification and valuation. *Ecosystem Services, 5*, 27–39. https://doi.org/10.1016/j.ecoser.2013.07.004.

Bagstad, K. J., Villa, F., Batker, D., Harrison-Cox, J., Voigt, B., & Johnson, G. W. (2014). From theoretical to actual ecosystem services: Mapping beneficiaries and spatial flows in ecosystem service assessments. *Ecology and Society, 19*(2), 64. https://doi.org/10.5751/ES-06523-190264.

Barbier, E. B. (2016). The protective service of mangrove ecosystems: A review of valuation methods. *Marine Pollution Bulletin, 109*(2), 676–681. https://doi.org/10.1016/j.marpolbul.2016.01.033.

Barbier, E. B., Hacker, S. D., Kennedy, C., Koch, E. W., Stier, A. C., & Silliman, B. R. (2011). The value of estuarine and coastal ecosystem services. *Ecological Monographs, 81*(2), 169–193.

Bayani, N., & Barthelemy, Y. (2016). Integrating ecosystems in risk assessments: Lessons from applying InVEST models in data-deficient countries. In F. G. Renaud, K. Sudmeier-Rieux, M. Estrella, & U. Nehren (Eds.), *Ecosystem-based disaster risk reduction and adaptation in practice* (pp. 227–254). Cham: Springer.

Boumans, R., Roman, J., Altman, I., & Kaufman, L. (2015). The Multiscale Integrated Model of Ecosystem Services (MIMES): Simulating the interactions of coupled human and natural systems. *Ecosystem Services, 12*, 30–41. https://doi.org/10.1016/j.ecoser.2015.01.004.

Bruins, R. J. F., Canfield, T. J., Duke, C., Kapustka, L., Nahlik, A. M., & Schafer, R. B. (2017). Using ecological production functions to link ecological processes to ecosystem services. *Integrated Environmental Assessment and Management, 13*(1), 52–61. https://doi.org/10.1002/ieam.1842.

Bryant, B. P., Borsuk, M. E., Hamel, P., Oleson, K. L. L., Schulp, C. J. E., & Willcock, S. (2018). Transparent and feasible uncertainty assessment adds value to applied ecosystem services modeling. *Ecosystem Services, 33*, 103–109. https://doi.org/10.1016/j.ecoser.2018.09.001.

Busch, D. S., & McElhany, P. (2016). Estimates of the direct effect of seawater pH on the survival rate of species groups in the California current ecosystem. *PloS One, 11*(8), e0160669.

Carpenter, S. R., Mooney, H. A., Agard, J., Capistrano, D., DeFries, R. S., Díaz, S., Dietz, T., Duraiappah, A. K., Oteng-Yeboah, A., & Pereira, H. M. (2009). Science for managing ecosystem services: Beyond the Millennium Ecosystem Assessment. *Proceedings of the National Academy of Sciences, 106*(5), 1305–1312.

Carr, E. W., Shirazi, Y., Parsons, G. R., Hoagland, P., & Sommerfield, C. K. (2018). Modeling the economic value of Blue Carbon in Delaware Estuary Wetlands: Historic estimates and future projections. *Journal of Environmental Management, 206*, 40–50. https://doi.org/10.1016/j.jenvman.2017.10.018.

Chan, K. M., & Ruckelshaus, M. (2010). Characterizing changes in marine ecosystem services. *F1000 Biology Reports* 2.

Cheung, W. L., Frölicher, T. L., Asch, R. G., Jones, M. C., Pinsky, M. L., Reygondeau, G., Rodgers, K. B., Rykaczewski, R. R., Sarmiento, J. L., Stock, C., & Watson, J. R. (2016). Building confidence in projections of the responses of living marine resources to climate change. *ICES Journal of Marine Science, 73*(5), 1283–1296.

Costanza, R., d'Arge, R., De Groot, R., Farber, S., Grasso, M., Hannon, B., Limburg, K., Naeem, S., O'Neill, R. V., Paruelo, J., Raskin, R. G., Sutton, P., & van den Belt, M. (1997). The value of the world's ecosystem services and natural capital. *Nature, 387*(6630), 253–260.

Costanza, R., de Groot, R., Sutton, P., Van der Ploeg, S., Anderson, S. J., Kubiszewski, I., Farber, S., & Turner, R. K. (2014). Changes in the global value of ecosystem services. *Global Environmental Change, 26*, 152–158.

Costanza, R., de Groot, R., Braat, L., Kubiszewski, I., Fioramonti, L., Sutton, P., Farber, S., & Grasso, M. (2017). Twenty years of ecosystem services: How far have we come and how far do we still need to go? *Ecosystem Services, 28*, 1–16.

Craft, C., Clough, J., Ehman, J., Joye, S., Park, R., Pennings, S., Guo, H., & Machmuller, M. (2009). Forecasting the effects of accelerated sea-level rise on tidal marsh ecosystem services. *Frontiers in Ecology and the Environment, 7*(2), 73–78.

Cranford, P. J., Kamermans, P., Krause, G., Mazurié, J., Buck, B. H., Dolmer, P., Fraser, D., Van Nieuwenhove, K., O'Beirn, F. X., & Sanchez-Mata, A. (2012). An ecosystem-based approach and management framework for the integrated evaluation of bivalve aquaculture impacts. *Aquaculture Environment Interactions, 2*(3), 193–213.

Daily, G. C., Polasky, S., Goldstein, J., Kareiva, P. M., Mooney, H. A., Pejchar, L., Ricketts, T. H., Salzman, J., & Shallenberger, R. (2009). Ecosystem services in decision making: Time to deliver. *Frontiers in Ecology and the Environment, 7*(1), 21–28.

DeAngelis, D. L., & Rose, K. A. (1992). Which individual-based based approach is most appropriate for a given problem. In D. L. DeAngelis & L. J. Gross (Eds.), *Individual-based models and approaches in ecology* (pp. 67–86). London: Chapman and Hall.

Elliott, M., & O'Higgins, T. G. (2020). From the DPSIR, the D(A)PSI(W)R(M) emerges... a butterfly-'protecting the natural stuff and delivering the human stuff'. In T. O'Higgins, M. Lago, & T. H. DeWitt (Eds.), *Ecosystem-based management, ecosystem services and aquatic biodiversity: Theory, tools and applications* (pp. 61–86). Amsterdam: Springer.

Forst, M. F. (2009). The convergence of integrated coastal zone management and the ecosystems approach. *Ocean & Coastal Management, 52*(6), 294–306.

Francesconi, W., Srinivasan, R., Pérez-Miñana, E., Willcock, S. P., & Quintero, M. (2016). Using the Soil and Water Assessment Tool (SWAT) to model ecosystem services: A systematic review. *Journal of Hydrology, 535*, 625–636.

Fulford, R. S., Peterson, M. S., & Grammer, P. O. (2011). An ecological model of the habitat mosaic in estuarine nursery areas: Part I-Interaction of dispersal theory and habitat variability in describing juvenile fish distributions. *Ecological Modelling, 222*, 3203–3215.

Fulford, R. S., Russell, M., & Rogers, J. E. (2016). Habitat restoration from an ecosystem goods and services perspective: Application of a spatially explicit individual-based model. *Estuaries and Coasts, 39*(6), 1801–1815.

Fulford, R. S., Heymans, S. J. J., & Wu, W. (2020). Mathematical modelling for ecosystem-based management (EBM) and ecosystem goods and services (EGS) assessment. In T. O'Higgins, M. Lago, & T. H. DeWitt (Eds.), *Ecosystem-based management ecosystem services and aquatic biodiversity: Theory, tools and applications* (pp. 275–290). Springer, Amsterdam.

Fulton, E. A., Link, J. S., Kaplan, I. C., Savina-Rolland, M., Johnson, P., Ainsworth, C., Horne, P., Gorton, R., Gamble, R. J., & Smith, A. D. M. (2011). Lessons in modelling and management of marine ecosystems: The Atlantis experience. *Fish and Fisheries, 12*(2), 171–188.

Fulton, E. A., Smith, A. D. M., Smith, D. C., & Johnson, P. (2014). An integrated approach is needed for ecosystem based fisheries management: Insights from ecosystem-level management strategy evaluation. *PloS One, 9*(1), e84242.

Gret-Regamey, A., Siren, E., Brunner, S. H., & Weibel, B. (2017). Review of decision support tools to operationalize the ecosystem services concept. *Ecosystem Services, 26*, 306–315. https://doi.org/10.1016/j.ecoser.2016.10.012.

Guannel, G., Ruggiero, P., Faries, J., Arkema, K., Pinsky, M., Gelfenbaum, G., Guerry, A., & Kim, C. K. (2015). Integrated modeling framework to quantify the coastal protection services supplied by vegetation. *Journal of Geophysical Research-Oceans, 120*(1), 324–345. https://doi.org/10.1002/2014jc009821.

Guerry, A. D., Ruckelshaus, M. H., Arkema, K. K., Bernhardt, J. R., Guannel, G., Kim, C. K., Marsik, M., et al. (2012). Modeling benefits from nature: Using ecosystem services to inform coastal and marine spatial planning. *International Journal of Biodiversity Science Ecosystem Services & Management, 8*(1–2), 107–121. https://doi.org/10.1080/21513732.2011.647835.

Hanson, C., Finisdore, J., & Ranganathan, J. (2012). *The Corporate Ecosystem Services Review: Guidelines for identifying business risks and opportunities arising from ecosystem change*. Washington, DC: World Resources Institute.

Harris, D. L., Rovere, A., Casella, E., Power, H., Canavesio, R., Collin, A., Pomeroy, A., Webster, J. M., & Parravicini, V. (2018). Coral reef structural complexity provides important coastal protection from waves under rising sea levels. *Science Advances, 4*(2). https://doi.org/10.1126/sciadv.aao4350.

Johnson, G. W., Bagstad, K. J., Snapp, R. R., & Villa, F. (2012). Service Path Attribution Networks (SPANs): A network flow approach to ecosystem service assessment. *International Journal of Agricultural and Environmental Information Systems, 3*, 54–71.

Kaplan, I. C., Horne, P. J., & Levin, P. S. (2012). Screening California current fishery management scenarios using the Atlantis end-to-end ecosystem model. *Progress in Oceanography, 102*, 5–18.

Karjalainen, T. P., Marttunen, M., Sarkki, S., & Rytkonen, A. M. (2013). Integrating ecosystem services into environmental impact assessment: An analytic-deliberative approach. *Environmental Impact Assessment Review, 40*, 54–64.

Kassakian, J., Jones, A., Martinich, J., & Hudgens, D. (2017). Managing for no net loss of ecological services: An approach for quantifying loss of coastal wetlands due to sea level rise. *Environmental Management, 59*(5), 736–751.

Kelble, C. R., Loomis, D. K., Lovelace, S., Nuttle, W. K., Ortner, P. B., Fletcher, P., Cook, G. S., Lorenz, J. J., & Boyer, J. N. (2013). The EBM-DPSER conceptual model: Integrating ecosystem services into the DPSIR framework. *Plos One, 8*(8), 12. https://doi.org/10.1371/journal.pone.0070766.

Kim, C. K., Toft, J. E., Papenfus, M., Verutes, G., Guerry, A. D., Ruckelshaus, M. H., Arkema, K. K., et al. (2012). Catching the right wave: Evaluating wave energy resources and potential compatibility with existing marine and coastal uses. *Plos One, 7*(11), 14. https://doi.org/10.1371/journal.pone.0047598.

Lewis, N. S., Fox, E. W., & DeWitt, T. H. (2019). Estimating the distribution of harvested estuarine bivalves with natural-history-based habitat suitability models. *Estuarine, Coastal and Shelf Science, 219*, 453–472.

Link, J. S., Fulton, E. A., & Gamble, R. J. (2010). The northeast US application of ATLANTIS: a full system model exploring marine ecosystem dynamics in a living marine resource management context. *Progress in Oceanography, 87*(1–4), 214–234.

Liquete, C., Piroddi, C., Drakou, E. G., Gurney, L., Katsanevakis, S., Charef, A., & Egoh, B. (2013). Current status and future prospects for the assessment of marine and coastal ecosystem services: A systematic review. *PloS One, 8*(7), e67737.

Little, S., Spencer, K. L., Schuttelaars, H. M., Millward, G. E., & Elliott, M. (2017). Unbounded boundaries and shifting baselines: Estuaries and coastal seas in a rapidly changing world. *Estuarine, Coastal and Shelf Science, 198*, 311–319. https://doi.org/10.1016/j.ecss.2017.10.010.

Mach, M. E., Martone, R. G., & Chan, K. M. A. (2015). Human impacts and ecosystem services: Insufficient research for trade-off evaluation. *Ecosystem Services, 16*, 112–120. https://doi.org/10.1016/j.ecoser.2015.10.018.

Marois, D. E., & Mitsch, W. J. (2015). Coastal protection from tsunamis and cyclones provided by mangrove wetlands–a review. *International Journal of Biodiversity Science, Ecosystem Services & Management, 11*(1), 71–83.

Marshall, K. N., Kaplan, I. C., Hodgson, E. E., Hermann, A., Busch, D. S., McElhany, P., Essington, T. E., Harvey, C. J., & Fulton, E. A. (2017). Risks of ocean acidification in the California current food web and fisheries: Ecosystem model projections. *Global Change Biology, 23*(4), 1525–1539.

Martínez-López, J., Bagstad, K. J., Balbi, S., Magrach, A., Voigt, B., Athanasiadis, I., Pascual, M., Willcock, S., & Villa, F. (2018). Towards globally customizable ecosystem service models. *Science of the Total Environment, 650*, 2325–2336. https://doi.org/10.1016/j.scitotenv.2018.09.371.

Mendoza, E., Oderiz, I., Martinez, M. L., & Silva, R. (2017). Measurements and modelling of small scale processes of vegetation preventing dune erosion. *Journal of Coastal Research*, 19–27. https://doi.org/10.2112/si77-003.1.

Millennium Ecosystem Assessment. (2005). Synthesis report. Island, Washington, DC.

Ochoa, V., & Urbina-Cordona, N. (2017). Tools for spatially modeling ecosystem services: Publication trends, conceptual reflections and future challenges. *Ecosystem Services, 26*, 155–169. https://doi.org/10.1016/j.ecoser.2017.06.011.

Oleson, K. L. L., Falinski, K. A., Lecky, J., Rowe, C., Kappel, C. V., Selkoe, K. A., & White, C. (2017). Upstream solutions to coral reef conservation: The payoffs of smart and cooperative decision-making. *Journal of Environmental Management, 191*, 8–18. https://doi.org/10.1016/j.jenvman.2016.12.067.

Pearson, S. G., Storlazzi, C. D., van Dongeren, A. R., Tissier, M. F. S., & Reniers, A. (2017). A Bayesian-based system to assess wave-driven flooding hazards on coral reef-lined coasts. *Journal of Geophysical Research-Oceans, 122*(12), 10099–10117. https://doi.org/10.1002/2017jc013204.

Peck, M. A., Arvanitidis, C., Butenschön, M., Canu, D. M., Chatzinikolaou, E., Cucco, A., Domenici, P., Fernandes, J. A., Gasche, L., & Huebert, K. B. (2016). Projecting changes in the distribution and productivity of living marine resources: A critical review of the suite of modelling approaches used in the large European project VECTORS. *Estuarine, Coastal and Shelf Science, 201*, 40–55.

Plagányi, É. E. (2007). Models for an ecosystem approach to fisheries. Food and Agriculture Org. (FAO) Fisheries Technical Paper No. 477.

Plagányi, É. E., Bell, J. D., Bustamante, R. H., Dambacher, J. M., Dennis, D. M., Dichmont, C. M., Dutra, L. X. C., Fulton, E. A., Hobday, A. J., & van Putten, E. I. (2011). Modelling climate-change effects on Australian and Pacific aquatic ecosystems: A review of analytical tools and management implications. *Marine and Freshwater Research, 62*(9), 1132–1147.

Pomeroy, A., Lowe, R., Symonds, G., Van Dongeren, A., & Moore, C. (2012). The dynamics of infragravity wave transformation over a fringing reef. *Journal of Geophysical Research-Oceans, 117*, 17. https://doi.org/10.1029/2012jc008310.

Quataert, E., Storlazzi, C., van Rooijen, A., Cheriton, O., & van Dongeren, A. (2015). The influence of coral reefs and climate change on wave-driven flooding of tropical coastlines. *Geophysical Research Letters, 42*(15), 6407–6415. https://doi.org/10.1002/2015gl064861.

Roelvink, D., Reniers, A., van Dongeren, A., de Vries, J. V., McCall, R., & Lescinski, J. (2009). Modelling storm impacts on beaches, dunes and barrier islands. *Coastal Engineering, 56* (11–12), 1133–1152. https://doi.org/10.1016/j.coastaleng.2009.08.006.

Rose, K. A., Allen, J. I., Artioli, Y., Barange, M., Blackford, J., Carlotti, F., Cropp, R., Daewel, U., Edwards, K., Flynn, K., Hill, S. L., HilleRisLambers, R., Huse, G., Mackinson, S., Megrey, B., Moll, A., Rivkin, R., Salihoglu, B., Schrum, C., Shannon, L., Shin, Y. J., Smith, S. L., Smith, C., Solidoro, C., John, M. S., & Zhou, M. (2010). End-to-end models for the analysis of marine ecosystems: Challenges, issues, and next steps. *Marine and Coastal Fisheries, 2*, 115–130.

Rose, K. A., Fiechter, J., Curchitser, E. N., Hedstrom, K., Bernal, M., Creekmore, S., Haynie, A., Ito, S., Lluch-Cota, S., Megrey, B. A., Edwards, C. A., Checkley, D., Koslow, T., McClatchie, S., Werner, F., MacCall, A., & Agostini, V. (2015). Demonstration of a fully-coupled end-to-end model for small pelagic fish using sardine and anchovy in the California Current. *Progress in Oceanography, 138*, 348–380.

Rosenberg, A. A., & McLeod, K. L. (2005). Implementing ecosystem-based approaches to management for the conservation of ecosystem services. *Marine Ecology Progress Series, 300*, 270–274.

Ruckelshaus, M., McKenzie, E., Tallis, H., Guerry, A., Daily, G., Kareiva, P., Polasky, S., et al. (2015). Notes from the field: Lessons learned from using ecosystem service approaches to inform real-world decisions. *Ecological Economics, 115*, 11–21. https://doi.org/10.1016/j.ecolecon.2013.07.009.

Rustigian, H. L., Santelmann, M. V., & Schumaker, N. H. (2003). Assessing the potential impacts of alternative landscape designs on amphibian population dynamics. *Landscape Ecology, 18*, 65–81.

Schumaker, N. H., & Brookes, A. (2018). HexSim: a modeling environment for ecology and conservation. *Landscape Ecology, 33*, 197–211.

Schumaker, N. H., Ernst, T., White, D., Baker, J., & Haggerty, P. (2004). Projecting wildlife responses to alternative future landscapes in Oregon's Willamette Basin. *Ecological Applications, 14*, 381–400.

Sharps, K., Masante, D., Thomas, A., Jackson, B., Redhead, J., May, L., Prosser, H., Cosby, B., Emmett, B., & Jones, L. (2017). Comparing strengths and weaknesses of three ecosystem services modelling tools in a diverse UK river catchment. *Science of the Total Environment, 584*, 118–130. https://doi.org/10.1016/j.scitotenv.2016.12.160.

Sheppard, C., Dixon, D. J., Gourlay, M., Sheppard, A., & Payet, R. (2005). Coral mortality increases wave energy reaching shores protected by reef flats: Examples from the Seychelles. *Estuarine, Coastal and Shelf Science, 64*(2–3), 223–234. https://doi.org/10.1016/j.ecss.2005.02.016.

Song, X.-P. (2018). Global estimates of ecosystem service value and change: Taking into account uncertainties in satellite-based land cover data. *Ecological Economics, 143*, 227–235.

Swaney, D. P., Humborg, C., Emeis, K., Kannen, A., Silvert, W., Tett, P., Pastres, R., et al. (2012). Five critical questions of scale for the coastal zone. *Estuarine, Coastal and Shelf Science, 96*, 9–21. https://doi.org/10.1016/j.ecss.2011.04.010.

Tallis, H., & Polasky, S. (2009). Mapping and valuing ecosystem services as an approach for conservation and natural-resource management. In R. S. Ostfeld & W. H. Schlesinger (Eds.), *Year in ecology and conservation biology* (pp. 265–283). Blackwell Publishing, Oxford: Annals of the New York Academy of Sciences.

Travers, M., Shin, Y. J., Jennings, S., & Cury, P. (2007). Towards end-to-end models for investigating the effects of climate and fishing in marine ecosystems. *Progress in Oceanography, 75*(4), 751–770.

Turner, K. G., Anderson, S., Gonzales-Chang, M., Costanza, R., Courville, S., Dalgaard, T., Dominati, E., et al. (2016). A review of methods, data, and models to assess changes in the value of ecosystem services from land degradation and restoration. *Ecological Modelling, 319*, 190–207. https://doi.org/10.1016/j.ecolmodel.2015.07.017.

Van Dongeren, A., Lowe, R., Pomeroy, A., Trang, D. M., Roelvink, D., Symonds, G., & Ranasinghe, R. (2013). Numerical modeling of low-frequency wave dynamics over a fringing coral reef. *Coastal Engineering, 73*, 178–190. https://doi.org/10.1016/j.coastaleng.2012.11.004.

Verutes, G. M., Arkema, K. K., Clarke-Samuels, C., Wood, S. A., Rosenthal, A., Rosado, S., Canto, M., Bood, N., & Ruckelshaus, M. (2017). Integrated planning that safeguards ecosystems and balances multiple objectives in coastal Belize. *International Journal of Biodiversity Science Ecosystem Services & Management, 13*(3), 1–17. https://doi.org/10.1080/21513732.2017.1345979.

Villa, F., Bagstad, K. J., Voigt, B., Johnson, G. W., Portela, R., Honzak, M., & Batker, D. (2014a). A methodology for adaptable and robust ecosystem services assessment. *PloS One, 9*(3), e91001.

Villa, F., Voigt, B., & Erickson, J. D. (2014b). New perspectives in ecosystem services science as instruments to understand environmental securities. *Philosophical Transactions of the Royal Society B, 369*, 20120286. https://doi.org/10.1098/rstb.2012.0286.

Weijerman, M., Fulton, E. A., Kaplan, I. C., Gorton, R., Leemans, R., Mooij, W. M., & Brainard, R. E. (2015). An integrated coral reef ecosystem model to support resource management under a changing climate. *PloS One, 10*(12), e0144165.

Weijerman, M., Link, J. S., Fulton, E. A., Olsen, E., Townsend, H., Gaichas, S., Hansen, C., Skern-Mauritzen, M., Kaplan, I. C., & Gamble, R. (2016). Atlantis Ecosystem Model Summit: Report from a workshop. *Ecological Modelling, 335*, 35–38.

Willcock, S., Martínez-López, J., Hooftman, D. A. P., Bagstad, K. J., Balbi, S., Marzo, A., Prato, C., Sciandrello, S., Signorello, G., Voigt, B., Villa, F., Bullock, J. M., & Athanasiadis, I. N. (2018). Machine learning for ecosystem services. *Ecosystem Services, 33*, 165–174. https://doi.org/10.1016/j.ecoser.2018.04.004.

3

Rapid Benefit Indicator Tools

Justin Bousquin and Marisa Mazzotta

Abstract Given the many interconnections between socio-economic and ecological systems, for Ecosystem-Based Management (EBM) to be effective, decision makers must consider metrics for both. Supply side tools and assessments characterize ecosystem condition, functioning, and potential to provide ecosystem goods and services (EGS). Demand side tools, including economic valuation, assess people's preferences for EGS and sometimes estimate the monetary amount people are willing to pay for a good or service. However, economic valuation is often omitted from assessments, due to lack of data or expertise; and economic valuation alone may not sufficiently capture all important aspects of some decisions. Benefit-relevant indicators have evolved as a way to measure the connection between goods or services that may be provided by an ecosystem, and people who may benefit from those services, while stopping short of valuation (Olander et al., Ecol Indic 85:1262–1272, 2018). Like economic valuation, benefit-relevant indicators can help assess trade-offs and compare alternative outcomes (National Ecosystem Services Partnership, Federal resource management and ecosystem services guidebook. National Ecosystem Services Partnership, Duke University, Durham, 2016). The Rapid Benefit Indicators (RBI) approach is an easy-to-use process for choosing a structured set of non-monetary benefit-relevant indicators for assessment (Mazzotta et al., Integr Environ Assess Manag 15:148–159, 2019). The RBI approach indicators are intended to be applied in conjunction with existing ecosystem service assessment approaches and tools, to connect changes in the availability of EGS to the locations where, and how, people benefit from those goods and services. Though developed for use with urban freshwater wetland restoration, the general RBI approach and indicator framework may be adapted and applied to other

J. Bousquin (✉)
US Environmental Protection Agency, Gulf Ecosystem Measurement and Modeling Division, Gulf Breeze, FL, USA
e-mail: Bousquin.Justin@epa.gov

M. Mazzotta
US Environmental Protection Agency, Atlantic Coastal Environmental Sciences Division, Narragansett, RI, USA

environmental changes or ecological systems. This chapter will detail the RBI approach and highlight how RBI tools can inform resource management decisions.

Lessons Learned

- When implementing EBM, it is not enough to simply maintain or restore the functioning of ecosystems; it is essential to also consider benefits from services that people want and need
- Socio-economic metrics must be linked to changes in ecosystems to be relevant to EBM policy and management questions
- Indicator-based methods inform decisions when direct measures of economic values are unavailable, overly complex, elicit resistance or are otherwise inadequate
- A structured process for selecting benefit indicators, such as the Rapid Benefit Indicators Approach, helps practitioners choose the right metrics
- Tools, such as the RBI checklist tool, the RBI spatial analysis tool and the RBI national catchment dataset, can make it easier for practitioners to evaluate benefits from services as part of their EBM approach

Needs to Advance EBM

- EBM methods and policy would benefit from more explicitly addressing and communicating benefits to people resulting from managing ecosystems
- EBM practitioners need tools that will allow them to evaluate the services and benefits provided by a wide array of ecosystems

1 Evaluating Benefits

Ecologists and economists have embraced the ecosystem services concept as an important support for understanding and managing social-ecological systems. Ecosystem-Based Management (EBM) seeks to protect, maintain or restore ecosystems and their functioning so that they can provide the services that people want and need (Piet et al. 2017; Delacámara et al. 2020). Decisionmakers using an EBM approach need to weigh trade-offs among the benefits and costs of different actions, and ecosystem service metrics can inform those trade-offs. Despite consensus around the importance of considering ecosystem services, it remains difficult for practitioners to choose the right metrics (Boyd et al. 2016; Olander et al. 2017).

There are metrics specific to each component of the ecosystem service framework (Fig. 1). Biophysical metrics are commonly used to model or monitor how an ecological system responds to human actions (the ecological outcomes). Ecosystem service metrics describe the ecosystem goods and services (EGS) produced by the system, based on its condition and resulting ability to perform functions needed to produce those EGS. To avoid double-counting of benefits to people, ecosystem service metrics should measure what is directly enjoyed, consumed or used by people, i.e. final ecosystem goods and services (FEGS; Boyd and Banzhaf 2007;

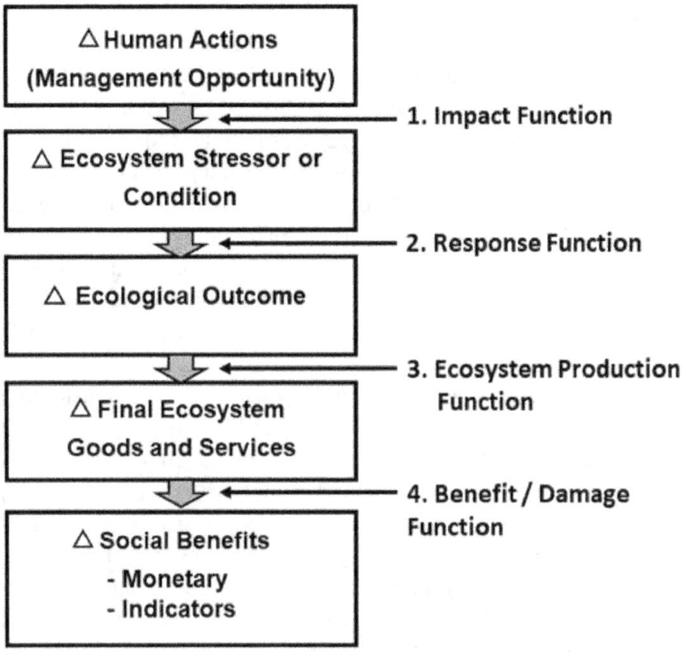

Fig. 1 Conceptual framework linking management actions to ecological change and to socio-economic benefits. (Adapted from Wainger and Mazzotta 2011)

Ringold et al. 2013). Changes in FEGS that result from human actions will lead to changes in human well-being, which may be measured in monetary terms or using indicators, as presented in this chapter.

There are many tools and metrics available to measure the functioning of ecological systems or their ability to produce ecosystem services, although these do not always measure the endpoints that are most appropriate for evaluating changes in human well-being. For example, there is often a bias towards, or over-reliance on, land use and land cover data (Tashier and Ringold 2019). Simply measuring an ecosystem's ability to produce goods and services does not provide evidence that those goods and services are used or enjoyed by people. Even when using the most precise metrics for ecological outcomes or ecosystem services, if these metrics cannot be linked to the socio-economic benefits that result from changes in the ecosystem, the assessment will be incomplete and risks being irrelevant to people. The closer metrics get to measuring the change in social benefits, the better those metrics can inform trade-offs.

Environmental decision makers, the people deciding between environmental management actions, often require or desire monetary measures of benefits to people. Economic valuation approaches monetize the value of ecosystem goods and services to people (Heal 2000b; Freeman et al. 2014). Valuation methods include the use of market prices, where available (e.g., for commercially-harvested

EGS such as fish or timber), but most EGS are not traded in markets and thus require the use of non-market valuation methods (National Research Council 2005; Champ et al. 2017). These include revealed preference methods that use people's behavior to infer values (e.g., the travel cost method, Parsons 2017); stated preference methods that use hypothetical questions or comparisons of choices to ask people to directly state their values (e.g. contingent valuation, Bateman and Willis 1999); and benefit transfer methods that apply values from existing studies to a new location and/or context (Johnston et al. 2015). Each of these methods is appropriate for different contexts and types of values. Whereas use values, where people directly interact with FEGS, may be measured by any of these methods; non-use values, where people do not directly interact with the service, can only be measured using stated preference methods or benefit transfer of stated preference studies. Each method has its advantages as well as shortcomings or pitfalls (Champ et al. 2017; Johnston et al. 2015; Freeman et al. 2014). These methods also tend to be resource-intensive, and location- and context-specific (Spash and Vatn 2006; Heal 2000a). Even when valuation studies are performed well, if the estimated values are not appropriately linked to available biophysical models or metrics, or if the appropriate biophysical metrics are not available, the estimated values will not be responsive to changes in ecosystems (Johnston et al. 2012; Schultz et al. 2012). Especially in the context of EBM, socio-economic metrics that can't be linked to changes in ecosystems are of limited relevance to policy or management questions.

In cases where assessments include metrics linking each component of the ecosystem services framework, decision makers may still face various difficulties in applying a comprehensive EGS assessment. Resource managers who must make decisions affecting EGS, such as state agencies, may not have in-house expertise to conduct an assessment, and may lack resources or support for commissioning an appropriate study. Olander et al. (2017) suggest that many methods are not responsive to policy and management changes, not generalizable enough to transfer from one context to another, or are too burdensome in terms of needed expertise, data or costs. Developing new metrics and assessment tools to fill gaps should involve decision makers to help ensure that their needs are met and that applications are feasible, with relevant outcomes (Ojo et al. 2018). However, tool developers must balance these factors against over-tailoring metrics and tools to be so location-specific that they are not transferable or overly burdensome to apply. For applied tools to be more widely employed, there often needs to be a shift in emphasis from accuracy to more general applicability and feasibility in terms of time, money, expertise and data availability. Practitioners have limited time and resources to learn new tools and to translate results to be relevant to their stakeholders; often, a less-precise but more easily-applied approach is sufficient for the types of decisions being made, especially when it will be used as part of a broader process of stakeholder engagement and adaptive management (Kline and Mazzotta 2012).

2 Non-monetary Benefit Indicators

Recognizing the limitations of monetary economic valuation, including uncertainties around value estimates but especially with regard to the expertise required and cost of primary studies, indicator-based studies offer an alternative (Boyd and Wainger 2002). Often study results are used primarily to stimulate public discussion or better direct further investigations (Thaler et al. 2014). In these cases, indicators may avoid some of the pitfalls and resistance that value estimates can elicit. Indicators can inform decisions when direct measures of economic value are unavailable, overly complex to estimate, or otherwise inadequate (Meadows 1998; Bossel 1999; Layke 2009). Desirable indicator variables have a strong relationship to the phenomena of interest yet are simple enough to be effectively monitored and/or modeled (Dale and Beyeler 2001).

Benefit indicators, in contrast to strictly biophysical indicators, provide information regarding the benefits and values of ecological changes to people. It is possible to use sound economic principles to formulate these indicators and capture the important aspects of value, while avoiding the burden of calculating dollar values (Mazzotta et al. 2019). Benefit-relevant indicators link biophysical outcomes to benefits for an identifiable group of people, in order to evaluate trade-offs and make decisions. An indicator is considered *relevant* when it captures something that directly alters beneficiaries' well-being, in units that are relevant to those beneficiaries (Olander et al. 2018).

The National Ecosystem Services Partnership (NESP) Guidebook presents an assessment framework that includes benefit-relevant indicators (NESP 2016). The guidebook suggests that conditions that influence values or preferences fall into five general categories (see also Wainger et al. 2001; Wainger et al. 2010):

1. Service Quality—quality of the service for its intended use
2. Capital and Labor—availability of capital and labor that complement ecological outputs in order to create goods and services
3. Number and characteristics of users
4. Reliability—reliability of the future stream of services
5. Scarcity and Substitutability—number of substitutes for the services provided

By incorporating these five conditions, benefit-relevant indicators complement ecological metrics with indicators more closely tied to what people value and prefer. Methods to measure benefit-relevant indicators can range from simple to complex and such metrics can be qualitative or quantitative. Though benefit-relevant indicators go a long way toward improving metrics, there is a great deal of flexibility and interpretation involved in defining the indicators for application. This level of flexibility makes benefit-relevant indicators broadly applicable; however, it also can result in inconsistency across applications, making them harder to compare or transfer across studies or locations with disparate contexts. While developing and applying benefit-relevant indicators is easier for decision makers than conducting economic valuation, it still requires economic or social science expertise.

3 Rapid Benefit Indicators (RBI) Approach

The Rapid Benefit Indicators (RBI) approach is an easy-to-use screening method for consistent and transparent site level assessment (Mazzotta et al. 2016, Mazzotta et al. 2019). The RBI approach is an extension of previous work on benefit indicators and parallels many benefit-relevant indicator concepts. The RBI approach uses a list of generic questions that capture important aspects of benefits and value to people, similar to the benefit-relevant indicators list of conditions. Practitioners answer each question to develop a set of non-monetary benefit indicators. Using a uniform indicator framework ensures greater consistency, while providing the latitude to select indicators best fitting the decision context. The RBI approach is intended to be used in conjunction with ecosystem service assessment approaches and tools to connect changes in ecosystems to changes in EGS and ultimately to benefits to people.

3.1 Five Step Rapid Benefit Indicator Process

The RBI indicators fit within a broader five-step process (Fig. 2) adapted from structured decision-making (Gregory et al. 2012) and in alignment with EBM (DeWitt et al. 2020). Each step in the process builds on the previous step. The **first step**, describing the decision context, includes identifying stakeholders and their objectives. Based on these objectives, the **second step** is to select the relevant ecosystem services and resulting benefits to assess, and to specify how those services and benefits are defined. Defining the ecosystem services and benefits is critical in order to be able to select appropriate indicators in step three. Identifying stakeholders, their values and how those values align with ecosystem services and benefits is essential and can be facilitated by FEGS approaches and tools (Sharpe et al. 2020).

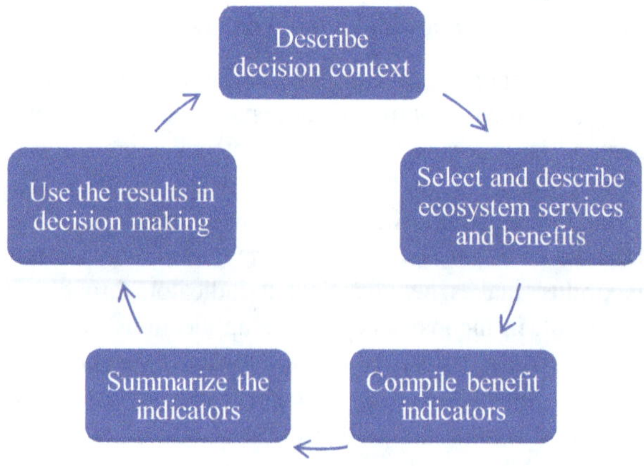

Fig. 2 The five-step process used to apply the RBI in decision-making (Mazzotta et al. 2019; adapted from Gregory et al. 2012)

The Rapid Benefit Indicators (RBI) approach Guidebook (Mazzotta et al. 2016) includes a case study where ecosystem services and benefits were selected from a list of those mentioned by resource managers during semi-structured interviews (Druschke and Hychka 2015). This list was further reduced to a set of ecosystem services that result in local benefits that are easily differentiated at a site scale. The RBI approach was designed to be applied at the site scale, which means that EGS that do not vary across sites do not need to be included in the assessment.

In **step three**, indicators are selected and compiled based on five questions, some with sub-questions. This is described in more detail below. In **step four**, compiled indicators are summarized to assist interpretation of results. **Step five** takes results summarized in step four and uses them in decision making.

Fostering adaptive management is a core part of EBM (Delacámara et al. 2020). To support adaptive management, the RBI decision process may be applied iteratively. After indicator results are used in decision making the practitioner may apply the same indicators over time to monitor both the results of actions and how changing conditions in the study area may suggest new priorities for action.

3.1.1 Rapid Benefit Indicator Questions (Step 3)

The five questions and their sub-questions are the core of the RBI approach. Some of these questions and sub-questions are optional, while others are required, as discussed further below.

Question 1: Can people benefit from an ecosystem service?

As an initial screening question, this helps determine whether people are able to benefit from an ecosystem service. It requires a site to currently meet, or to meet after restoration, three criteria:

1. It produces a final ecosystem good or service,
2. There are people who will benefit from the EGS, and
3. Complementary inputs required for benefits to reach people, if any, are available.

Sites that do not meet these criteria do not result in a benefit and require no further assessment for that EGS. A site fails the first criterion if it is not large enough or does not have high enough functioning to produce services of the required quantity or quality. For example, a site may be too fragmented to provide habitat for a bird species of interest to bird watchers. A site fails the second criterion if general demand for the service is lacking (see Russell et al. 2020 for similar requirements of Natural Capital Accounting). This is a more superficial precursor to question 2, assessing how many people benefit, and stops short of going into spatial relationships or quantification. For example, nearby flooding after recent storms is strong enough evidence of demand for flood-reduction services, without assessing where flooding occurred in relation to sites. For some ecosystem services, complementary inputs or conditions must be available for enjoyment of the service. A site fails the third criterion if such inputs or conditions are necessary but not available. Examples include infrastructure allowing physical access, or lack of institutional constraints such as regulatory harvest limits (NESP 2016; Olander et al. 2018).

Question 2: How many people benefit?

This question focuses on quantifying the pool of people who can benefit from an EGS at each site. In quantifying economic values, often the aggregate value of EGS is more sensitive to the number who benefit than the magnitude of individual values (Bateman et al. 2006). Thus, the number of beneficiaries provides an indication of the total benefits from an ecological change. The RBI uses a spatial approach to count beneficiaries, defining how services flow to people, defining the area where people may benefit, and quantifying the number of people who could benefit within the defined area. Services are generally produced in-situ and then either services flow to areas where people can access them, people travel to the site to access services, or both. RBI uses three general categories (Fig. 3) to characterize these spatial relationships (Fisher et al. 2009; Bagstad et al. 2013):

(a) Services are generated and must be enjoyed on site or within a geographic area (Fig. 3a), for example, through recreational uses of a site. To benefit, people must be located at or travel to that site or geographic area. When evaluating these services, the pool of beneficiaries depends on how far people will travel to the site.
(b) Services are generated on site and flow in all directions to a surrounding area (Fig. 3b), for example, birds or pollinators that use the site for habitat and move through nearby areas where people can benefit. People within the area where services flow will be able to benefit. When evaluating these services, it is important to consider how far services travel, whether those travel paths are blocked in any direction, and how those travel paths overlap with people who might benefit.
(c) Services are generated on site and flow in a single, or restricted, direction to a surrounding area (Fig. 3c), for example, downstream flood risk reduction from a wetland. People within the area where services flow will be able to benefit. This is true for services that flow downstream, such as water retention or purification. When evaluating these services, it is important to consider how far services

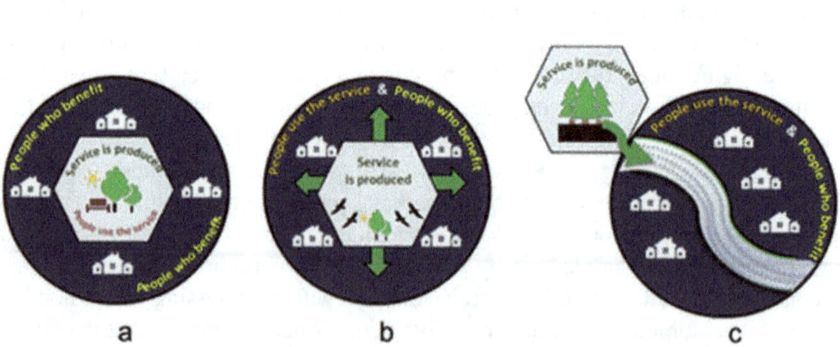

Fig. 3 The three categories RBI uses to characterize spatial relationships between where services are produced and where people access them (Mazzotta et al. 2019)

travel, whether anything impedes or assists that flow, in what direction, and how the travel path overlaps with people who might benefit.

Nonuse values for EGS that are important to people although they do not directly interact with them are a special case in which people do not have to be in any particular spatial relationship to the service, but simply need to be aware of the service and value it. These services can benefit people at varying distances without direct contact with people. For example, people may value the fact that a site provides habitat for a rare or unique species, although they do not see or interact with that species. The RBI approach is not tailored to the complexities of evaluating nonuse services, which are beyond the scope of this method. However, if a site is particularly rare or unique, or contains rare species, nonuse values may be significant and should be noted and evaluated using other approaches (Wainger et al. 2018; Richardson and Loomis 2009).

The spatial extent of relevant areas for assessment will vary based on local conditions and attributes of the ecosystem, landscape, and beneficiaries (Vajjhala et al. 2008). Distance decay may impact the service, causing service quantity or quality to decrease with distance from the source (Fisher et al. 2009; Bagstad et al. 2013). Services may also decrease in quantity or quality when they encounter "sinks," features on the landscape that absorb, degrade or deplete the service or the conduit transporting the service (Bagstad et al. 2013). Like services, benefits may experience distance decay, where people's values diminish with increasing distance from the area where services are accessed (Hanley et al. 2003; Bateman et al. 2006; Campbell et al. 2007; Campbell et al. 2008). One way to account for decreasing benefits with distance is to divide the number of beneficiaries into beneficiary pools based on distance bands and assigning lower weight to farther bands to help account for the decay in value. For example, when evaluating a recreational service, the number of beneficiaries within walking distance and driving distance may be estimated separately, with lower weight applied to those farther away.

Question 3: How much are people likely to benefit?

This question assesses the magnitude of benefits. How much people benefit from an ecosystem service is assessed by indicators answering four questions. Some of these questions may not be relevant to every ecosystem service, and those that are relevant may have one or more indicators. These questions are based on core concepts of economic theory of supply, demand, and value (Freeman et al. 2014; Nicholson and Snyder 2012). Economic theory posits that each of these factors, all else equal, will increase or decrease a person's value for a good or service through their effects on the demand function (the function that relates price or willingness to pay to quantity and quality of a good or service).

a. What is the quality of the service?

People benefit more from higher quality ecosystem services. Existing tools that measure the functioning of ecological systems or their ability to produce ecosystem services (e.g. Lewis et al. 2020; Culhane et al. 2020) may help inform assessment of the quality of ecosystem services provided by a site.

Fig. 4 In this illustration, the service areas shown in the blue and yellow circles represent recreational parks, and the location of individual beneficiaries is indicated by the house icons in each circle. The house within the green overlap area has access to two parks, where other homes depicted each can easily access only one park. Therefore, services provided by the parks are less scarce for the house in the green area and an additional park will have a lower incremental value for people in that location, than for those living in the other homes depicted, who have fewer substitutes (Yee et al. 2017)

b. Are there substitutes for the service or is the service scarce?

In general, people benefit less from each additional unit of a good or service they could receive. In other words, for a given beneficiary, the more units of a service that are available, the lower the value of more of that service or additional sources of that service (Yee et al. 2017; Fig. 4). Thus, fewer substitutes or greater scarcity lead to higher value, all else equal. The scarcity of ecosystem services already available to beneficiaries is informed by estimating the number of similar ecosystems and/or technological substitutes already providing those services. An increase in substitutes indicates lower value, so the direction of influence is reversed for this indicator as compared to the others.

c. What is the quality of complements to the EGS?

For some EGS, complements may be required for people to benefit, or may enhance benefits. For example, without a boat launch, people may not be able to access a waterway to benefit from recreational boating; and a higher quality boat launch will enhance benefits of boating. Thus, people benefit more when the quality of complementary inputs, other goods and services used with the ecosystem service, is higher. This is only important for services that are enhanced by complementary factors.

d. How strong are people's preferences?

People's strength of preferences influences their economic value (willingness to pay) for improvements in a good or service. In general, those with stronger preferences will benefit more from a given improvement in a good or service. For example, a more avid birder may have a higher value for improved bird habitat and resultant birding opportunities than someone who is less interested in birding. Characteristics of the beneficiaries that influence preferences are specific to the local context and the

EGS in question and may involve complex interactions among factors, and thus may be difficult to account for. In practice, this factor may often be omitted from assessment due to lack of data, or may be incorporated through more qualitative approaches such as opportunities for public comment.

Question 4: What are the social equity implications?

Determining social equity implications examines the population receiving the benefits and evaluates whether they are socio-economically disadvantaged. Benefits can be more important for vulnerable populations or people facing environmental concerns. These groups tend to have fewer resources to access ecosystem services yet may rely on them more (Norman et al. 2012). Similar to quantification of how many people benefit (Question 2), evaluating social equity implications involves characterizing the spatial relationship between where services are produced and where people access them. However, instead of quantifying the number of people who benefit, what is important for social equity is characterizing attributes of the people who benefit that make that population socio-economically disadvantaged.

Question 5: How reliable are benefits expected to be over time?

Determining the reliability of benefits over time explores the probability that some change will occur over time to inhibit the production of services or flow of benefits to people. When benefits are provided reliably over a longer period the total value of those benefits is greater. For example, benefits of a restored coastal marsh will diminish over time if the marsh becomes submerged due to sea-level rise. Features that impact the reliable delivery of services over time need to be site-specific but not necessarily benefit-specific.

3.1.2 Using Answers to Rapid Benefit Indicator Questions in Decision Making (Steps 4 & 5)

After developing a set of indicators and quantifying those indicators by answering the five generic questions, the next step, step four, is to summarize the indicator metrics. Summarizing metrics into a table can help when making comparisons across sites. These summary tables are analogous to the consequence tables used in structured decision making (see Gregory et al. 2012 for examples), where metrics evaluate the performance of each site. All tools developed to facilitate application of the RBI approach (described below) provide a summary table for this step. This summary table does not rank sites quantitatively or aggregate metrics.

The last step in the process, step five, is to evaluate the metrics in the summary table in light of the decision(s) to be made. There are many ways to weigh the different metrics and the resulting trade-offs when choosing among different actions for different sites. The indicator metrics may be used as is in disaggregated form as a basis for discussions, to inform participatory or consensus-type decisions; or they may be aggregated using Multi-Criteria Decision Analysis (MCDA) methods (Belton and Stewart 2002; Gregory et al. 2012). MCDA methods have successfully been used to aggregate RBI metrics (Martin and Mazzotta 2018a, b; Martin et al. 2018).

4 Tools for Applying the Rapid Benefit Indicator Approach

The RBI approach was originally developed for application to urban freshwater wetland restoration. The general approach and indicator framework will work with other types of environmental changes or within different ecological systems. Indicators for five benefits of urban freshwater wetlands (Fig. 5) have been previously developed and integrated into tools that help users more easily apply the Rapid Benefit Indicators approach.

The rest of this chapter focuses on three tools developed to assist in applying the RBI approach:

1. Checklist Tool
2. Spatial Analysis Toolset
3. National Catchment Dataset

These tools are available for download for those practitioners who may want to use them in their application, or use them as models that may be modified for other services and contexts. Each tool has unique aspects that make it more or less useful to specific audiences and applications (Table 1). For example, two of the tools produce a color-coded summary report for indicator results (Fig. 6).

Ecosystem Service	How people benefit
Flood water regulation	**Reduced Flood Risk:** The risks from floods to people and structures are reduced.
Scenic landscapes	**Scenic Views:** People can enjoy scenic views.
Learning opportunities	**Environmental Education:** People can benefit from studying nature or from enhanced connection to nature.
Recreational opportunities	**Recreation:** People can enjoy recreation
Birds	**Bird Watching:** People can watch or hear birds.

Fig. 5 The five ecosystem services and benefits that have previously developed indicators for freshwater wetlands restoration sites (Mazzotta et al. 2019)

Table 1 Differences between tools

Tool name	Software required	Custom benefits	Summary report	Guidance	Main advantage
Checklist tool	Excel/PDF	Y	Y	High	Document process
Spatial analysis toolset	ArcGIS	Limited	Y	Medium	Automated
National Catchment Dataset	GIS	N	N	Low	No data required

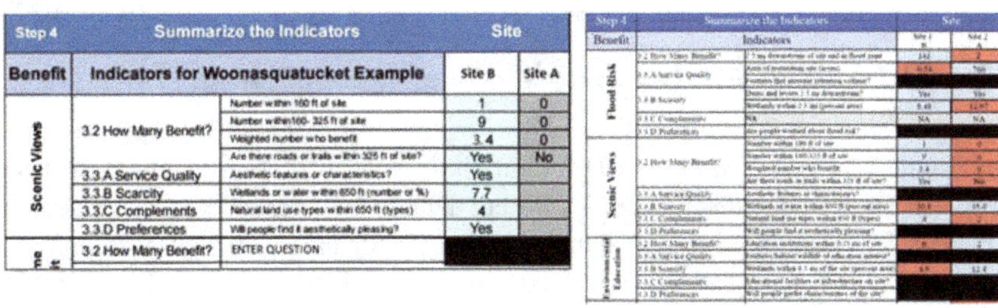

Fig. 6 Partial example color-coded summary reports produced by the Checklist Tool (Left) and Spatial Analysis Toolset (Right)

4.1 RBI Checklist Tool

The *Rapid Benefit Indicators (RBI) checklist tool*[1] is for recording results of manually-conducted analysis. Users can develop indicators using a variety of resources including paper maps, online maps, stakeholder engagement, or site visits. The checklist tool helps users go through the assessment process and provides space to document data sources, assumptions and other supplemental information.

The checklist tool is available in a pdf format as a low-tech solution that users can fill out by hand and print. The first two pages document the decision context and scope the ecosystem services and benefits being assessed. The next five pages of the pdf are each specific to one of the five ecosystem services and benefits that have previously developed indicators (Fig. 5). The last page is for summarizing results. In some cases, data entered in earlier pages of the pdf are automatically copied to their section in the summary table.

The macro-enabled Excel checklist tool has added functionality that walks the tool user through the step by step decision process. The Excel checklist tool uses responses entered in each form to dictate the next entry. For example, if a site fails to

[1]https://cfpub.epa.gov/si/si_public_record_Report.cfm?Lab=NHEERL&dirEntryId=331110.

meet criteria in Question 1, the tool skips forms for Questions 2–5 for that site. The Excel checklist has built-in forms for the five ecosystem services and benefits with previously developed indicators (Fig. 5). The Excel checklist can also document and assess newly developed indicators for different benefits, new ecological systems or different ecological changes using the five-question framework from the RBI approach. When site assessment is complete, the Excel checklist tool automatically summarizes indicators in a color-coded summary table (Fig. 6), where red indicates worse, blue indicates better, and gray indicates NA/neutral, relative to the average for all sites (for quantitative indicators) or based on characteristics increasing or decreasing benefits (for yes/no indicators). This summary does not rank sites.

4.2 RBI Spatial Analysis Tools

The *Rapid Benefit Indicators (RBI) spatial analysis tools*[2] compile indicators based on spatial data (Bousquin et al. 2017). Where datasets are available, this significantly expedites analysis. However, spatially-derived indicators are not adequate for answering all of the questions in the RBI approach. This tool is specific to the five ecosystem services and benefits with previously developed indicators (Fig. 5). This tool requires some familiarity with Geographic Information Systems (GIS). Results are summarized by site in table format, either spatially in the site dataset attribute table or as a pdf report (Fig. 6). Although the results are summarized spatially, the tool does not symbolize the results on a map or produce map products directly.

The *Rapid Benefit Indicators (RBI) spatial analysis tools* are an ArcGIS Python toolbox. This means users must have ESRI's desktop software, ArcMap® or ArcCatalog®, to open and use the tools (ESRI 2011). The toolbox does not require installation; users simply point ArcMap or ArcCatalog to it and then interact with it just like other toolboxes. Being written as a Python toolbox makes the tools more transparent and adaptable. Users who are familiar with the Python language can open the code and see all input handling and processes. This makes it easier to update inputs or processing to fit newly developed indicators.

The spatial analysis toolset includes seven individual tools (Table 2). The main tool, the *Full Indicator Assessment Tool*, runs a complete analysis of any or all of the five existing ecosystem services and their benefits. This is the fastest way to assess multiple ecosystem services and benefits. The additional tools in the toolset perform partial analysis. *Part* tools perform a variety of functions, including downloading data (e.g. *Flood Data Download Tool*), performing partial analysis using user updated parameters (e.g. *Social Equity of Benefits Tool* where the default buffer distance can be altered), or performing a specific part of the analysis that could be transferable to newly developed indicators (e.g. the *Presence/Absence to Yes/No Tool* can test for the presence of other spatial features near a site that could serve as indicators).

---

[2]https://cfpub.epa.gov/si/si_public_record_Report.cfm?Lab=NHEERL&dirEntryId=338471.

Table 2 Individual tools in the spatial analysis toolset

Tool name	Purpose
Full indicator assessment tool	Assess any/all of five benefit indicators
Benefit reliability tool	Assess benefit reliability (Question 5)
Flood data download tool	Download hydrologic data for flood risk indicators
Flood risk reduction tool	Assess reduced flood risk benefit indicators
Presence/absence to yes/no tool	Assess presence/absence for custom benefit indicators
Report generation tool	Summarize indicator results
Social equity of benefits tool	Assess benefit social equity implications (Question 4)

Table 3 National spatial datasets to use as default inputs

Data layer name	Source	Description
Restoration site polygons	EnviroAtlas[a]	Potentially restorable wetlands on agricultural land
Population raster	EnviroAtlas[1]	Dasymetric population
Address points	OpenStreetMap	OSM buildings[b]
Flood zone polygons	FEMA,[c] EnviroAtlas[1]	Flood zones Estimated Flood Plains
Dams/levees	USGS	National Hydrography Dataset
Wetland polygons	EnviroAtlas[1]	National Wetlands Inventory
Catchments	NHDPlusV2[d]	Catchments
Conservation lands	EnviroAtlas[1]	Protected Areas Database of the US (PADUS)
Landuse/Greenspace polygons	NLCD[e] NOAA[f]	National Land Cover Database (NLCD) Coastal change analysis program (C-CAP)
Trails	OpenStreetMap[g]	HikeBikeMap
Roads	TIGER/line,[h] OpenStreetMap[7]	Roads Roads
Educational institution points	DHS[i]	Homeland infrastructure foundation—Public schools, private schools
Bus stops	OpenStreetMap[7]	Buses
Social vulnerability	SVI,[j] EJScreen[k]	Social vulnerability index Demographic indicators

[a]https://enviroatlas.epa.gov/enviroatlas/interactivemap/
[b]https://osmbuildings.org/
[c]https://msc.fema.gov/portal/search
[d]http://www.horizon-systems.com/NHDPlus/NHDPlusV2_data.php
[e]https://www.mrlc.gov/data
[f]https://coast.noaa.gov/digitalcoast/data/ccapregional.html
[g]https://www.openstreetmap.org
[h]https://www.census.gov/geo/maps-data/data/tiger-line.html
[i]https://hifld-geoplatform.opendata.arcgis.com/
[j]https://coast.noaa.gov/digitalcoast/data/sovi.html
[k]https://ejscreen.epa.gov/mapper/

Although the way spatial analysis tools analyze data is predefined, there is flexibility in what the user can load as input datasets. Most of the inputs have nationally consistent datasets that can serve as a default (Table 3). This helps ensure the tools will not require data collection in most places. In many cases there are better alternatives to the national data. For example, national datasets may be incomplete (e.g. FEMA Flood zones; Bousquin and Hychka 2019), may lead to less precise results (e.g. EnviroAtlas raster population data when compared to address points; Bousquin et al. 2015), or may have higher resolution local datasets (e.g. statewide land use datasets like the one used in the Woonasquatucket watershed, RI case study; Martin et al. 2018).

4.3 RBI National Catchment Dataset

The *Rapid Benefit Indicators (RBI) catchment dataset*[3] allows a user to compare sites in different catchments based on a catchment characterization previously performed using a sub-set of indicator metrics (Bousquin and Hychka 2019). Rather than being a tool to help users analyze their data, this is a national dataset of results for two indicators of reduced flood risk benefits. The first indicator answers question two, *how many people benefit*. The second indicator answers the *is the service scarce* part of question three, *how much are people likely to benefit*. Using this dataset provides the least flexibility, as indicators and data inputs are predetermined.

Using this dataset requires some GIS experience, as comparing the data across restoration sites requires several steps. First, users can download these results as a comma separated values (csv) file for their region of interest. Next users will need to download the NHDPlusV2 catchment shapefile for their region. This catchment shapefile is the same one the *spatial analysis tools* use (Table 3). The csv file contains a "COMID" field to join the data to the NHDPlusV2 catchment shapefile's "FEATUREID" field. After joining the csv to the shapefile, GIS software (e.g. ArcGIS, QGIS, grass, R, etc.) can be used to visualize data by catchment. Catchment results can then be overlaid and compared to locations for restoration sites (Fig. 7). Comparisons can be visual, using site coordinates or imagery, or spatial, overlaying and summarizing indicators for a shapefile of sites.

Each catchment has 15 columns or fields. Most fields represent similar information but aggregated in different ways or determined based on different data. Each row presents values for a specific catchment. The dataset does not have the spatial resolution to differentiate sites within the same catchment.

For the first indicator, *how many people benefit*, beneficiaries are people in flood-prone areas downstream of a wetland restoration (Fig. 7). If restored, these people are in proximity to receive reduced flood-risk as a result. We recommend using values in either the "EA_pct_d" or "fld_pct_d" fields. These values represent the sum of the population in flood-prone areas in that catchment and catchments five km downstream. The difference between the two fields is in the data used to identify flood-prone areas. In the "EA_pct_d" field, an EnviroAtlas modeled layer defines flood-prone areas (Woznicki et al. 2019). In the "fld_pct_d" field, FEMA inland flood zones define flood-prone areas (FEMA 2018). Both estimates overlay the

[3]https://catalog.data.gov/dataset/nhdplusv2-catchment-rbi-data.

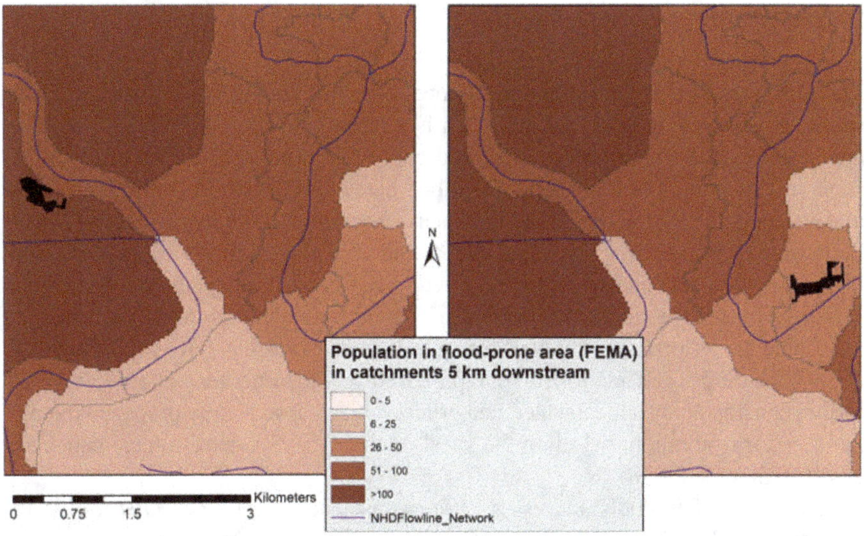

Fig. 7 Once catchments results from the csv file are joined to a spatial catchments dataset sites can be overlaid and compared based on characterizations for the catchment they fall within. This example shows a site (black) in a catchment with >100 people in FEMA flood-prone areas downstream of the site (Left) compared to a nearby site (black) in a catchment with 33 people in FEMA flood-prone areas downstream of the site (Right)

flood-prone areas with the same dasymetric 2010 population data and use the same methods to combine estimates from downstream catchments. Where there are more people in flood-prone areas downstream there is more demand for increased flood-reduction benefits. Thus, it is a maximizing criterion, and a higher value makes a catchment higher priority for restoration.

For the second indicator, *is the service scarce*, existing wetlands may already be providing flood-risk reduction services for the same beneficiaries. We recommend using values in the field "wet_pct", which represent the percent of that catchment that is currently wetlands. In catchments with abundant wetlands already providing flood-reduction services, added units of this service have less value than in catchments where wetlands providing flood-reduction services are scarce. Thus, it is a minimizing criterion, and a lower value makes a catchment higher priority for restoration.

The two indicator field values may need manipulation to inform decision making. Regions have physiographic differences that impact wetland suitability, flood-risk, and population density. As a result, the distribution of values within different regions is diverse. For example, a 10% difference in wetlands between catchments in west Texas is drastic but would be minor in coastal Minnesota. To account for this, Bousquin and Hychka (2019) suggest binning the indicators into discrete categories based on the distribution of regional values, and show a method using four quartiles, dividing catchment values into four categories of equal number. Different discretizations are better suited depending on thresholds, decision context and overarching objectives. Many GIS applications aid users in choosing discrete categories when visualizing data.

5 Summary

The RBI approach helps formalize a process for developing benefit indicators for site-level assessment. The RBI approach is a rapid screening assessment, meant to reduce the burden in expertise, data, and cost placed on practitioners using socio-economic metrics. The three tools described here were developed to help decision-makers apply the RBI approach indicators in a consistent and transparent way. These tools were developed for a subset of EGS and applied within the context of urban freshwater wetlands and could be modified or reproduced for other ecosystems and services. In a case-study application of the spatial analysis tool, we screened 65 candidate restoration sites within a watershed, using the RBI combined with MCDA methods, and identified four preferred restoration sites. These sites were further considered by a local watershed group, which subsequently proposed three of the sites for restoration, based on the sites' potential EGS values to local beneficiaries (Martin et al. 2018).

Each of the RBI tools addresses different user needs. The *checklist tool* is for recording results of manually-conducted assessments or for developing new indicators for other types of environmental changes or within different ecological systems. The *spatial analysis tool* is for users with some GIS expertise who have geospatial data to apply the existing rapid benefit indicators for urban freshwater wetland restoration sites. The *NHDPlus national catchment dataset* is for users only interested in assessing flood-protection benefits. Users of this dataset must be willing to accept reduced flexibility in exchange for not having to perform analysis on multiple input data sets.

We have presented a framework for consistently developing a set of benefit indicators, along with a set of tools for applying this framework. Alongside ecosystem service metrics, benefit indicators help ecological and social interactions be better considered within an Ecosystem-Based Management framework. When stakeholder values are used to select ecosystem services to assess and are reflected in benefits assessments selection trade-offs are better informed and social-ecological systems are better connected in Ecosystem-Based management decision-making.

Disclaimer This chapter has been subjected to Agency review and has been approved for publication. The views expressed in this paper are those of the author(s) and do not necessarily reflect the views or policies of the U.S. Environmental Protection Agency.

References

Bagstad, K., Johnson, G., Voigt, B., & Villa, F. (2013). Spatial dynamics of ecosystem service flows: A comprehensive approach to quantifying actual services. *Ecosystem Services, 4*, 117–125. https://doi.org/10.1016/j.ecoser.2012.07.012.

Bateman, I., & Willis, K. (1999). *Valuing environmental preferences: Theory and practice of contingent valuation method*. Oxford: Oxford University Press on Demand.

Bateman, I., Day, B., Georgiou, S., & Lake, I. (2006). The aggregation of environmental benefit values: Welfare measures, distance decay and total WTP. *Ecological Economics, 60*, 450–460. https://doi.org/10.1016/j.ecolecon.2006.04.003.

Belton, V., & Stewart, T. (2002). *Multiple criteria decision analysis: An integrated approach* (372 p). Boston, MA: Kluwer Academic Publishers.

Bossel, H. (1999). *Indicators for sustainable development: Theory, methods, applications. A report to the Balaton group* (123 p). Winnipeg, MB: International Institute for Sustainable Development.

Bousquin, J., & Hychka, K. (2019). A geospatial assessment of flood vulnerability reduction by freshwater wetlands – a benefit indicators approach. *Frontiers in Environmental Sciences, 54*, 1–14.

Bousquin, J., Hychka, K., & Mazzotta, M. (2015). *Benefit indicators for flood regulation services of wetlands: A modeling approach.* Narragansett, RI: US Environmental Protection Agency, Office of Research and Development, National Health and Environmental Effects Research Laboratory, Atlantic Ecology Division. EPA/600/R-15/191.

Bousquin, J., Mazzotta, M., & Berry, W. (2017). *Rapid benefit indicator (RBI) spatial analysis toolset and manual.* Gulf Breeze, FL: US EPA Office of Research and Development, National Health and Environmental Effects Research Laboratory, Gulf Ecology Division. https://www.epa.gov/water-research/rapid-benefit-indicators-rbi-approach.

Boyd, J., & Banzhaf, S. (2007). What are ecosystem services? The need for standardized environmental accounting units. *Ecological Economics, 63*(2–3), 616–626. https://doi.org/10.1016/j.ecolecon.2007.01.002.

Boyd, J., & Wainger, L. (2002). Landscape indicators of ecosystem service benefits. *American Journal of Agricultural Economics, 84*(5), 1371–1378.

Boyd, J., Ringold, P., Krupnick, A., Johnson, R., Weber, M., & Hall, K. (2016). Ecosystem services indicators: Improving the linkage between biophysical and economic analyses. *International Review of Environment and Resource Economics, 8*, 359–443. https://doi.org/10.2139/ssrn.2662053.

Campbell, D., Hutchinson, W., & Scarpa, R. (2007). Using choice experiments to explore the spatial distribution of willingness to pay for rural landscape improvements. *Environmental Planning A, 41*(1), 97–111. https://doi.org/10.1068/a4038.

Campbell, D., Scarpa, R., & Hutchinson, W. (2008). Assessing the spatial dependence of welfare estimates obtained from discrete choice experiments. *Letters in Spatial Resource Sciences, 1*, 117–126. https://doi.org/10.1007/s12076-008-0012-6.

Champ, P. A., Boyle, K. J., & Brown, T. C. (Eds.). (2017). *A primer on nonmarket valuation* (2nd ed.). Dordrecht: Springer.

Culhane, F. E., Robinson, L. A., & Lillebø, A. I. (2020). Approaches for estimating the supply of ecosystem services: Concepts for ecosystem-based management in coastal and marine environments. In T. O'Higgins, M. Lago, & T. H. DeWitt (Eds.) *Ecosystem-based management, ecosystem services and aquatic biodiversity: Theory, tools and applications* (pp. 105–126). Amsterdam: Springer.

Dale, V., & Beyeler, S. (2001). Challenges in the development and use of ecological indicators. *Ecological Indicators, 1*, 3–10. https://doi.org/10.1016/S1470-160X(01)00003-6.

Delacámara, G., O'Higgins, T., Lago, M., & Langhans, S. (2020). Ecosystem-based management: moving from concept to practice. In T. O'Higgins, M. Lago, & T. H. DeWitt (Eds.), *Ecosystem-based management, ecosystem services and aquatic biodiversity: Theory, tools and applications* (pp. 39–60). Amsterdam: Springer.

DeWitt, T. H., Berry, W. J., Canfield, T. J., Fulford, R. S., Harwell, M. C., Hoffman, J. C., Johnston, J. M., Newcomer-Johnson, T. A., Ringold, P. L., Russel, M. J., Sharpe, L. A., & Yee, S. J. H. (2020). The final ecosystem goods and services (FEGS) approach: A beneficiary-centric method to support ecosystem-based management. In T. O'Higgins, M. Lago, & T. H. DeWitt (Eds.), *Ecosystem-based management, ecosystem services and aquatic biodiversity: Theory, tools and applications* (pp. 127–148). Amsterdam: Springer.

Druschke, C. G., & Hychka, K. C. (2015). Manager perspectives on communication and public engagement in ecological restoration project success. *Ecology and Society, 20*(1), 58.

ESRI. (2011). *ArcGIS Desktop: Release 10*. Redlands, CA: Environmental Systems Research Institute.

FEMA. (2018). FEMA flood map service center portal. Website and web map. Retrieved from https://msc.fema.gov/portal/search.

Fisher, B., Turner, R., & Morling, P. (2009). Defining and classifying ecosystem services for decision making. *Ecological Economics, 68*(3), 643–653. https://doi.org/10.1016/j.ecolecon.2008.09.014.

Freeman, A., Herriges, J., & Kling, C. (2014). *The measurement of environmental and resource values: Theory and methods* (3rd ed.). New York: RFF Press.

Gregory, R., Failing, L., Harstone, M., Long, G., McDaniels, T., & Ohlson, D. (2012). *Structured decision making: A practical guide to environmental management choices* (312 p). Chichester, UK: Wiley-Blackwell.

Hanley, N., Schläpfer, F., & Spurgeon, J. (2003). Aggregating the benefits of environmental improvements: Distance-decay functions for use and non-use values. *Journal of Environmental Management, 68*, 297–304. https://doi.org/10.1016/S0301-4797(03)00084-7.

Heal, G. (2000a). Valuing ecosystem services. *Ecosystems, 3*(1), 24–30. https://doi.org/10.1007/s100210000006.

Heal, G. (2000b). *Nature and the marketplace: Capturing the value of ecosystem services.* Washington, DC: Island Press.

Johnston, R., Schultz, E., Segerson, K., Besedin, E., & Ramachandran, M. (2012). Enhancing the content validity of stated preference valuation: The structure and function of ecological indicators. *Land Economics, 88*(1), 102–120. https://doi.org/10.3368/le.88.1.102.

Johnston, R., Rolfe, J., Rosenberger, R., & Brouwer, R. (2015). *Benefit transfer of environmental and resource values* (Vol. 14). New York: Springer.

Kline, J., & Mazzotta, M. (2012). *Evaluating tradeoffs among ecosystem services in the management of public lands.* Gen. Tech. Rep. PNW-GTR-865. Portland, OR: US Department of Agriculture, Forest Service, Pacific Northwest Research Station. 48 p, 865.

Layke, C. (2009). *Measuring nature's benefits: A preliminary roadmap for improving ecosystem service indicators* (36 p). Washington, DC: World Resources Institute.

Lewis, N. S., Marois, D. E., Littles, C. J., & Fulford, R. S. (2020). Projecting changes to coastal and estuarine ecosystem goods and services- models and tools. In T. O'Higgins, M. Lago, & T. H. DeWitt (Eds.). *Ecosystem-based management, ecosystem services and aquatic biodiversity: Theory, tools and applications* (pp. 235–254). Amsterdam: Springer.

Martin, D., & Mazzotta, M. (2018a). Non-monetary valuation using multi-criteria decision analysis: Sensitivity of additive aggregation methods to scaling and compensation assumptions. *Ecosystem Services, 29*, 13–22. https://doi.org/10.1016/j.ecoser.2017.10.022.

Martin, D. M., & Mazzotta, M. (2018b). Non-monetary valuation using multi-criteria decision analysis: Using a strength-of-evidence approach to inform choices among alternatives. *Ecosystem Services, 33*, 124–133. https://doi.org/10.1016/j.ecoser.2018.06.001.

Martin, D., Mazzotta, M., & Bousquin, J. (2018). Combining ecosystem services assessment with structured decision making to support ecological restoration planning. *Environ Manag, 62*(3), 608–618. https://doi.org/10.1007/s00267-018-1038-1.

Mazzotta, M., Bousquin, J., Ojo, C., Hychka, K., Gottschalk Druschke, C., Berry, W., & McKinney, R. (2016). *Assessing the benefits of wetland restoration: A rapid benefit indicators approach for decision makers.* Narragansett, RI: USEPA, Office of Research and Development. EPA/600/R-16/084.

Mazzotta, M., Bousquin, J., Berry, W., Ojo, C., McKinney, R., Hyckha, K., & Druschke, C. G. (2019). Evaluating the ecosystem services and benefits of wetland restoration by use of the rapid benefit indicators approach. *Integrated Environmental Assessment and Management, 15*(1), 148–159. https://doi.org/10.1002/ieam.4101.

Meadows, D. (1998). *Indicators and information for sustainable development* (78 p). Hartland Four Corners, VT: The Sustainablity Institute.

National Ecosystem Services Partnership. (2016). *Federal Resource Management and ecosystem services guidebook* (2nd ed.). Durham: National Ecosystem Services Partnership, Duke University. https://nespguidebook.com.

National Research Council. (2005). *Valuing ecosystem services: Toward better environmental decision-making*. Washington, DC: National Academies Press.

Nicholson, W., & Snyder, C. (2012). *Microeconomic theory: Basic principles and extensions.* Nelson Education.

Norman, L., Villarreal, M., Lara-Valencia, F., Yuan, Y., Nie, W., Wilson, S., Amaya, G., & Sleeter, R. (2012). Mapping socio-environmentally vulnerable populations access and exposure to ecosystem services at the US–Mexico borderlands. *Applied Geography, 34*, 413–424. https://doi.org/10.1016/j.apgeog.2012.01.006.

Ojo, C., Mulvaney, K., Mazzotta, M., & Berry, W. (2018). A marketing plan for scientists: Building effective products and connecting with stakeholders in meaningful ways. *Solutions, 9*(2), 1.

Olander, L., Polasky, S., Kagan, J. S., Johnston, R. J., Wainger, L., Saah, D., Maguire, L., Boyd, J., & Yoskowitz, D. (2017). So you want your research to be relevant? Building the bridge between ecosystem services research and practice. *Ecosystem Services, 26*, 170–182. https://doi.org/10.1016/j.ecoser.2017.06.003.

Olander, L. P., Johnston, R. J., Tallis, H., Kagan, J., Maguire, L. A., Polasky, S., Urban, D., Boyd, J., Wainger, L., & Palmer, M. (2018). Benefit relevant indicators: Ecosystem services measures that link ecological and social outcomes. *Ecological Indicators, 85*, 1262–1272. https://doi.org/10.1016/j.ecolind.2017.12.001.

Parsons, G. (2017). The travel cost model. In P. A. Champ, K. J. Boyle, & T. C. Brown (Eds.), *A primer on non-market valuation* (2nd ed., pp. 187–234). Dordrecht: Springer.

Piet, G., Delacámara, G., Gómez, C. M., Lago, M., Rouillard, J., Martin, R., & van Duinen, R. (2017). Making ecosystem-based management operational. Deliverable 8.1, European Union's Horizon 2020 Framework Programme for Research and Innovation grant agreement No. 642317.

Richardson, L., & Loomis, J. (2009). The total economic value of threatened, endangered and rare species: An updated meta-analysis. *Ecological Economics, 68*, 1535–1548.

Ringold, P., Boyd, J., Landers, D., & Weber, M. (2013). What data should we collect? A framework for identifying indicators of ecosystem contributions to human well-being. *Front Ecol Environ, 11*(2), 98–105. https://doi.org/10.1890/110156.

Russell, M. J., Rhodes, C., Sinha, R. K., Van Houtven, G., Warnell, G., & Harwell, M. C. (2020). Ecosystem-based management and natural capital accounting. In T. O'Higgins, M. Lago, & T. H. DeWitt (Eds.), *Ecosystem-based management, ecosystem services and aquatic biodiversity: Theory, tools and applications* (pp. 149–164). Amsterdam: Springer.

Schultz, E., Johnston, R., Segerson, K., & Besedin, E. (2012). Integrating ecology and economics for restoration: Using ecological indicators in valuation of ecosystem services. *Restoration Ecology, 20*(3), 304–310. https://doi.org/10.1111/j.1526-100X.2011.00854.x.

Sharpe, L., Hernandez, C., & Jackson, C. (2020). Prioritizing stakeholders, beneficiaries and environmental attributes: A tool for ecosystem-based management. In T. O'Higgins, M. Lago, & T. H. DeWitt (Eds.), *Ecosystem-based management, ecosystem services and aquatic biodiversity: Theory, tools and applications* (pp. 189–212). Amsterdam: Springer.

Spash, C. L., & Vatn, A. (2006). Transferring environmental value estimates: Issues and alternatives. *Ecological Economics, 60*(2), 379–388. https://doi.org/10.1016/j.ecolecon.2006.06.010.

Tashier, A., & Ringold, R. (2019). A critical review of available ecosystem services data according to the final ecosystem goods and services framework. *Ecosphere, 10*, 3. https://doi.org/10.1002/ecs2.2665.

Thaler, T., Boteler, B., Dworak, T., Holen, S., & Lago, M. (2014). Investigating the use of environmental benefits in the policy decision process: A qualitative study focusing on the EU water policy. *Journal of Environmental Planning and Management, 57*(10), 1515–1530.

Vajjhala, S., John, A., & Evans, D. (2008). *Determining the extent of market and extent of resource for stated preference survey design using mapping methods* (43 p). Washington, DC: Resources for the Future.

Wainger, L., & Mazzotta, M. (2011). Realizing the potential of ecosystem services: A framework for relating ecological changes to economic benefits. *Environmental Management, 48*, 710. https://doi.org/10.1007/s00267-011-9726-0.

Wainger, L., King, D., Salzman, J., & Boyd, J. (2001). Wetland value indicators for scoring mitigation trades. *Stanford Environment Law Journal, 20*(2), 413–477.

Wainger, L., King, D., Mack, R., Price, E., & Maslin, T. (2010). Can the concept of ecosystem services be practically applied to improve natural resource management decisions? *Ecol Econ, 69*, 978–987. https://doi.org/10.1016/j.ecolecon.2009.12.011.

Wainger, L., Helcoski, R., Farge, K., Espinola, B., & Green, G. (2018). Evidence of a shared value for nature. *Ecological Economy, 154*, 107–116. https://doi.org/10.1016/j.ecolecon.2018.07.025.

Woznicki, S., Baynes, J., Panlasigui, S., Mehaffey, M., & Neale, A. (2019). Development of a spatially complete floodplain map of the conterminous United States using random forest. *Science of the Total Environment, 647*, 942–953. Accessed from EnviroAtlas: ftp://newftp.epa.gov/epadatacommons/ORD/EnviroAtlas/Estimated_floodplain_CONUS.zip.

Yee, S., Bousquin, J., Bruins, R., Canfield, T., DeWitt, T., de Jesús Crespo, R., Dyson, B., Fulford, R., Harwell, M., Hoffman, J., & Littles, C. (2017). *Practical strategies for integrating final ecosystem goods and services into community decision-making*. Washington, DC: US Environmental Protection Agency, Office of Research and Development.

4

Models and Mapping Tools to Inform Resilience Planning after Disasters

Mark Myer and John M. Johnston

Abstract In the aftermath of Superstorm Sandy in 2012, recovery and rebuilding efforts focused on resilience and diversified infrastructure that included consideration of the benefits that healthy ecosystems provide. County governments on Long Island identified a need for tools to map coastal and estuarine areas that may provide ecosystem services. Current methods of ecosystem service mapping often rely on complicated statistical models, labor-intensive site validation, or proprietary data. We examined a method of fast ecosystem services mapping that relies on publicly-available data, includes stakeholder input, and uses ArcGIS software that is ubiquitous in municipal planning. This chapter provides an example of ecosystem service mapping that generates easily explained visualizations suitable for non-scientific audiences with tools already available to municipal planning departments. We explain how to define indicators of benefit presence, obtain data, and create maps using examples from a collaboration with Nassau County, Long Island, New York.

Lessons Learned

- Some situations do not require an effort-intensive modeling approach to ecosystem services mapping—for these, a quick estimate serves the purpose.
- Consulting with stakeholders at every step of the process is essential. As researchers, our ideas of which ecosystems and benefits are important may not match theirs.
- It is possible to map the areas that may provide ecosystem services using publicly available data with a combination of expert consultation and inductive reasoning.

M. Myer
City of New Orleans Mosquito Termite and Rodent Control Board, New Orleans, LA, USA

J. M. Johnston (✉)
US EPA Office Research and Development, Center for Environmental Measurement and Modeling, Athens, GA, USA
e-mail: johnston.johnm@epa.gov

Needs to Advance EBM

- Despite several contemporaneous efforts, there is no agreed-upon standard to rigorously define ecosystem services. This subjectivity introduces uncertainty into service mapping.
- Local partners may not immediately perceive utility in ecosystem-based management, underscoring the need for communication and outreach that enumerates its advantages over established paradigms.

1 Hurricane Sandy and the South Shore of Nassau County

Superstorm Sandy made landfall in New Jersey and New York on October 29, 2012, leaving behind a wake of destruction along the U.S. eastern seaboard. The storm caused billions of dollars in damage, killed at least 147 people in the United States (Blake et al. 2013) and dozens more overseas, and it left millions temporarily without electricity and fuel (Diakakis et al. 2015). In Nassau County, Long Island, - New York, voluntary evacuations were announced for the south shore's storm surge area in anticipation of extensive damage. On impact, Sandy's storm surge was nearly 14 ft above mean low tide, causing inundation of coastal areas and shoreline changes from erosion and accretion of sand and sediment (Hapke et al. 2013). Disruption to south shore bays from an influx of salt water and sediment was widespread, including a reduction in eelgrass (*Zostera marina*), which serves as a crucial habitat for local shellfish (Tinoco 2017). In the aftermath of the storm, New York City and the surrounding communities committed to rebuilding and adding infrastructure in ways that increased resilience to future natural disasters, with consideration of green infrastructure methods of reducing stormflow, increasing infiltration, and reducing nutrient runoff to improve ecosystem services (Interboro Team 2014; The City of New York 2013). As part of the Federal response to the disaster, the U.S. Environmental Protection Agency's Office of Research and Development and Region 2 (including New York, New Jersey and Puerto Rico) worked with the Federal Emergency Management Agency, New York Department of State, county policy makers, and others to identify projects that diversify the built and green infrastructure portfolio while increasing community resilience to natural disasters.

Hempstead Bay is in the western part of Long Island's south shore embayment, extending approximately from Far Rockaway in Queens to Massapequa in southeastern Nassau County (Fig. 1). Long Beach forms the barrier between the Bay and the Atlantic Ocean on the cityward side, with Jones Beach continuing the chain of barrier islands to the east. The area supports a diversity of ecosystems, including freshwater streams, brackish streams, tidal wetlands and marshes, seagrass and open bay, and barrier islands. The Bay is integral to the lifestyle enjoyed by residents, both by providing ecosystem services directly and indirectly supporting others. Shellfishing for clams, oysters, and scallops is both a recreational activity and part of the local economy, with bay scallops alone contributing millions of dollars per year (Peconic Estuary Program 2015). Sailing and recreational boating are enjoyed

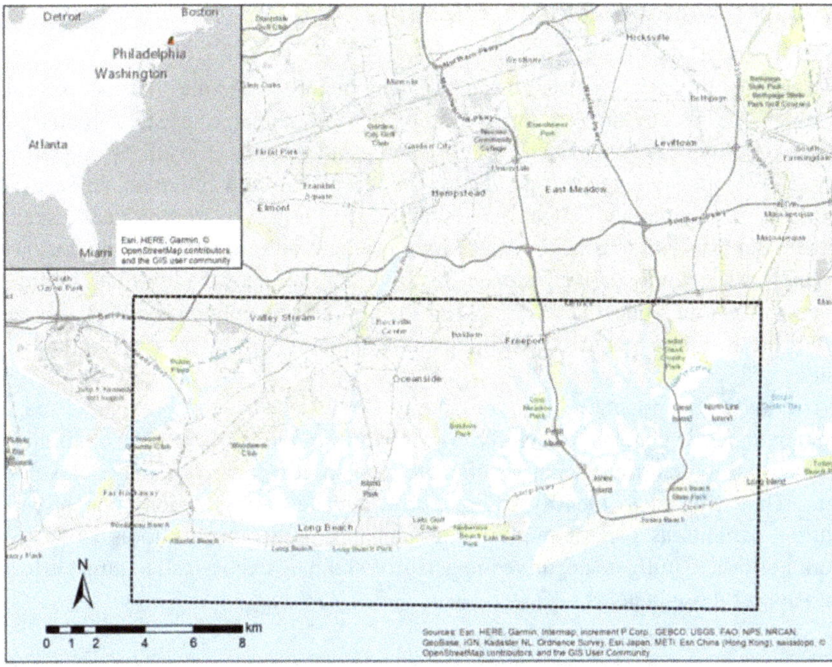

Fig. 1 Study Area: South shore of Nassau County in Long Island, NY. Hempstead Bay is located between the south shore and Long Beach, in the dotted box

in the warmer seasons. Several rare and endangered birds including the piping plover (*Charadrius melodus*) and yellow-crowned night heron (*Nyctanassa violacea*) reside in the area, which makes it a destination for birdwatchers (Cohen et al. 2006; Hodgman et al. 2015). Aquatic vegetation, both emergent and submerged, attenuates wave energy and decreases inundation from storm surge (Paul et al. 2012). The Bay ecosystem is recovering but faces pressures from development and pollution. The south shore experienced decades of stressors from development activities resulting in loss of submerged aquatic vegetation and coastal wetlands and increased nutrient loadings that impair coastal ecosystems (Hartig et al. 2002). Local commercial shellfish landings are far from historic highs because of habitat loss. Though storms are a normal dynamic of coastal ecosystems, areas stressed by human activity have less resilience, limiting their ability to recover from disturbance (Carpenter et al. 2001). The maps and methods described here were provided to the county planning office to help communicate the presence of and potential for enhancement of ecosystem services. Increasing awareness of local ecosystems and their services, especially through the use of intuitive maps, was a first step in building understanding of the relationships between wetland vegetation and coastal resilience and fisheries production.

2 Background on Nature's Benefits

We consider the term "nature's benefits"—the benefits of nature that people care about—to be synonymous with ecosystem services. Final ecosystem goods and services are provided or created by ecosystems and directly enjoyed or utilized by people. We acknowledge the distinction between final and intermediate services (Lamothe and Sutherland 2018; DeWitt et al. 2020); however, we don't address this further. Our goal was to help Nassau County visualize and communicate the location of benefits that people derive from the Bay, therefore we used the informal, intuitive concept of nature's benefits. We avoid confusing the concept by clarifying that any part of the Bay that is utilized or enjoyed and was not built by people is a nature's benefit.

The local government of Nassau County (Fig. 1) managers were interested in mapping nature's benefits to identify priority areas of the Bay that provide multiple services and to determine what benefits are located near areas of planned development. They wanted to identify, characterize, and describe locations that provide nature's benefits as part of public communication and outreach tools as part of overall efforts to mitigate negative impacts from human activity and to target efforts that support those benefits.

3 Geographic Information Systems—Utility of Arc ModelBuilder

The desired product for Nassau County was an illustrative map, suitable for inclusion in a handout or a poster for public-facing communication, that could be quickly produced without a long, data-gathering process. Inspired by the "service-providing area" maps of Angradi et al. (2016) used to visually characterize geographical areas that provide targeted ecosystem services, these are intended to quickly convey where benefits are likely to occur. Although they are simplifications of the natural world, a major strength is that they may not require fieldwork or an expensive monitoring program to generate. Nature's benefit maps outline the geographical areas that have the potential to provide ecosystem services, conveying those benefits through the use of a few, easily described indicators that outline the boundaries/presence of a given benefit. We considered cost, simplicity, and end-user ease-of-use as drivers of the overall design goals, and we targeted county and municipal planners (and their communications staff) as the intended users.

To meet these design goals, we used the ModelBuilder semi-automated map-building feature available in ESRI ArcGIS, the most popular and widely used Geographic Information System (GIS). GIS is used to work with spatial data and create maps and is considered an essential tool in municipal planning, environmental science, ecology, environmental economics, and many other disciplines. ModelBuilder is a visual representation of GIS operations in ArcGIS as a directed

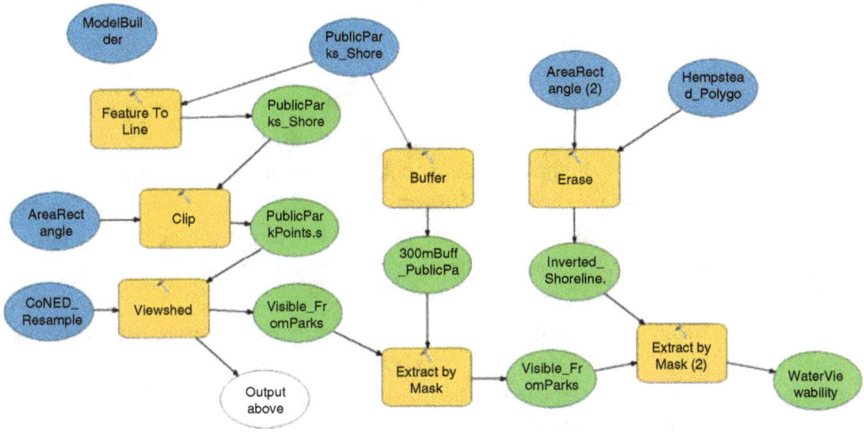

Fig. 2 A ModelBuilder directed graph (i.e., a GIS workflow) used to create a map of areas of the shore suitable for recreational viewing of aquatic animals. A blue oval is a map file, a yellow box is a GIS operation, and a green or clear oval is an output file

graph (Fig. 2). Shapes and colors are used to represent a map file, a table of input data, a mapping operation, or an output file. Following the arrows shows the GIS workflow as steps from input to output. For example, in Fig. 2, the input file PublicParks_Shore is a shapefile that contains areas of public parkland adjacent to the ocean and is represented as a blue oval. Following the two arrows from the PublicParks_Shore input, the map operations performed on the shapefile are *Feature To* Line, which turns the shape into a line, and *Buffer*, which outputs a polygon of the area a given distance from its input. Map operations are represented as orange rectangles. The output is represented as a green oval that can be used in other map operations, creating a continuous workflow. A ModelBuilder workflow can be saved as a portable file and shared for reuse and editing by other users with ArcGIS software. Benefits maps can be delivered with ModelBuilder files as a compressed archive with all required input data in a single package.

4 Steps to Generate a Nature's Benefits Map

Generating a nature's benefit map begins with local knowledge to determine which benefits are present and amenable to mapping. It is important to consult with residents, representatives of businesses that are associated with the local landscape (farming, fishing, or ecotourism for example), political representatives, and researchers including ecologists, hydrologists, and geologists. Consulting as many experts as possible strengthens the impact of the product and fosters inclusion and ownership. Soliciting the values and preferences of the various stakeholders also helps rank the priority order of the potential list of benefits (Sharpe et al. 2020). Meetings, surveys, or personal correspondence can all be used to determine which benefits are most important to a community. It may be necessary to brainstorm a preliminary list of benefits to jumpstart conversations, especially among audiences that have not been exposed to ecosystem services concepts. It is also possible to highlight the benefits of nature that may be overlooked or underappreciated by including these in a preliminary list for stakeholder consultation.

Amenability to benefits mapping means that a particular benefit should be present in a fixed location and there should be information available to determine indicators of its presence. Reliable indicator data must be available to determine the location of benefits so that GIS operations can be performed to identify their location across the county (i.e., area of interest). Benefits must occur in a discrete location that can be visualized on a map to be considered amenable. If a benefit rarely exists or occurs almost everywhere, it is unlikely to be a good choice for nature's benefits mapping as it may not help inform differences between management alternatives for a particular decision. For example, reduction of bay nitrogen pollution is a benefit related to the presence of denitrifying bacteria (Christensen et al. 1987). However, microscopic bacteria can potentially occur everywhere in the Bay, and their location and abundance are also in flux, so it is difficult or impossible to map this benefit.

5 Indicator Selection

Because many benefits cannot be directly observed or quantified, we use indicators to estimate their presence. In our case indicators are environmental (i.e., habitat) characteristics known to occur with the presence of a plant or animal species. We define an indicator as a mappable (fixed, measurable) quantity that spatially co-occurs with the benefit. As such, nature's benefits maps indicate where a benefit may be present but is not guaranteed. In other words, the presence of the indicator is necessary but not sufficient to ensure the presence of the benefit. Reliable indicators are crucial because a map based on faulty assumptions will be misleading or incorrect. The guiding question to ask is "what are the one or two characteristics that are almost always present when this benefit is provided?"

To illustrate indicator selection, we use the examples of hard clam gathering and shore fishing. For the hard clam example, the benefit is those clams that are harvestable, so metrics are needed to convert this benefit into rules for mapping the extent of this benefit. Because clam collectors can only reach so far underwater, even if they are using a specialized tool, one indicator will be water depth less than 2 m (about 6 ft) at mean tide, which will represent the area where a collector could reasonably reach the bottom of the Bay at low tide (Wells 1957). The second indicator represents areas where hard clams are likely to live. We researched the ecology of the hard clam and found they tend to live in areas with a sand or mud bottom (Wells 1957; Walker and Tenore 1984). Therefore, our indicators for hard clam collecting are areas of the Bay with a sand or mud bottom in two meters or less depth at mean tide, because that is where hard clams that people can reach are most likely to occur. It's important to note that only clams that people can harvest are considered a nature's benefit: if they can't be collected, they aren't considered a benefit. Because of this, most nature's benefit maps include at least one accessibility indicator.

Similarly, for shore-based fishing, indicators should reflect the potential of the public to catch fish while standing on a shore. The first indicator will be those areas identified as shoreline; the area must be adjacent to water, because an angler needs to be on the shore to catch fish. Our second indicator for shore-based fishing is public accessibility, for example state or local parks. Finally, an angler can reasonably cast a line about 30 m (around 90 ft) at the most, so benefits locations will be within 30 m of shoreline. Combining these indicators results in areas that provide the benefit of shore-based fishing.

6 Mapping Indicators

Once indicators that provide the mapping boundaries of a given nature's benefit are chosen, the next step is to find spatial data to represent the indicator, download it preferably from a publicly available source, and load it into a GIS platform. Indicators can be in almost any geodata format, from elevation and land cover rasters to wetland and soil type polygons and bathymetry contours. In the eight examples created for Nassau County, publicly available indicator data were used, avoiding the need for the use of proprietary or privileged information.

In the example of hard clam gathering, our first indicator was an indicator of accessibility, as water depth two meters or less at mean tide. A raster elevation map showing water depth of tidal zones is the indicator, and we used the Coastal National Elevation Database Project (United States Geological Survey 2019). The website contains a download link for the Topobathymetric Digital Elevation Model (https://gis.ny.gov/elevation/NYC-topobathymetric-DEM.htm), which shows the data availability for coastal U.S. waters. As an example of ensuring that indicator data are current and not outdated, we used a special report on the website that detailed how the topography and bathymetry model was adjusted after Hurricane Sandy (Stronko 2013). The second indicator was sand bottom type. Bay bottom substrate type was provided by the United States Fish and Wildlife Service's National Wetlands Inventory (NWI) (United States Fish and Wildlife Service 2019). This dataset includes wetland and estuarine ecosystems for the coastal United States. We obtained the Wetlands and Deepwater Code Diagram from the NWI to provide the bottom substrate of estuaries and bays (Fig. 3).

For shore-based fishing, obtaining all of the indicators was more challenging. For the first indicator, an internet search for "United States shoreline polygons" directed us to the National Oceanic and Atmospheric Administration's (NOAA) Office for Coastal Management shoreline website (National Oceanic and Atmospheric Administration 2019) where we ultimately chose the NOAA Composite Shoreline dataset, which the site says is for high-resolution cartographic work. We then clipped it to our study area using GIS to reflect the south Hempstead Bay area.

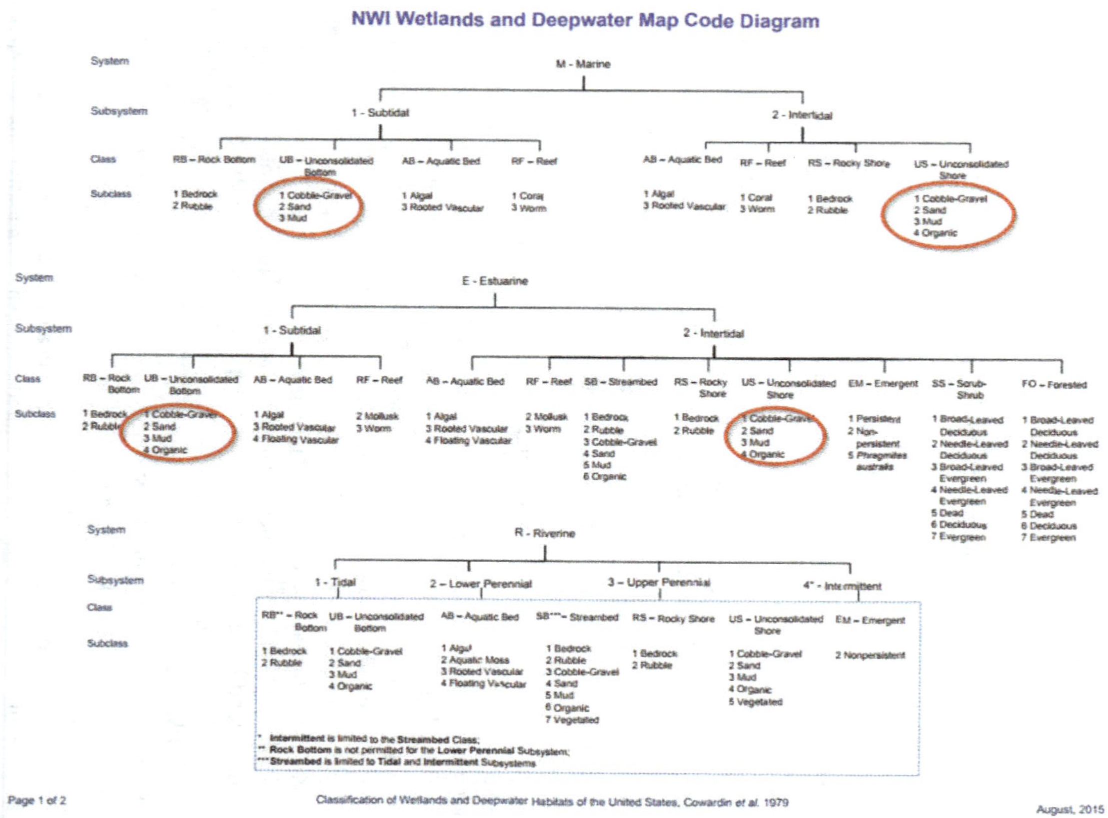

Fig. 3 The National Wetlands Inventory contains indicators of aquatic bottom types (cobble-gravel, sand, mud, organic). Sand bottom type, important for hard clams, is identified with red ovals

Fig. 4 Screen capture of ArcMap's Table of Contents for Nassau County shoreline fishing benefits mapping. Public parks indicator was created from other data

The second indicator, publicly-accessible areas, didn't exist for our area of interest, therefore, we created it. The ESRI World Topographic Map is provided with an ArcGIS Online subscription, and we traced the boundaries of all the areas labeled as public parks or beaches along the shore using the Editor tool to create a new polygon file (Fig. 4). Unlike other indicator data, these were not from a U.S. government source. Even though we did not use the original map, instead we referred to it as a guide to make a new polygon file, it is important to examine the type of license to ensure proper use and to credit the original source on any documents as demonstrated in Figs. 5(a and b).

The third indicator for shore-based fishing was any area of publicly-accessible shoreline within 30 m of shore. We used distance buffering within ArcGIS ModelBuilder to include only areas within 30 m, and there was no need to download another indicator dataset.

7 Nature's Benefits in Nassau County

Our collaboration with Nassau County started with an initial list of 20 nature's benefits. Soliciting input from the anticipated end users of the maps was essential in delivering a useful product. Some of our initial choices of benefits, such as SCUBA diving and guided boat touring, were ruled out as requiring excessively subjective judgments to choose indicators. Others, like waterfowl hunting and seal habitat, were discarded because they were deemed less important. After two rounds of deliberation, we chose the following benefits: bay scallop habitat, hard clam collecting, shore fishing, offshore striped bass fishing, summer flounder fishing, vegetative wave attenuation, aquatic animal viewscapes, and yellow-crowned night heron habitat.

Line fishing from the shore, by a single angler. Private property is not included in this service, and areas shown represent areas that are accessible from public land.

Ecosystem Service Type	Cultural and Provisioning (a non-material benefit that contributes to the development and cultural advancement of people, and a product obtained from an ecosystem).
Present	Areas of shoreline that are within 30 meters of a publicly-accessible park, beach, or recreational area.
Absent	Areas of shoreline that are greater than 30 meters from a publicly-accessible park, beach, or recreational area.
Rationale	The service of shore-caught fish is present within 30 meters of publicly-accessible shore points, which represents the maximum distance that an angler could reasonably cast a line.
Liiting Factor	Accessibility of shoreline to anglers. Areas of the shore that are privately owned were not included as public resources.
Data Sources	Locations of publicly-accessible parks, beaches, and recreational areas: Author's polygon drawing based on Google Maps and ESRI World Topographical Map. Shoreline: National Oceanic and Atmospheric Administration (NOAA) Office for Coastal Management Composite Shoreline
For Further Reading	List of public access to New York and Long Island marine waters for beach fishing and boat launches with contact information: http://www.dec.ny.gov/outdoor/7901.html

Fig. 5 (**a** and **b**) Nature's Benefits handouts for shore fishing and hard clam collecting

This stakeholder collaboration effort resulted in eight nature's benefit maps that illustrated a variety of benefits important to people who live on the Bay. We created 1-page handouts for each nature's benefit map (e.g., Figs. 5(a and b)), explained the indicators used to create it, and provided background information on the benefit along with further reading. Data sources were included with the example handouts, and a selection of the peer-reviewed literature that guided our choice of indicators

b

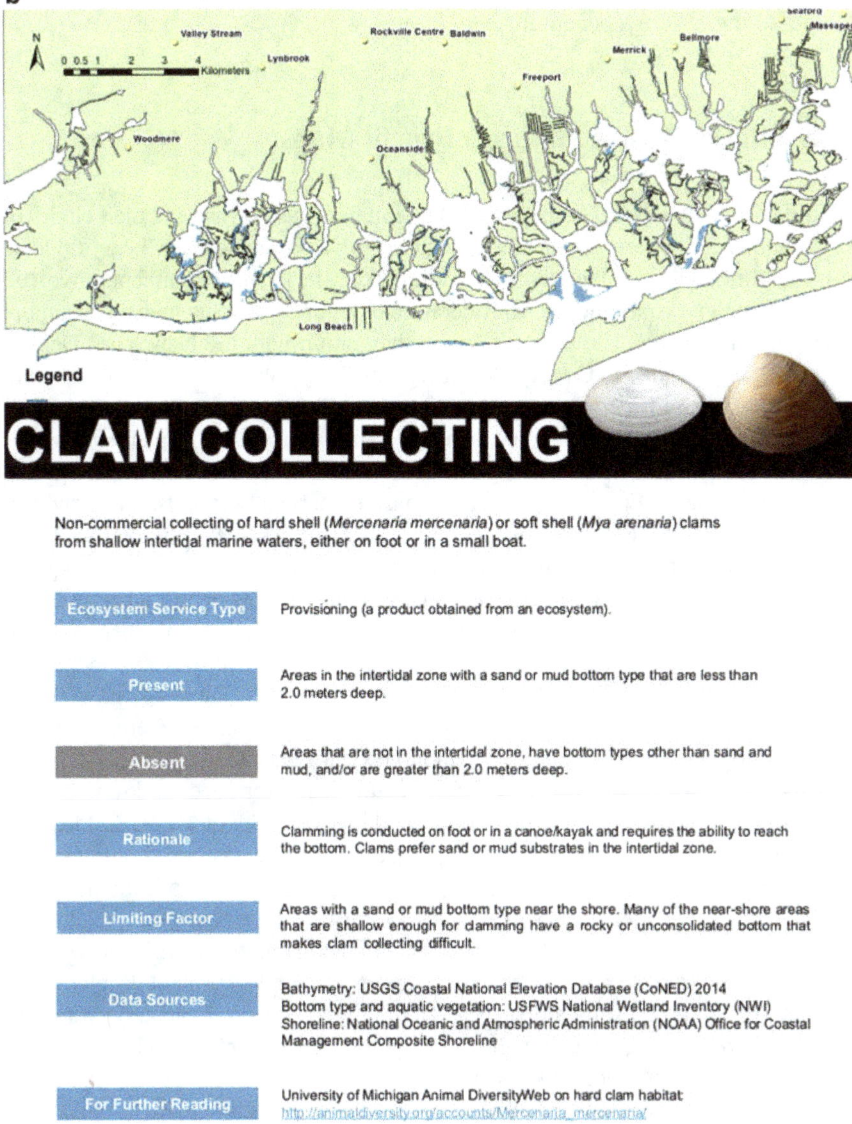

Fig. 5 (continued)

was listed in a “For Further Reading” section. We provided a comprehensive package to Nassau County that included the eight benefit handouts, all input files, Arc ModelBuilder files, a Further Reading document that included peer-reviewed literature supporting indicator choices, and two guidance documents. A summary was also provided with an illustrated step-by-step manual that guides a user through indicator election, creating new maps, and navigating the GIS software for making changes to the Nature's benefit maps.

8 Synthesis—Using Nature's Benefit Maps

The utility of nature's benefit maps is in their intuitive ease-of-use and their clarity of presentation. They are especially effective for communicating benefits to a non-scientific audience, including decision makers in local governments involved with planning and zoning where consideration of nature's benefits may reach a more sustainable and resilient solution. In our collaboration with the New York Department of State and Nassau County Department of Public Works, two use cases were prominent: (1) from a regional perspective, identifying areas of nature's benefit hotspots (i.e., multiple overlapping benefits); and (2) from an implementation or management perspective, identifying the most limiting factor determining the spatial extent of the benefit. Ecosystems, such as coastal emergent wetlands, that provide a high density of benefits can be managed for preservation or actively improved. For vegetative wave attenuation the limiting factor was the extent of emergent aquatic vegetation, rather than submerged aquatic vegetation or the extent of wetland, intertidal, and aquatic zones. Knowing the limiting factor allows planners to communicate how best to increase the amount of a nature's benefit, possibly increasing its spatial extent. Decision makers and the public both wanted to know where benefits were located and how to manage to potentially increase benefits. Nature's benefits mapping provides both information needs for decision support.

A moderate degree of skill with GIS is required. A user needs to know how to arrange files in the proper directories for ArcGIS, open and edit ModelBuilder, and must have some familiarity with GIS operations like clipping to be able to fully utilize the tool. ArcGIS is required to utilize ModelBuilder and replicate the examples here, but we decided this was acceptable because ArcGIS is in widespread use in municipal planning departments, such as the Nassau County government, who are the primary end-users. This approach is broadly transferable to other GIS platforms, including open-source software such as QGIS (https://qgis.org/en/site/). Maps are considered provisional, indicating the potential for benefits, unless fieldwork is done to confirm benefit presence and absence.

End-user guidance and feedback at each step (sometimes described as being "co-developed") was invaluable in delivering products the county could use. The handouts served as templates for other benefits, and we recommend that nature's benefit maps be used in the initial stages of planning, including developing and evaluating potential alternative scenarios for a given management effort. These modeling and mapping tools have applications in health impact assessments, environmental impact assessments, and municipal planning and zoning. Our emphasis was on making best use of publicly available data and translating these clearly for ease-of-use, including end-user modification and extension to serve a range of interests and needs.

Acknowledgements We would like to acknowledge the following individuals for their invaluable contributions to numerous project discussions, including Rabi Kieber (USEPA Region 2), Jonathan Halfon (FEMA), Elizabeth Codner-Smith (TNC), Anthony Dvarskas (Stony Brook), Paul Beyer (NYDOS), Muluken Muche and Florence Fulk (USEPA), Barry Pendergrass (NYDOS), Carolyn LaBarbiera (NYDOS), Sean Sallie and Joseph Cuomo (Nassau County Dept. of Public Works), Nadia Seeteram (ORISE) and Bennett Brooks (Consensus Building Institute). Although this document has been reviewed in accordance with Environmental Protection Agency (EPA) policy and approved for publication, it may not necessarily reflect official agency policy. Mention of trade names or commercial products does not constitute endorsement or recommendation for use. This research was supported in part by an appointment to the ORISE Fellowship Program at the U.S. EPA, Office of Research and Development, Athens, Georgia, administered by the Oak Ridge Institute for Science and Education through Interagency Agreement No. DW8992298301 between the U.S. Department of Energy and the U.S. EPA.

Disclaimer This chapter has been subjected to Agency review and has been approved for publication. The views expressed in this paper are those of the author(s) and do not necessarily reflect the views or policies of the U.S. Environmental Protection Agency.

References

Angradi, T. R., Launspach, J. J., Bolgrien, D. W., Bellinger, B. J., Starry, M. A., Hoffman, J. C., Trebitz, A. S., Sierszen, M. E., & Hollenhorst, T. P. (2016). Mapping ecosystem service indicators in a Great Lakes estuarine area of concern. *Journal of Great Lakes Research, 42* (3), 717–727.

Blake, E. S., Kimberlain, T. B., Berg, R. J., Cangialosi, J. P., & Beven, J. L. I. (2013). *Tropical cyclone report Hurricane Sandy (AL182012)*. National Hurricane Center.

Carpenter, S., Walker, B., Anderies, J. M., & Abel, N. (2001). From metaphor to measurement: Resilience of what to what? *Ecosystems, 4*(8), 765–781.

Christensen, J. P., Murray, J. W., Devol, A. H., & Codispoti, L. A. (1987). Denitrification in continental shelf sediments has major impact on the oceanic nitrogen budget. *Global Biogeochemical Cycles, 1*(2), 97–116.

Cohen, J. B., Fraser, J. D., & Catlin, D. H. (2006). Survival and site fidelity of piping plovers on Long Island, New York. *Journal of Field Ornithology, 77*(4), 409–417.

DeWitt, T. H., Berry, W. J., Canfield, T. J., Fulford, R. S., Harwell, M. C., Hoffman, J. C., Johnston, J. M., Newcomer-Johnson, T. A., Ringold, P. L., Russel, M. J., Sharpe, L. A., & Yee, S. J. H. (2020). The final ecosystem goods and services (FEGS) approach: A beneficiary-centric method to support ecosystem-based management. In T. O'Higgins, M. Lago, & T. H. DeWitt (Eds.), *Ecosystem-based management, ecosystem services and aquatic biodiversity: Theory, tools and applications* (pp. 127–148). Amsterdam: Springer.

Diakakis, M., Deligiannakis, G., Katsetsiadou, K., & Lekkas, E. (2015). Hurricane Sandy mortality in the Caribbean and continental North America. *Disaster Prevention and Management: An International Journal, 24*(1), 132–148.

Hapke, C., Brenner, O., Hehre, R., & Reynolds, B. (2013). *Coastal change from Hurricane Sandy and the 2012–13 winter storm season—Fire Island, New York*. (Open-File Report 2013–1231). U.S. Geological Survey. Retrieved from http://pubs.usgs.gov/of/2013/1231/.

Hartig, E. K., Gornitz, V., Kolker, A., Mushacke, F., & Fallon, D. (2002). Anthropogenic and climate-change impacts on salt marshes of Jamaica Bay, New York City. *Wetlands, 22*(1), 71–89.

Hodgman, T. P., Elphick, C. S., Olsen, B. J., Shriver, W. G., Correll, M. D., Field, C. R., Ruskin, K. J., & Wiest, W. A. (2015). *The conservation of tidal marsh birds: Guiding action at the intersection of our changing land and seascapes*. Saltmarsh Habitat & Avian Research Program.

Interboro Team. (2014). *Living with the Bay: A comprehensive regional resiliency plan for Nassau County's South shore*. Retrieved from https://www.hud.gov/sites/documents/INTERBORO_IP_BRIEFING_BOOK.PDF.

Lamothe, K. A., & Sutherland, I. J. (2018). Intermediate ecosystem services: The origin and meanings behind an unsettled concept. *International Journal of Biodiversity Science, Ecosystem Services & Management, 14*(1), 179–187.

National Oceanic and Atmospheric Administration. (2019). NOAA Shoreline Website. Retrieved from https://shoreline.noaa.gov/.

Paul, M., Bouma, T. J., & Amos, C. L. (2012). Wave attenuation by submerged vegetation: Combining the effect of organism traits and tidal current. *Marine Ecology Progress Series 444*:31–41

Peconic Estuary Program. (2015). Restoring the Peconic Bay Scallop. Retrieved from https://www.peconicestuary.org/peconic-bay-scallop/.

Sharpe, L., Hernandez, C., & Jackson, C. (2020). Prioritizing stakeholders, beneficiaries and environmental attributes: A tool for ecosystem-based management. In T. O'Higgins, M. Lago, & T. H. DeWitt (Eds.), *Ecosystem-based management, ecosystem services and aquatic biodiversity: Theory, tools and applications* (pp. 189–212). Amsterdam: Springer.

Stronko, J. (2013). *Hurricane Sandy science plan—Coastal impact assessments: U.S. Geological Survey Fact Sheet 2013–3090*. Retrieved from https://pubs.usgs.gov/fs/2013/3090/.

The City of New York. (2013). A stronger, more resilient New York. PlaNYC.

Tinoco, A. (2017). *Effects of Hurricane Sandy on Great South Bay, Long Island: Assessing water quality, seagrass and associated nekton communities* (M.S. Marine and Atmospheric Science). Brook, NY: Stony Brook University Stony.

United States Fish and Wildlife Service. (2019). National spatial data infrastructure-wetlands layer. Retrieved from https://www.fws.gov/wetlands/data/NSDI-Wetlands-Layer.html.

United States Geological Survey. (2019). USGS EROS Archive—Digital Elevation—Coastal National Elevation Database (CoNED) Project—Topobathymetric Digital Elevation Model (TBDEM). Retrieved from https://www.usgs.gov/centers/eros/science/usgs-eros-archive-digital-elevation-coastal-national-elevation-database-coned?qt-science_center_objects=0#qt-science_center_objects.

Walker, R., & Tenore, K. (1984). The distribution and production of the hard clam, *Mercenaria mercenaria*, in Wassaw Sound, Georgia. *Estuaries, 7*(1), 19–27.

Wells, H. (1957). Abundance of the hard clam *Mercenaria mercenaria* in relation to environmental factors. *Ecology, 38*(1), 123–128. https://doi.org/10.2307/1932134.

5

Ecosystem-Based Management: Moving from Concept to Practice

Gonzalo Delacámara, Timothy G. O'Higgins, Manuel Lago, and Simone Langhans

Abstract Ecosystem-Based Management (EBM), intended to restore, enhance and/or protect the resilience of ecosystems, is gaining momentum. It is often argued, though, that some of the difficulties to provide practical guidance to conduct EBM stems from the lack of a clear definition. EBM emphasises on factoring in complex linkages in social-ecological systems; dealing with adequate scales (both time and space wise); promoting adaptive management of complex and dynamic systems; and adopting integrated assessment and management frameworks. This chapter shows, on one side, challenges to build consensus on a definition that is both conceptually and theoretically sound as well as practicable; on the other, the enabling factors that make EBM actually happen.

Lessons Learned

- Ecosystem-Based Management (EBM) moves away from a limited (partial) consideration of natural systems and society as separate entities.
- Society will face increasing levels of uncertainty—EBM is a more flexible approach that allows for the identification of some of those uncertainties, so

G. Delacámara (✉)
IMDEA Water Institute, Alcalá de Henares, Madrid, Spain
e-mail: gonzalo.delacamara@imdea.org

T. G. O'Higgins
Marine and Renewable Energy Ireland (MaREI), Environmental Research Institute, University College Cork (UCC), Cork, Ireland
e-mail: tim.ohiggins@ucc.ie

M. Lago
Ecologic Institute, Berlin, Germany
e-mail: Manuel.lago@ecologic.eu

S. Langhans
Department of Zoology, University of Otago, Dunedin, New Zealand

Basque Centre for Climate Change (BC3), Leioa, Spain
e-mail: simone.langhans@bc3research.org

that they can be considered in decision-making, hence allowing society to adapt to error.

- EBM is a recognition of the need to "juggling all balls" at once. Since reality is complex, responses will also have to be so.
- EBM is an integrated approach in many different ways: acknowledging for the recognition of linkages between society and ecosystems, between different types of ecosystems, throughout time and space, looking at things from multiple standpoints.
- Enhanced governance is critical to make EBM happen. This is not just about transparency, accountability and meaningful social participation. It is also about using the right incentives, cooperating, coordinating decisions between sectors, improving our knowledge and information base, uptaking innovations, etc.

Needs to Advance EBM

- We need to better understand the complexity of the social, behavioural side of social-ecological systems, to match our understanding of the ecological side.

1 Defining Ecosystem-Based Management: Minding Mice at a Crossroads or Not Quite?

Ecosystem-Based Management (EBM) has gained increased popularity in recent years; yet, it has been argued that lack of consensus on its definition is precluding progress in terms of design and practical implementation of such approaches. EBM emphasises on considering ecosystem connections; dealing with appropriate spatial and temporal scales; fostering adaptive management of complex social-ecological systems; managing those systems in an integrated way; accounting for the dynamic nature of ecosystems and society, etc. There is no single agreed-upon definition for EBM, which sometimes is even referred to as the ecosystem approach (EA) itself. What seems common to all definitions, though, is the acknowledgement of the complexity and interspecies relationships within ecological systems, although broader governance elements are also of great importance in most definitions. This chapter does not develop a linguistic, nominal investigation of different definitions of EBM but rather aims at adding conceptual clarity while, at the same time, focusing on enabling factors for their effective uptake in decision- and policy-making for biodiversity and aquatic ecosystem services conservation.

We often demand clear-cut definitions: obvious, without any need of proof, unambiguous, unequivocal. However, demanding rigidly defined areas of doubt and uncertainty is a challenging exercise. Grasping at straws to some extent. Social-ecological systems such as aquatic ecosystems are highly adaptive and complex. How could one expect their management to be defined in an indisputable way if characterised by adaptability and complexity (Preiser et al. 2018)? Nonetheless, this is far from being a futile exercise since it also provides useful insights for the practical implementation of these management approaches (see Robinson and Culhane 2020 for further details on complexity, EBM and a framework to assess linkages).

1.1 *The Many-Sided Definition of Ecosystem-Based Management*

According to the outcomes of the AQUACROSS research project,[1] Ecosystem-Based Management (EBM) can be understood (Gómez et al. 2016) as

> any management or policy option intended to restore, enhance and/or protect the resilience of the ecosystem.

As such, EBM would stand for any course of action intended to improve the ability of an ecosystem to remain within critical thresholds, to respond to change, and/or to transform, so that a new equilibrium or a new evolutionary path follow (ibid.).

When applied to aquatic ecosystems, EBM can be said to set the foundations for effective and widely applicable management concepts and practices. Yet, one may think that EBM actually lacks a definition (and a universal 'grammar'), which may hinder implementation (Long et al. 2015, 2017; Willaert et al. 2019); that EBM requires extensive data and sophisticated modelling (Addison et al. 2019); that EBM is linked to naïve attempts to describe complex and adaptive systems, squeezing the universe between our fingers; and, on more practical grounds, that there are neither enough resources to deliver EBM approaches (Curtice et al. 2012) nor a clear mandate and institutional setup for EBM in prevailing legislation (Nilsson and Bohman 2015; Link et al. 2019). Harwell (2020) presents a non-exhaustive overview of the broad suite of U.S. federal environmental laws and regulations linked to ecosystem services and a survey of the legal scholar literature on the topic in the U.S. environmental law. The combined outcome of all these premises, whether utterly true or false, is definitely limiting the ability of EBM to deliver, let alone in aquatic ecosystems as shown in Langhans et al. (2018). As a matter of fact, EBM is often defined using combinations of underlying principles (Long et al. 2017). EBM aims at ensuring that choices do not negatively affect ecosystem functions, so that the delivery of aquatic ecosystem services and benefits stemming from them can be

[1]AQUACROSS was a Horizon 2020 project aimed to support EU efforts to protect aquatic biodiversity and ensure the provision of aquatic ecosystem services. As such, AQUACROSS sought to advance knowledge and application of ecosystem-based management (EBM) for aquatic ecosystems to support the timely achievement of the EU 2020 Biodiversity Strategy targets. AQUACROSS has received funding from the European Union's Horizon 2020 Programme for Research, Technological Development and Demonstration under Grant Agreement no. 642317.

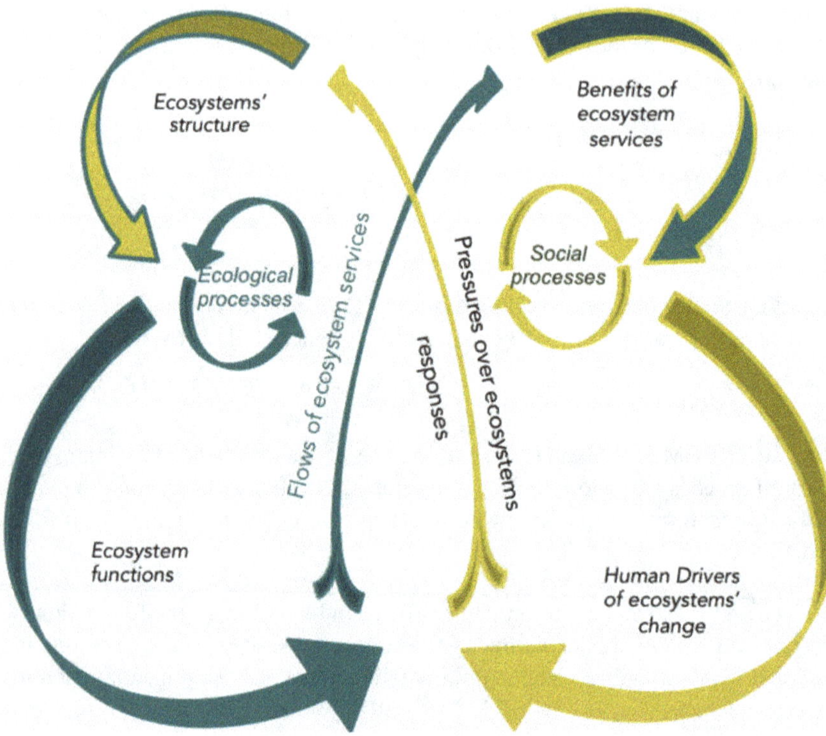

Fig. 1 Social-ecological systems as interlinked, complex, adaptive systems. Source: Gómez et al. (2016)

sustained in the long term. This makes EBM highly relevant to maintain and restore connections between social and ecological systems (Keesstra et al. 2018).

Social-ecological systems (SES) are therefore complex and adaptive systems that should be analysed in a holistic, integrated way (see Ostrom 2009). On the one hand, through the notion of SES, one can analyse the detrimental consequences over ecosystems that result from the satisfaction of multiple demands of services provided by nature and to society. As in Fig. 1 (see below), this could be seen as the demand side of the "butterfly" used to represent a SES. It shows how the demand and use of naturally provided services is an outcome of social processes, including markets and governing institutions, and determined by multiple factors (such as population and economic growth, climate change, technological progress, etc.). These demands of services result in pressures over ecosystems and further changes in their structure. On the other hand, one can analyse the potential of ecosystems to continue delivering ecosystem services on which human life, the social system and the ecological system itself depend upon, and how this all affects human wellbeing. This is the supply side, which if analysed can shed light on the functioning of ecosystems and how changes, induced by human actions, are linked to human wellbeing and sustainability.

EBM is also a way to address uncertainty and variability in ecological systems (Link et al. 2012) that, by definition are dynamic (DeFries and Nagendra 2017), in an attempt to embrace change, to learn from the past and to design and implement adaptive policies throughout the management process (Schultz et al. 2015). In other words, in the absence of sound governance frameworks and inclusive models (covering a wide range of social actors), EBM seems deemed flawed, as in the case of relevant sectoral policies such as the EU Common Agricultural Policy (Pe'er et al. 2019) or the Common Fishery Policy (Lado 2016). On the contrary, if properly designed and implemented, EBM rather shows the flexibility to identify uncertainties and incorporate those into decisions (through exploration of alternatives), and to adapt to error or uncertainty through an iterative review of progress toward the goal for which an EBM-based plan was developed.

By explicitly factoring in the full range of ecological and social interactions and processes necessary to sustain ecosystem structure, functions and services, EBM seems to offer great expectations for aquatic ecosystems management (Tallis et al. 2010; Langhans et al. 2018, 2019). One of the potential advantages of EBM is that, in principle, it should go well beyond conventional scientific and policy practice.

All dimensions of this multi-layered definition of EBM foster integration across ecological and social systems (Collins et al. 2011), explicitly address the need for sustainable patterns of resource use (Grehan et al. 2009), acknowledge interspecies relationships (Long et al. 2015), and call for the consideration of ecological thresholds and other environmental limits (Samhouri et al. 2017; Möllmann et al. 2015). Basing management decisions on the ecosystem entails that planning needs to be adapted to the dynamics of the whole ecosystem to at least preserve, if not to enhance, their potential to delivering services and benefits society depends upon.

Albeit difficult, describing is always simpler than analysing; defining may also be easier than understanding. Stefan Zweig, in *Chess Story* (Zweig 2011) said:

> We are happy when people/things conform and unhappy when they don't. People and events don't disappoint us, our models of reality do. It is my model of reality that determines my happiness or disappointments.

It is critical to note that all definitions of EBM have some common features but mainly differ in their view of the connections between ecosystems and society, which has clear implications in terms of modelling (see Fulford et al. 2020 for a detailed discussion on mathematical modelling for EBM and ecosystem services).

1.2 Disambiguation of the Concept

There may be good reasons to believe that EBM does not add too much value if compared with the so-called ecosystem approach (EA), such as introduced by the Secretariat of the Convention on Biological Diversity (CBD 2008) (see Enright and Boteler 2020 for a deep review of the implicit origins of the approach and its subsequent adoption in marine International Law). As a matter of fact, a number of authors still use the latter expression rather than the former (Bennett et al. 2015; McIntyre 2019). According to CBD (ibid.)

> the ecosystem approach is a strategy for the integrated management of land, water and living resources that promotes conservation and sustainable use in an equitable way. Thus, the application of the ecosystem approach will help to reach a balance of the three objectives of the Convention: conservation; sustainable use; and the fair and equitable sharing of the benefits arising out of the utilization of genetic resources.

The EA can be considered as a management approach that takes the ecosystem itself as the relevant scale (time and space wise, but also on organisational grounds) but this can only be inferred from some of its 12 principles. For instance, principle 5 states:

> conservation of ecosystem structure and functioning, in order to maintain ecosystem services, should be a priority target of the ecosystem approach.

Principle 6 in turn adds:

> ecosystems must be managed within the limits of their functioning.

Finally, principle 7 stresses upon the fact that

> the ecosystem approach should be undertaken at the appropriate spatial and temporal scales.

Beyond linguistic flourishes, the EA could certainly be considered an ecosystem approach to management, hence quite close to the notion of EBM, although other authors add some nuances (Kirkfeldt 2019). Do fine distinctions between these terms essentially matter? Not quite, within this context... One may actually argue that the EA does not preclude any other conservation approaches applied to aquatic ecosystems; hence, it may be compatible and consistent with such approaches, such as EBM (if considered a beast of a different coat), integrated water resources management (IWRM), integrated coastal zone management (ICZM), etc. In fact, these approaches may support the implementation of the EA in different biomes (Gómez et al. 2016).

As above, EBM emphasises on considering ecosystem connections (something that is actually coherent with the EA), dealing with adequate spatial and temporal scales, promoting adaptive management, deepening integration, accounting for the dynamic nature of ecosystems, etc. As in Sect. 1.1, there is no single agreed-upon, canonical definition for EBM, which sometimes is even referred to as the EA itself (Kirkfeldt, op. cit.).

Unlike the Common Classification of Ecosystem Services (CICES) (Haines-Young and Potschin 2018), for example, which was created ad hoc to homogenise a typology, not more than an accounting system, just as many other typologies, indicator sets, mapping exercises, databases, etc., this book is meant to shed light on available conceptual and management approaches, rather than to create a new 'taxonomy' of EBM definitions.

Further to the notions of EA and EBM, not more than a decade ago practitioners such as the International Union for the Conservation of Nature (Cohen-Shacham et al. 2016) and the World Bank (MacKinnon et al. 2008; Browder et al. 2019; Erin Gray et al. 2019) did coin the term Nature Based Solutions (NBS) (Davies and Lafortezza 2019), which has pervaded global discussions on water resources management since the release of the World Water Development Report 2018 (UNESCO 2018).

Some of those NBS, for instance, aim at restoring and conserving aquatic ecosystems via natural means. Those NBS can take the form of green infrastructures intended to maintain and enhance landscape, soil cover, and groundwater sources in order to improve their natural traits, the ecological services and other abiotic components they deliver, and to favour climate change adaptation and reduced vulnerability to extreme events and water quality degradation. The distinctive character of NBS would then have to do with their single purpose (i.e. conserving aquatic ecosystems), but also with a specific set of means (i.e. natural ones).

This example clearly underlines some distinctive features of NBS. On the one hand, not every solution that is effective on environmental grounds (e.g., increasing water recharge rates in groundwater bodies) is a NBS. On the other hand, NBS are interventions over ecosystems: they use natural processes rather than replacing nature, hence emulating ecological functions. It is also clear that, as in the case of the example above, the intended environmental outcome (i.e. recharging an aquifer) is not the end but the means that makes these NBS relevant for EBM of those aquatic ecosystems. Last but not least, it should also be mentioned that NBS are not simply means to reset a spoiled ecosystem to its pristine state, which is often not a feasible endeavour (if possible at all), but rather to adapt prevailing developments so as to enhance or to recover ecosystem functions that were either fragile or have been lost altogether.

As in EC (2015), NBS are

> living solutions inspired by, continuously supported by and using nature, which are designed to address various societal challenges in a resource-efficient and adaptable manner and to provide simultaneously economic, social, and environmental benefits.

Maes and Jacobs (2017), in turn, add further conceptual clarity when defining NBS as

> any transition to a use of ecosystem services with decreased input of non-renewable natural capital and increased investment in renewable natural processes.

Nesshöver et al. (2017) stress, though, that NBS need to be developed and discussed in relation to existing concepts to clarify their added value.

Along these lines, Integrated Water Resources Management (IWRM) can arguably be presented as an empirical concept more than a theoretical one (Smith and Clausen 2015). It was also built up from the actual experience of practitioners. The concept has now long been used (since the first global water conference, the United Nations –UN– Water Conference, in Mar del Plata, Argentina, on March 14th, 1977), even if practice is still limited to some areas of the world such as the European Union or Australia, to name a few. However, it was not until the UN Conference on Environment and Development (Rio, Brazil, 1992) that the concept was refined and became the object of wide discussions as to its practical implications. GWP (2000) states:

> IWRM is a process which promotes the co-ordinated development and management of water, land and related resources, in order to maximize the resultant economic and social welfare in an equitable manner without compromising the sustainability of vital ecosystems.

IWRM could therefore be seen as a wider notion, which could benefit from the design and implementation of EBM approaches.

Quite the same applies to Integrated Coastal Zone Management (ICZM). EC (2012), concerning the implementation of ICZM echoed the previous recognition of EC (2000) that coastal management lacked vision and was based on a very limited understanding of processes and dynamics in coastal ecosystems, as a result of the gap between scientific research and data collection, and end-users (see also Lewis et al. 2020 for a detailed review on the complexity and disparity of results from applying the existing coastal and estuarine ecosystem goods and services models and tools).

To cover the aquatic ecosystems continuum, Marine Spatial Planning (MSP, see Pauli 2010), has often been mentioned as an ecosystem-based approach. Through analysing and allocating parts of marine ecosystems to specific uses in order to meet ecological, economic, and social objectives, MSP integrates across economic sectors and among agencies, delivers adaptive approaches, and is focused on longer time frames. Le Tissier (2020), provides an insightful review on the relationship and contradictions regarding the juxtaposition of EBM and ICM and MSP.

IWRM and ICZM, let alone MSP, are all processes that entail a new fashion of governance, one that goes well beyond conventional and restrictive views based on integrity, stakeholder engagement, transparency and accountability or mere institutional reforms, and progresses towards mastering complexity, thus considering the redesign of incentives, horizontal and vertical coordination of sectoral policies, promoting innovation at different levels, etc. (such as in the OECD Water Governance Programme, for instance: OECD 2011).

Further to these conceptual approaches, a number of policy initiatives have sparked the development of a wide range of assessment frameworks and decision support systems for EBM (Harvey et al. 2017; Link and Browman 2017; Alexander and Haward 2019; Lago et al. 2019). Yet, a major challenge remains as to the design of an operational framework that links, in an efficient way, the assessment of biodiversity and ecological processes and their meaningful consideration in decision-making processes (and not just in policy making), despite recent progress (Langhans et al. 2018; Gómez et al. 2017; Do Yun et al. 2017). EBM faces both conceptual and functional limitations, in particular with regards to the lack of explicit acquaintance of the ecosystem services concept (Jordan et al. 2012), critical to connect ecological analyses with social wellbeing; a certain bias towards ecological dimensions rather than a balanced approach embedding social-ecological processes (Berkes 2012), which would enhance a more holistic understanding of relevant dynamics and feedback loops. Trade-offs, uncertainties, and non-linearities inherent in the management of aquatic ecosystems are somehow downplayed, when not overlooked at all (Curtin and Prellezo 2010; Culhane et al. 2020).

2 Distinctive Features of Ecosystem-Based Management of Aquatic Ecosystems

Despite a certain amount of uncertainty and lack of conceptual clarity, there has been some evolution in the policy use of EBM (Gelcich et al. 2018). There is an actual policy need for a consolidated, practical definition of the term that addresses its different dimensions: understanding complex and adaptive social-ecological systems, resilience thinking (Curtin and Parker 2014), accounting for ecosystem functions and services, favouring multi-functional responses, etc.

EBM of aquatic ecosystems should therefore be assessed, designed and implemented so as to solve specific problems, and these solutions are often likely to maximise contributions to the resilience of social-ecological systems as a whole. Understanding the enabling conditions (see Sect. 3.3 of this chapter) that foster cooperative behaviour, or the factors that hinder it, is of chief importance for the design and implementation of EBM approaches. Progress in evolutionary theory (Di Marco et al. 2019; Wasser 2013) and game theory (Arfanuzzaman and Syed 2018; Punt et al. 2014) have been considered as particularly auspicious.

Identifying biodiversity policy challenges and assessing different alternatives to address them is far from being just a purely scientific or political endeavour. This daunting task requires both a conducive policy process and harnessing scientific knowledge to face stakeholders throughout with the outcomes of their own choices and to support them across the whole policy cycle. There is no such thing as good science or good policy: the former should address societal challenges; the latter, in turn, should be based on evidence.

Due to the complexities involved in aquatic social-ecological systems, it is clear there is neither a design of a one-size-fits-all EBM approach nor just one EBM implementation path. Additionally, it is critical to understand that more science may not necessarily close the existing knowledge gaps. Rather, each individual situation may need to be considered in its institutional and political setting and requires consideration of site-specific trade-offs.

Under an EBM approach science is neither only intended to inform nor to make technically feasible and sound decisions. It is rather a means to build a credible knowledge base through dialogue between stakeholders and scientists. Sharpe et al. (2020) provide more information on the benefits of Structure Decision Making–SDM and Decision Support Tools–DST in engaging with stakeholders and supporting decision makers; Lewis et al. (2020) also shed light on a community-based decision support framework and tool for quantifying trade-offs in ecosystem goods and services.

It is often argued that EBM approaches are characterised by their contribution to ecological integrity, biodiversity, resilience and (in some versions, as in Granek et al. 2010 for coastal ecosystems) to ecosystem services delivery; by their use of scientific knowledge and appropriate spatial scales (Qiu et al. 2018), their acknowledgement of social-ecological connections, stakeholder engagement and accountability (Nunan et al. 2018); by transdisciplinarity and integrated management (Pires et al.

2018; Carmen et al. 2018) and by their adaptiveness (Long et al. 2015). However, unlike common wisdom, EBM does not exclusively show those features and several approaches that are not based on the ecosystem may well do so.

What could be said to be specific of EBM for aquatic ecosystems?

- *Holistic*. EBM gives consideration to ecological and social factors, which demands interdisciplinary knowledge and prominence to water governance and the relationship among aquatic species as well as with their abiotic environment. EBM protects the integrity of aquatic ecosystems as a means to preserve a complementary array of ecosystem services as well as to preserve aquatic biodiversity in its own right. EBM thus acknowledges social-ecological interactions (necessarily including terrestrial ecosystems too) and seeks inclusive policy-making processes that favour transparency, accountability and provide a better framework so as to make different stakeholders understand why they make the decisions they make.
- *Multi-functional*. Unlike more conventional approaches to biodiversity and ecosystem conservation that focus on single benefits, EBM is characterised by multiple functions and benefits, thus being able to strike the balance, at once, between different policy domains. To put it in a different way, EBM aims at maximising the joint value of all aquatic ecosystem services and abiotic components, rather than focusing only on the delivery of single services, such as drinking water, water for irrigation, natural assimilation capacity of pollution, etc.).
- *Based on evidence*. Just like other management approaches, EBM is based on scientific knowledge. This is but solemnising what is obvious, though. What makes EBM stand out is the kind of scientific knowledge that is mobilised, as well as the way in which this is factored into decision-making processes. The role of science in EBM is two-fold: on one side, it is intended to be relevant for policy —and decision-making—; on the other, it is essential for the credibility of social knowledge and for the legitimacy of policy decisions it intends to inform and improve. Needless to say, in addition to scientific knowledge, traditional knowledge on ecosystem management (Joa et al. 2018) also provides critical insights.
- *Mindful of the spatial scale (and ecological organisational levels)*. Management based on the aquatic ecosystem is much more challenging and sophisticated than managing single water bodies (or even watersheds). This might imply devolution to local communities, but may also require action at higher levels through, for example, transboundary or even global cooperation. EBM decisions are to take place at the appropriate level and scale, taking into account aquatic ecosystem boundaries but also complex interlinkages and adaptive processes. Actually, EBM keeps a close relationship with the notion of meta-ecosystem (Loreau et al. 2003). This

(continued)

concept provides a powerful theoretical tool to ascertain the emergent properties that arise from spatial coupling of local ecosystems, such as global source-sink dynamics, diversity-productivity patterns, stabilisation of ecosystem processes, and indirect interactions at landscape or regional scales.

- *Across the water continuum*. Not only complex interactions within the social-ecological system are considered, but also the relationships among aquatic realms (i.e. freshwater ecosystems, coastal ecosystems and transitional waters, marine ecosystems), including unknown and unpredictable effects.
- *Calling for policy coordination*. By definition, EBM calls for (horizontal — across sectoral policies— and vertical —in multi-level governance systems, across different levels of government within the same sectoral policy —) policy coordination. As EBM requires cooperative agreements and collective action to share the range of aquatic ecosystem services obtained across different stakeholders and policy domains, and by seeking to balance ecological and social concerns, as above, these approaches open new opportunities of concurrently pursuing different policy objectives.
- *Adaptive and dynamic*. EBM is a way of adaptive management. Aquatic ecosystem processes and functions are complex and variable. Through acknowledging that there are no optimal solutions and that the future is uncertain, EBM seeks to build or strengthen adaptation capacities by restoring critical aquatic ecosystems and supporting social abilities to respond to a range of possible future scenarios. Low-hanging fruits of management interventions in the short term should be weighed against longer-term benefits of alternative actions. Uncertain events may alter long-term goals or show new, alternate lines of action. As a result of that, not only those aims but also management approaches to reach them should be regularly revisited, hence making monitoring critical so that indications of potential opportunities or difficulties are spotted sufficiently in advance.

3 Enabling Factors for the Effective Uptake of Ecosystem-Based Management

Beyond contributions to conceptual clarity, what seems more important is what makes EBM feasible, i.e. the enabling conditions for this kind of management approaches to happen in practice. Other chapters in this book (quote) will address some of those practical dimensions more specifically. In this chapter we discuss these factors from a conceptual point of view.

3.1 Moving Away from Conventional Praxis in Ecosystem Management

It is very often the case that management toward a single goal often underestimates consequences or dependencies on other parts of ecosystems. For instance, traditional approaches to biodiversity conservation tend to focus on single pressures and specific impacts, flagship umbrella species (Kalinkat et al. 2017), hotspots (Marchese 2015), lack of integration of traditional and scientific knowledge (Sutherland et al. 2014). This is not to say that there is not utility in species-centred conservation approaches. Surrogate species remain popular as they provide useful and necessary shortcuts for some conservation programmes (Caro 2010). As a matter of fact, it is not only a question of the object of conservation programmes but rather about management approaches themselves. Those programmes achieve measurable outcomes (Herzon et al. 2018); however, positive results in the conservation of emblematic species may come at the expense of degrading resilience and increasing vulnerability.

From a management and policy perspective, conventional approaches are based on sectoral (and frequently conflicting) policies. On the contrary, EBM offers a promise of making the multiple co-benefits linked to the enhancement of the ecosystem overall visible. EBM opens new possibilities to meet different policy objectives simultaneously and quite often through cooperative approaches and policy synergies. The so-called nexus perspective is a good example of this (see Venghaus and Hake 2018, for an overview of nexus thinking in EU policies, or Zhang et al. 2018, for a specific application to hydropower generation).

Traditional ecosystem management approaches also tend to maximise the delivery of some ecosystem services (with a bias towards productive ones, such as drinking water, water for irrigation. . .) whilst impairing the ecosystem's capacity to deliver other services, including those linked to self-regulation and support. Traditional management has overreached in altering ecosystems for a single purpose. Emerging EBM, in contrast, finds opportunities in benefits stemming from the restoration of natural features. See, for instance, Sklar et al. (2019), where coastal risk reduction (i.e. storm protection through wave attenuation) and resilience (i.e. flood storage compensation schemes) come in the form of protection, enhancement and restoration of natural features such as mangroves, coral reefs, seagrass beds, sand dunes, inter-tidal and sub-tidal wetlands, mudflats, floodplains, salt marshes, etc.

An additional feature of conventional approaches is the neglect (to a very different extent) of the inherent uncertainties of social-ecological systems and the adoption of mostly deterministic approaches to future challenges when modelling the consequences of future scenarios. EBM necessarily separates from optimality (see Heal et al. 2001 for a seminal discussion on challenges posed by the transition from optimality to sustainability) and, through acknowledging irreducible uncertainties, emphasises on the relevance of building adaptability. The idea of building adaptation capacities, by the way, is far from being just one of restoring critical ecosystems (or ecosystem functions and processes), but also of strengthening social abilities (i.e. investing in social capital) to respond in a robust way to a range of possible futures.

In summary, EBM underlines the trade-offs between focusing on one single species, pressure, impact, sector, or ecosystem service, and adopting a more complex approach. In other words, EBM offers the possibility to discern between planning for the long-term or get caught in a series of lock-ins: technological, analytical, and institutional (Lukasiewicz et al. 2016), which clearly hinder adaptability.

3.2 Adaptive Governance of Aquatic Ecosystems

Schultz et al. (2015) defined governance as "the structures and processes by which people in societies make decisions and share power, creating the conditions for ordered rule and collective actions, or institutions of social coordination" (p. 7369). It is precisely the recognition of the above-mentioned lock-ins that supports the idea that ecosystems have to be governed. This implies, among other things, overcoming (impact) remedial, reactive, mostly unplanned, ad hoc approaches to ecosystem management in favour of pre-emptive, proactive, planned, collective, and coordinated decisions. To put it in a different way, EBM provides an opportunity for adaptive governance leading to higher social-ecological resilience (as Boyd et al. 2015 suggested).

There is a binding constraint for collective action, though: EBM is a leap in the dark or, at best, a chimera in the absence of a strong and enabling institutional setup (Börgström et al. 2015). As DeFries and Nagendra (2017) put it, ecosystem management is very much a "wicked problem", with no straightforward solution. Focusing on ecosystems rather than on single species or functions or processes requires defining specific spatial, temporal and organisational scales of ecosystem services provision. This, of course, has implications in turn for the appropriate scale and structure of ecosystem management institutions.

"Good governance" has gained momentum in the discussion about EBM (Long et al. 2015; Soma et al. 2015; Schultz et al. 2015; Gunderson et al. 2016; Bodin 2017; Bundy et al. 2017; Link et al. 2019). As Grindle (2010) pointed out,

> good governance is a good idea. We would all be better off, and citizens of many developing countries would be much better off, if public life were conducted within institutions that were fair, judicious, transparent, accountable, participatory, responsive, well-managed, and efficient [. . .] Who, after all can reasonably defend bad governance? [. . .].

Good governance, though, carries a similar virus to EBM: the appeal of the idea somehow outpaces its capacity to deliver. It is sure-fire that EBM demands transparency, inclusiveness, a good knowledge base, the appropriate spatial scale, policy coordination, etc. Yet, aren't these defining characteristics of good public policy overall?

An effective uptake of EBM requires adapting prevailing institutions and policy-making processes and jumping the fence to address challenges of a political nature.

Firstly, it is important to stress upon the idea that EBM is a means to an end. This entails that the objectives of EBM need to be defined and this requires the identification of what set of ecosystem services may be sustainably delivered as well as their relative importance. Beyond what Irvine et al. (2016) call shared values, it is a fact that there is a number of asymmetries (information, preferences, etc.) that explain why different individuals value ecosystem services in a different way (see Harwell et al. 2020 to deepen on the role of Strategic Communication in EBM for overcoming information asymmetries). Key to achieving mutually beneficial situations for biodiversity and ecosystem services is identifying those differences and trade-offs between users (Cavender-Bares et al. 2015, and also DeWittt et al. 2020 for a discussion on beneficiary-centric orientation supporting EBM). Indeed, what is special about EBM is that it gives prevalence to this collective choice, which makes everyone accountable for her or his own decisions.

Secondly, balancing trade-offs between ecosystem services and users requires finding the best way to meet pre-defined environmental objectives. This can be done through a number of alternatives. Choices between alternative courses of action are a collective endeavour, which also entails trade-offs: between short-term opportunity costs and long-term benefits, mitigated pressures and weaker provision of productive services or enhanced long-term water security, tackling critical events or managing risks and resilience, etc.

Foremost, seizing opportunities stemming from EBM demands a meaningful change of our mind-set. Whilst traditional measures are designed to respond to a particular challenge, EBM is linked, as above, to multiple ancillary benefits and may simultaneously contribute to various policy objectives, well beyond their intended outcome. The essence of EBM is precisely to be able to reap off the benefits of synergies across different ecological and policy realms. Nonetheless, current assessment methodologies such as optimisation models or cost-effectiveness analysis tend to be limited to account for multiple benefits, hence biasing decisions against EBM solutions, but also for uncertainty (Borgström et al. 2015), consideration of local trade-offs and disservices (Traoré et al. 2018), integration of ecosystem services in land and landscape planning and management (Turkelboom et al. 2018; Grêt-Regamey et al. 2017; Arkema et al. 2015), non-linearities (Grêt-Regamey et al. 2014), and non-convexities (Hyytiäinen et al. 2015).

4 Conclusion

Most of us do not internalize that a social-ecological system is a complex entity, dynamic and adaptive both on ecological and social grounds. Virtually all readers would be willing to say they are perfectly aware of that assertion; some could even explain what it connotes beyond its primary or more literal meaning. Yet, most of us would overlook that evidence when making choices affecting ecosystems. Similar to our brain which has one side (left) that helps us think logically and organize thoughts to build sentences, and another one (right) that helps us experience emotions and interpret nonverbal cues, our way of approaching the management of social-ecological systems responds in many ways to that dualism. As Kahneman (2011) would put it: we think fast and slow.

On the one hand, we feel driven to make decisions related to survival (i.e. withdrawing water from ecosystems for irrigation agriculture or drinking water supply). On the other, another part of us leads us to connection, complexity, and relationships. The key to progress in aquatic biodiversity and ecosystem conservation is to contribute, within the context of EBM, to both dimensions operating altogether. For the ecosystem to be genuinely healthy, all (specialized) elements of the social-ecological system must be integrated. That is, each element of the system has to perform its individual function while operating as part of the whole. Integration is neither more (but not less) than that: to bond different elements, already combined de facto (although sometimes they may have been disconnected, the reason why some interventions are aimed at restoring connectivity), so that the whole system works properly. Integration, inherent in EBM approaches, coordinates and balances those individual parts.

In recent decades, scientists from different disciplines have managed to develop assessment frameworks and even technologies that allow us to better understand the causal relationships between biological diversity and ecosystem services, between different types of ecosystem services, or between ecological processes and functions and social well-being (precisely through the notion of ecosystem services). However, none of them seems to guarantee that trends in biodiversity loss in aquatic ecosystems are reversed or that they are managed in a more integrated way. We have more information about aquatic ecosystems, somehow we know them better; yet, this does not always ensure that we understand how these complex and adaptive social-ecological systems do work.

Following the analogy regarding our brain, as above, one of the most relevant findings in scientific literature is the plasticity and ductile nature of our brain. The same applies to social-ecological systems to some extent: that is, they are systems that always change and not only at certain stages of their evolution. What modifies those systems? In essence, a series of experiences does. Depending on decisions and pressures on aquatic ecosystems, different parts of the ecosystem are activated and the way they do so to respond (as is also the case in social systems) determines the health of these socio-ecological systems and their level of biodiversity. Different connections then become more sophisticated and the system is reconfigured. There is some good news in that evidence: no social-ecological system, provided certain thresholds are not irreversibly exceeded, is a slave to each of these decisions and responses (both human and ecological). Thus, EBM approaches can intervene so that ecosystems become healthier and we result happier. The genetic diversity of aquatic ecosystems at stake plays an important role in how to respond to multiple pressures. However, as decisive or more than that genetic component is everything that happens in these social-ecological systems. In other words, beyond the basic 'architecture' of ecosystems, it is our choices that truly determine whether social-ecological systems evolve in a resilient and integrated way or not.

The integration inherent to the EBM concept precisely consists in facilitating the above-mentioned reconfigurations of social-ecological systems, in favouring the connections between the different parts of them. When these links are cooperative, the elements that integrate are created and strengthened.

References

Addison, P. F., Bull, J. W., & Milner-Gulland, E. J. (2019). Using conservation science to advance corporate biodiversity accountability. *Conservation Biology, 33*(2), 307–318.

Alexander, K. A., & Haward, M. (2019). The human side of marine ecosystem-based management (EBM): 'Sectoral interplay' as a challenge to implementing EBM. *Marine Policy, 101*, 33–38.

Arfanuzzaman, M., & Syed, M. A. (2018). Water demand and ecosystem nexus in the transboundary river basin: A zero-sum game. *Environment, Development and Sustainability, 20*(2), 963–974.

Arkema, K. K., Verutes, G. M., Wood, S. A., Clarke-Samuels, C., Rosado, S., Canto, M., & Faries, J. (2015). Embedding ecosystem services in coastal planning leads to better outcomes for people and nature. *Proceedings of the National Academy of Sciences, 112*(24), 7390–7395.

Bennett, E. M., Cramer, W., Begossi, A., Cundill, G., Díaz, S., Egoh, B. N., Geijzendorffer, R., et al. (2015). Linking biodiversity, ecosystem services, and human well-being: Three challenges for designing research for sustainability. *Current Opinion in Environmental Sustainability, 14*, 76–85.

Berkes, F. (2012). Implementing ecosystem-based management: Evolution or revolution? *Fish and Fisheries, 13*(4), 465–476.

Bodin, Ö. (2017). Collaborative environmental governance: Achieving collective action in social-ecological systems. *Science, 357*(6352), eaan1114.

Borgström, S., Bodin, Ö., Sandström, A., & Crona, B. (2015). Developing an analytical framework for assessing progress toward ecosystem-based management. *Ambio, 44*(3), 357–369.

Boyd, E., Nykvist, B., Borgström, S., & Stacewicz, I. A. (2015). Anticipatory governance for social-ecological resilience. *Ambio, 44*(1), 149–161.

Browder, G. J., Ozment, S., Rehberger, I., Gartner, T., & Lange, G.-M. (2019). *Integrating green and gray: Creating next generation infrastructure*. Washington, DC: World Bank Group and World Resources Institute. Retrieved from http://documents.worldbank.org/curated/en/680391553111128576/Integrating-Green-and-Gray-Creating-Next-Generation-Infrastructure.

Bundy, A., Chuenpagdee, R., Boldt, J. L., de Fatima Borges, M., Camara, M. L., Coll, M., Diallo, I., Fox, C., Fulton, E. A., Gazihan, A., Jarre, A., Jouffre, D., Kleisner, K. M., Knight, B., Link, J., Matiku, P. P., Masski, H., Moutopoulos, D. K., Piroddi, C., Raid, T., Sobrino, I., Tam, J., Thiao, D., Torres, M. A., Tsagarakis, K., van der Meeren, G. I., & Shin, Y.-J. (2017). Strong fisheries management and governance positively impact ecosystem status. *Fish and Fisheries, 18*(3), 412–439.

Carmen, E., Watt, A., Carvalho, L., Dick, J., Fazey, I., Garcia-Blanco, G., Grizzetti, B., Hauck, J., Izakovicova, Z., Kopperoinen, L., Liquete, C., Odeea, D., Steingröver, E., & Young, J. (2018). Knowledge needs for the operationalisation of the concept of ecosystem services. *Ecosystem Services, 29*, 441–451.

Caro, T. (2010). *Conservation by proxy: Indicator, umbrella, keystone, flagship, and other surrogate species*. Washington, DC: Island Press.

Cavender-Bares, J., Polasky, S., King, E., & Balvanera, P. (2015). A sustainability framework for assessing trade-offs in ecosystem services. *Ecology and Society, 20*(1), 17.

Cohen-Shacham, E., Walters, G., Janzen, C., & Maginnis, S. (2016). *Nature-based solutions to address global societal challenges*. Gland, Switzerland: IUCN. Retrieved from https://www.

iucn.org/sites/dev/files/content/documents/nature-based_solutions_to_address_global_societal_challenges.pdf.

Collins, S. L., Carpenter, S. R., Swinton, S. M., Orenstein, D. E., Childers, D. L., Gragson, T. L., Grimm, N. B., Grove, M. J., Harlan, S. L., Kaye, J. P., Knapp, A. K., Kofinas, G. P., Magnuson, J. J., McDowell, W. H., Melack, J. M., Ogden, L. A., Robertson, G. P., Smith, M. D., Whitmer, A. C., & Knapp, A. K. (2011). An integrated conceptual framework for long-term social–ecological research. *Frontiers in Ecology and the Environment, 9*(6), 351–357.

Convention on Biological Diversity (CBD). (2008). UNEP/CBD/COP/DEC/IX/7 9 October 2008.

Culhane, F. E., Robinson, L. A., & Lillebø, A. I. (2020). Approaches for estimating the supply of ecosystem services: Concepts for ecosystem-based management in coastal and marine environments. In T. O'Higgins, M. Lago, & T. H. DeWitt (Eds.), *Ecosystem-based management, ecosystem services and aquatic biodiversity: Theory, tools and applications* (pp. 105–126). Amsterdam: Springer.

Curtice, C., Dunn, D. C., Roberts, J. J., Carr, S. D., & Halpin, P. N. (2012). Why ecosystem-based management may fail without changes to tool development and financing. *Bioscience, 62*, 508–515.

Curtin, CG., & Parker, J.P. (2014). Foundations of resilience thinking. Conservation Biology, 28 (4), 912–923.

Curtin, R., & Prellezo, R. (2010). Understanding marine ecosystem based management: A literature review. *Marine Policy, 34*(5), 821–830.

Davies, C., & Lafortezza, R. (2019). Transitional path to the adoption of nature-based solutions. *Land Use Policy, 80*, 406–409.

DeFries, R., & Nagendra, H. (2017). Ecosystem management as a wicked problem. *Science, 356* (6335), 265–270.

DeWitt, T. H., Berry, W. J., Canfield, T. J., Fulford, R. S., Harwell, M. C., Hoffman, J. C., Johnston, J. M., Newcomer-Johnson, T. A., Ringold, P. L., Russel, M. J., Sharpe, L. A., & Yee, S. J. H. (2020). The final ecosystem goods and services (FEGS) approach: A beneficiary-centric method to support ecosystem-based management. In T. O'Higgins, M. Lago, & T. H. DeWitt (Eds.), *Ecosystem-based management, ecosystem services and aquatic biodiversity: Theory, tools and applications* (pp. 127–148). Amsterdam: Springer.

Di Marco, M., Harwood, T.D., Hoskins, A.J., Ware, C., Hill, S.L., Ferrier, S. (2019). Projecting impacts of global climate and land-use scenarios on plant biodiversity using compositional-turnover modelling. *Global Change Biology*.

Do Yun, S., Hutniczak, B., Abbott, J. K., & Fenichel, E. P. (2017). Ecosystem-based management and the wealth of ecosystems. *Proceedings of the National Academy of Sciences, 114*(25), 6539–6544.

EC. (2000). Directive 2000/60/EC of the European Parliament and of the council establishing a framework for the community action in the field of water policy. *OJ L 327, 22.12.*2000, 1–73. Brussels: European Commission.

EC. (2012). COM (2012) 491 final, Progress of the EU's integrated maritime policy, SWD(2012) 255 final. Brussels: European Commission.

EC. (2015). *Towards an EU research and innovation policy agenda for nature-based solutions & re-Naturing cities*. Final Report of the Horizon 2020 Expert Group on 'Nature-Based Solutions and Re-Naturing Cities'. Directorate-General for Research and Innovation Climate Action, Environment, Resource Efficiency and Raw Materials. Brussels: European Commission. Retrieved from https://ec.europa.eu/newsroom/horizon2020/document.cfm?doc_id=10195.

Enright, S. R., & Boteler, B. (2020). The ecosystem approach in international law. In T. O'Higgins, M. Lago, & T. H. DeWitt (Eds.), *Ecosystem-based management, ecosystem services and aquatic biodiversity: Theory, tools, and applications* (pp. 333–352). Amsterdam: Springer.

Erin Gray, E., Ozment, S., Altamirano, J. C., Feltran-Barbieri, R., & Morales, A. G. (2019). *Green-gray assessment: How to assess the costs and benefits of green infrastructure for water supply systems* (Working paper). Washington, DC: World Resources Institute.

Fulford, R. S., Heymans, S. J. J., & Wu, W. (2020). Mathematical modelling for ecosystem-based management (EBM) and ecosystem goods and services (EGS) assessment. In T. O'Higgins, M. Lago, & T. H. DeWitt (Eds.), *Ecosystem-based management, ecosystem services and aquatic biodiversity: Theory, tools and applications* (pp. 275–290). Amsterdam: Springer.

Gelcich, S., Reyes-Mendy, F., Arriagada, R., & Castillo, B. (2018). Assessing the implementation of marine ecosystem based management into national policies: Insights from agenda setting and policy responses. *Marine Policy, 92*, 40–47.

Gómez, C. M., Delacámara, G., Arévalo-Torres, J., Barbière, J., Barbosa, A. L., Boteler, B., Culhane, F., Daam, M., Gosselin, M.-P., Hein, T., Iglesias-Campos, A., Jähnig, S., Lago, M., Langhans, S., Martínez-López, J., Nogueira, A., Lillebø, A., O'Higgins, T., Piet, G., Pletterbauer, F., Pusch, M., Reichert, P., Robinson, L., Rouillard, J., & Schlüter, M. (2016). The AQUACROSS innovative concept. Deliverable 3.1. AQUACROSS, European Union's Horizon 2020 Framework Programme for Research and Innovation Grant Agreement No. 642317. Technical Report (February 19th, 2016). European Union (H2020 FP Grant Agreement)-AQUACROSS. Retrieved from https://aquacross.eu/sites/default/files/D3.1%20Innovative%20Concept.pdf.

Gómez, C. M., Delacámara G., Jähnig, S., Mattheiss, V., Langhans, S., Domisch, S.,Hermoso, V., Piet, G., Martínez-López, J., Lago, M., Boteler, B., Rouillard, J., Abhold, K., Reichert, P., Schuwirth, N., Hein, T., Pletterbauer, F., Funk, A., Nogueira, A., Lillebø, A., Daam, M., Teixeira, H., Robinson, L., Culhane, F., Schlüter, M., Martin, R., Iglesias-Campos, A., Barbosa, A..L., Arévalo-Torres, J., & O'Higgins, T. (2017). Developing the AQUACROSS assessment framework. Deliverable 3.2, AQUACROSS, European Union's horizon 2020 framework Programme for research and innovation Grant agreement no. 642317. Technical report. European Union (H2020 FP Grant agreement)-AQUACROSS. Retrieved from https://aquacross.eu/sites/default/files/D3.2_Assessment%20Framework.13012017.pdf.

Granek, E. F., Polasky, S., Kappel, C. V., Stoms, D. M., Reed, D. J., Primavera, E., Koch, W., Kennedy, C., Cramer, L. A., Hacker, S. D., Perillo, G. M. E., Aswani, S., Silliman, B., Bael, D., Muthiga, N., Barbier, E. B., & Wolanski, E. (2010). Ecosystem services as a common language for coastal ecosystem-based management. *Conservation Biology, 24*, 207–216.

Grehan, A. J., van den Hove, S., Armstrong, C. W., Long, R., van Rensburg, T., Gunn, V., Mikkelsen, E., ben de Mol, B., & Hain, S. (2009). HERMES: Promoting ecosystem-based management and the sustainable use and governance of deep-water resources. *Oceanography, 22*(1), 154–165.

Grêt-Regamey, A., Rabe, S. E., Crespo, R., Lautenbach, S., Ryffel, A., & Schlup, B. (2014). On the importance of non-linear relationships between landscape patterns and the sustainable provision of ecosystem services. *Landscape Ecology, 29*(2), 201–212.

Grêt-Regamey, A., Altwegg, J., Sirén, E. A., van Strien, M. J., & Weibel, B. (2017). Integrating ecosystem services into spatial planning – A spatial decision support tool. *Landscape and Urban Planning, 165*, 206–219.

Grindle, M. (2010). *Good governance: The inflation of an idea*. Faculty Research Working Paper Series, RWP 10–023, Harvard Kennedy School. Cambridge, Harvard University. Retrieved from https://dash.harvard.edu/handle/1/4448993.

Gunderson, A. R., Armstrong, E. J., & Stillman, J. H. (2016). Multiple stressors in a changing world: The need for an improved perspective on physiological responses to the dynamic marine environment. *Annual Review of Marine Science, 8*, 357–378.

GWP. (2000). *Integrated water resources management*. TAC Background Paper No. 4. Stockholm: Global Water Partnership (GWP). Retrieved from https://www.gwp.org/globalassets/global/toolbox/publications/background-papers/04-integrated-water-resources-management-2000-english.pdf.

Harvey, C. J., Kelble, C. R., & Schwing, F. B. (2017). Implementing "the IEA": Using integrated ecosystem assessment frameworks, programs, and applications in support of operationalizing ecosystem-based management. *ICES Journal of Marine Science, 74*(1), 398–405.

Haines-Young, R., & Potschin, M. B. (2018). Common international classification of ecosystem services (CICES) V5. 1 and guidance on the application of the revised structure. European Environment Agency (EEA). Retrieved from https://cices.eu/content/uploads/sites/8/2018/01/Guidance-V51-01012018.pdf.

Harwell, D. R. (2020). Ecosystem services in U.S. environmental law and governance for the ecosystem-based management practitioner. In T. O'Higgins, M. Lago, & T. H. DeWitt (Eds.), *Ecosystem-based management, ecosystem services and aquatic biodiversity: Theory, tools, and applications* (pp. 373–402). Amsterdam: Springer.

Harwell, M., Molleda, J. L., Jackson, C. A., & Sharpe, L. (2020). Establishing a common framework for strategic communications in ecosystem-based management the natural sciences. In T. O'Higgins, M. Lago, & T. H. DeWitt (Eds.), *Ecosystem-based management, ecosystem services and aquatic biodiversity: Theory, tools, and applications* (pp. 165–188). Amsterdam: Springer.

Heal, G., Daily, G. C., Ehrlich, P. R., & Salzman, J. (2001). Protecting natural capital through ecosystem service districts. *Stanford Environmental Law Journal, 20*, 333.

Herzon, I., Birge, T., Allen, B., Povellato, A., Vanni, F., Hart, K., Radleyd, G., Tucker, G., Keenleyside, C., Oppermann, R., Underwood, E., Poux, X., Beaufoy, G., & Pražan, J. (2018). Time to look for evidence: Results-based approach to biodiversity conservation on farmland in Europe. *Land Use Policy, 71*, 347–354.

Hyytiäinen, K., Ahlvik, L., Ahtiainen, H., Artell, J., Huhtala, A., & Dahlbo, K. (2015). Policy goals for improved water quality in the Baltic Sea: When do the benefits outweigh the costs? *Environmental and Resource Economics, 61*(2), 217–241.

Irvine, K., O'Brien, N., Ravenscroft, L., Cooper, N., Everard, M., Fazey, I., Reed, M. S., & Kenter, J. O. (2016). Ecosystem services and the idea of shared values. *Ecosystem Services, 21*, 184–193.

Joa, B., Winkel, G., & Primmer, E. (2018). The unknown known – A review of local ecological knowledge in relation to forest biodiversity conservation. *Land Use Policy, 79*, 520–530.

Jordan, S. J., O'Higgins, T., & Dittmar, J. A. (2012). Ecosystem services of coastal habitats and fisheries: Multiscale ecological and economic models in support of ecosystem-based management. *Marine and Coastal Fisheries, 4*(1), 573–586.

Kahneman, D. (2011). *Thinking, fast and slow*. Macmillan.

Kalinkat, G., Cabral, J. S., Darwall, W., Ficetola, G. F., Fisher, J. L., Giling, D. P., Gosselin, M.-P., Grossart, H. P., Jähnig, S. C., Jeschke, J. M., Knopf, K., Larsen, S., Onandia, G., Pätzig, M., Saul, W.-C., Singer, G., Sperfeld, E., & Jarić, I. (2017). Flagship umbrella species needed for the conservation of overlooked aquatic biodiversity. *Conservation Biology, 31*(2), 481–485.

Keesstra, S., Nunes, J., Novara, A., Finger, D., Avelar, D., Kalantari, Z., & Cerdà, A. (2018). The superior effect of nature based solutions in land management for enhancing ecosystem services. *Science of the Total Environment, 610*, 997–1009.

Kirkfeldt, T. S. (2019). An ocean of concepts: Why choosing between ecosystem-based management, ecosystem-based approach and ecosystem approach makes a difference. *Marine Policy, 106*, 103541.

Lado, E. P. (2016). *The common fisheries policy: The quest for sustainability*. New York: Wiley.

Lago, M., Boteler, B., Rouillard, J., Abhold, K., Jähnig, S., Iglesias-Campos, A., Delacámara, G., Piet, G., Hein, T., Nogueira, A., Lillebø, A., Strosser, P., Robinson, L., De Wever, A., O'Higgins, T., Schlüter, M., Török, L., Reichert, P., van Ham, C., Villa, F., & McDonald, H. (2019). Introducing the H2020 AQUACROSS project: Knowledge, assessment, and management for AQUAtic biodiversity and ecosystem services aCROSS EU policies. *Science of the Total Environment, 652*, 320–329.

Langhans, S. D., Domisch, S., Balbi, S., Delacámara, G., Hermoso, V., Kuemmerlen, M., Martin, R., Martínez-López, J., Vermeiren, P., Ferdinando Villa, F., & Jähnig, S. C. (2018). Combining eight research areas to foster the uptake of ecosystem-based management in fresh waters. *Aquatic Conservation: Marine and Freshwater Ecosystems, 29*(7), 1161–1173.

Langhans, S. D., Jähnig, S. C., Lago, M., Schmidt-Kloiber, A., & Hein, T. (2019). The potential of ecosystem-based management to integrate biodiversity conservation and ecosystem service provision in aquatic ecosystems. *Science of the Total Environment, 672*, 1017–1020.

Le Tissier, M. (2020). Unravelling the relationship between ecosystem-based management, integrated coastal zone management and marine spatial planning. In T. O'Higgins, M. Lago, & T. H. DeWitt (Eds.), *Ecosystem-based management, ecosystem services and aquatic biodiversity : Theory, tools, and applications* (pp. 403–416). Amsterdam: Springer.

Lewis, N. S., Marois, D. E., Littles, C. J., & Fulford, R. S. (2020). Projecting changes to coastal and estuarine ecosystem goods and services – Models and tools. In T. O'Higgins, M. Lago, & T. H. DeWitt (Eds.), *Ecosystem-based management, ecosystem services and aquatic biodiversity: Theory, tools, and applications* (pp. 235–254). Amsterdam: Springer.

Link, J. S., & Browman, H. I. (2017). Operationalizing and implementing ecosystem-based management. *ICES Journal of Marine Science, 74*(1), 379–381.

Link, J. S., Ihde, T. F., Harvey, C. J., Gaichas, S. K., Field, J. C., Brodziak, J. K. T., Towsend, H. M., & Peterman, R. M. (2012). Dealing with uncertainty in ecosystem models: The paradox of use for living marine resource management. *Progress in Oceanography, 102*, 102–114.

Link, J. S., Dickey-Collas, M., Rudd, M., McLaughlin, R., Macdonald, N. M., Thiele, T., & Rae, M. (2019). Clarifying mandates for marine ecosystem-based management. *ICES Journal of Marine Science, 76*(1), 41–44.

Long, R. D., Charles, A., & Stephenson, R. L. (2015). Key principles of marine ecosystem-based management. *Marine Policy, 57*, 53–60.

Long, R. D., Charles, A., & Stephenson, R. L. (2017). Key principles of ecosystem-based management: The fishermen's perspective. *Fish and Fisheries, 18*(2), 244–253.

Loreau, M., Mouquet, N., & Holt, R. D. (2003). Meta-ecosystems: A theoretical framework for a spatial ecosystem ecology. *Ecology Letters, 6*(8), 673–679.

Lukasiewicz, A., Pittock, J., & Finlayson, M. (2016). Institutional challenges of adopting ecosystem-based adaptation to climate change. *Regional Environmental Change, 16*(2), 487–499.

MacKinnon, K., Sobrevila, C., & Hickey, V. (2008). *Biodiversity, climate change, and adaptation: Nature-based solutions from the World Bank portfolio*. Washington, DC: World Bank. Retrieved from http://documents.worldbank.org/curated/en/149141468320661795/Biodiversity-climate-change-and-adaptation-nature-based-solutions-from-the-World-Bank-portfolio.

Maes, J., & Jacobs, S. (2017). Nature-based solutions for Europe's sustainable development. *Conservation Letters, 10*(1), 121–124.

Marchese, C. (2015). Biodiversity hotspots: A shortcut for a more complicated concept. *Global Ecology and Conservation, 3*, 297–309.

McIntyre, O. (2019). Environmental protection and the ecosystem approach. In S. C. McCaffrey, C. Leb, & R. T. Denoon (Eds.), *Research handbook on international water law* (pp. 126–146). Cheltenham: Edward Elgar Publishing.

Möllmann, C., Folke, C., Edwards, M., & Conversi, A. (2015). Marine regime shifts around the globe: Theory, drivers and impacts. *Philosophical Transactions of the Royal Society, B370*, 20130260.

Nesshöver, C., Assmuth, T., Irvine, K. N., Rusch, G. M., Waylen, K. A., Delbaere, B., et al. (2017). The science, policy and practice of nature-based solutions: An interdisciplinary perspective. *Science of the Total Environment, 579*, 1215–1227.

Nilsson, A. K., & Bohman, B. (2015). Legal prerequisites for ecosystem-based management in the Baltic Sea area: The example of eutrophication. *Ambio, 44*(3), 370–380.

Nunan, F., Menton, M., McDermott, C., & Schreckenberg, K. (2018). Governing for ecosystem health and human wellbeing. In K. Schreckenberg, G. Mace, & M. Poudyal (Eds.), *Ecosystem services and poverty alleviation* (pp. 159–173). Oxon: Routledge.

OECD. (2011). *Water governance in OECD Countries: A multi-level approach*. Paris: OECD Publishing. Retrieved from https://doi.org/10.1787/9789264119284-en.

Ostrom, E. (2009). A general framework for analyzing sustainability of social-ecological systems. *Science, 325*(5939), 419–422.

G. M., Martínez, G., Palomo, I., Dick, J., Casaera, J., van Dijk, J., Priess, J. A., Langemeyer, J., Mustajoki, J., Kopperoinen, L., Baptist, M. J., Peri, P. L., Mukhopadhyay, R., Aszalós, R., Roy, S. B., Luque, S., & Rusch, V. (2018). When we cannot have it all: Ecosystem services trade-offs in the context of spatial planning. *Ecosystem Services, 29*, 566–578.

ESCO. (2018). *United Nations world water assessment programme*. The United Nations world water development Report 2018: Nature-based solutions for water. Paris: UNESCO. Retrieved from https://unesdoc.unesco.org/ark:/48223/pf0000261424.

ghaus, S., & Hake, J. F. (2018). Nexus thinking in current EU policies–the interdependencies among food, energy and water resources. *Environmental Science & Policy, 90*, 183–192.

sser, S.P. (Ed.). (2013). Evolutionary theory and processes: Modern horizons: Papers in honour of Eviatar Nevo. Berlin: Springer.

laert, T., Garcia-Alegre, A. G., Queiroga, H., Cunha e Sá, M. A., & Lillebø, A. I. (2019). Measuring vulnerability of marine and coastal habitats' potential to deliver ecosystem services: Complex Atlantic region as case study. *Frontiers in Marine Science, 6*, 199.

ng, C., Chen, X., Li, Y., Ding, W., & Fu, G. (2018). Water-energy-food nexus: Concepts, questions and methodologies. *Journal of Cleaner Production, 195*, 625–639.

eig, S. (2011). *Chess story*. New York: New York Review of Books.

Pauli, G. A. (2010). *The blue economy: 10 years, 100 innovations, 100 milli* Paradigm Publications.

Pe'er, G., Zinngrebe, Y., Moreira, F., Sirami, C., Schindler, S., Müller, R., B D., Bezák, P., Bonn, A., Hansjürgens, B., Lomba, A., Möckel, S., Pass Schmidt, J., & Lakner, S. (2019). A greener path for the EU commo *Science, 365*(6452), 449–451.

Pires, A. P., Srivastava, D. S., & Farjalla, V. F. (2018). Is biodiversity able from climate change? What we know and what we don't. *Bioscience, 68*

Preiser, R., Biggs, R., De Vos, A., & Folke, C. (2018). Social-ecological adaptive systems: Organizing principles for advancing research meth *Ecology and Society, 23*(4), 46.

Punt, M. J., Weikard, H. P., & van Ierland, E. C. (2014). Game theory and m The effects of conservation autarky in a multiple-use environment. In P. Kumar, & T. Dedeurwaerdere (Eds.), *Handbook on the economics o and biodeversity* (pp. 264–277). Cheltenham: Edward Elgar.

Qiu, J., Carpenter, S. R., Booth, E. G., Motew, M., Zipper, S. C., Kucharik, C. & Turner, M. G. (2018). Understanding relationships among ecosystem se scales and over time. *Environmental Research Letters, 13*(5), 054020.

Robinson, L. A., & Culhane, F. E. (2020). Linkage frameworks: An explorati systems in ecosystem-based management. In T. O'Higgins, M. Lago, & T *Ecosystem-based management, ecosystem services and aquatic biodiversity applications* (pp. 213–234). Amsterdam: Springer.

Samhouri, J. F., Andrews, K. S., Fay, G., Harvey, C. J., Hazen, E. L., Henness K., Hunsicker, M. E., Large, S. I., Marshall, K. N., Stier, A. C., Tam, J. (2017). Defining ecosystem thresholds for human activities and environme California current. *Ecosphere, 8*(6), e01860.

Schultz, L., Folke, C., Österblom, H., & Olsson, P. (2015). Adaptive gov management, and natural capital. *Proceedings of the National Academy of* 7369–7374.

Sharpe, L., Hernandez, C., & Jackson, C. (2020). Prioritizing stakeholders environmental attributes: A tool for ecosystem-based management. In T. O & T. H. DeWitt (Eds.), *Ecosystem-based management, ecosystem services versity: Theory, tools and applications* (pp. 189–212). Amsterdam: Springe

Sklar, F. H., Meeder, J. F., Troxler, T. G., Dreschel, T., Davis, S. E., & Ruiz, Everglades: At the forefront of transition. In E. Wolanski, J. W. Da R. Ramachandran (Eds.), *Coasts and estuaries* (pp. 277–292). Amsterdam:

Smith, M., & Clausen, T. (2015). *Integrated water resource management: A* Marseille: World Water Council (WWC).

Soma, K., van Tatenhove, J., & van Leeuwen, J. (2015). Marine governance in a Regionalization, integration and cooperation for ecosystem-based manag *Coastal Management, 117*, 4–13.

Sutherland, W. J., Gardner, T. A., Haider, L. J., & Dicks, L. V. (2014). Ho traditional knowledge be effectively incorporated into international assessme 1–2.

Tallis, H., Levin, P. S., Ruckelshaus, M., Lester, S. E., McLeod, K. L., Fluharty, B. S. (2010). The many faces of ecosystem-based management: Making the pr in real places. *Marine Policy, 34*(2), 340–348.

Traoré, S., Salles, J.M., Tidball, M. (2018). Ecosystem services, ecosystem economic dynamics: Is it always worth to conserve natural capital? Retrie pdfs.semanticscholar.org/aeb2/d9f6876c0b32c2bfa6597662272bae8a1c93.pdf

Turkelboom, F., Leone, M., Jacobs, S., Kelemen, E., García-Llorente, M., Baró, M., Barton, D. N., Berry, P., Stange, E., Thoonen, M., Kalóczkaik, A., Vadin A. J., Czúczk, B., Röckmann, C., Wurbs, D., Odee, D., Preda, E., Gómez-Bagg

6

Remediation to Restoration to Revitalization: Engaging Communities to Support Ecosystem-Based Management and Improve Human Wellbeing at Clean-Up Sites

Kathleen C. Williams and Joel C. Hoffman

Abstract Remediation to Restoration to Revitalization (R2R2R) is a framework to identify ecological and policy-based relationships between large-scale aquatic sediment remediation projects, subsequent habitat restoration projects, and waterfront revitalization. A defining feature of R2R2R is that it possesses three essential feedback loops: a translational ecology feedback loop, an adaptive management feedback loop, and a project management feedback loop. The R2R2R framework builds on Ecosystem-Based Management (EBM) theory by addressing the role of humans through these feedback loops, and by recognizing the ability of communities to learn and make choices that improve the environment through translational science. In this framework, translating ecological changes from remediation and restoration projects to public benefits (e.g., swimmable water, potential for urban greenspace) using the concept of ecosystem services is critical to support decision-making. In practice, community perceptions and uses of the remediated and restored ecosystem or habitat are central to EBM. We use the Great Lakes Area of Concern program to illustrate how R2R2R exemplifies EBM for large, complex sediment remediation and aquatic habitat restoration projects.

Lessons Learned

- The Remediation to Restoration to Revitalization (R2R2R) framework is integrative of diverse interests through ongoing opportunities for engagement and a synthesis of input to inform research and project alternatives
- Consideration of translational ecology and adaptive management, in addition to the project, create distinct opportunities for engagement with the community, stakeholders, and project implementers

K. C. Williams · J. C. Hoffman (✉)
Great Lakes Toxicology and Ecology Division, United States Environmental Protection Agency, Office of Research and Development, Center for Computational Toxicology and Exposure, Duluth, MN, USA
e-mail: hoffman.joel@epa.gov

- Health Impact Assessment can create science-based, community-relevant, and decision context relevant recommendations

Needs to Advance EBM

- Identify relationships between a positive change in environmental stressors, such as sediment contamination and habitat degradation, and improvements in human health or quality of life
- Case studies inclusive of a broad range of environmental management contexts that contribute to our social capacity for inclusive, equitable decision-making in social-ecological systems

1 Introduction: The R2R2R Framework

Remediation to Restoration to Revitalization (R2R2R) is a framework to identify ecological and policy-based relationships between large-scale sediment remediation projects, subsequent habitat restoration projects, and community revitalization. Ecological outcomes of remediation and restoration may be defined in terms of either ecological quality (e.g., a sediment quality target) or quantity (e.g., acres of wetland restored; Krantzberg 2003). Revitalization outcomes promote human-wellbeing, including social equity, while protecting or improving natural capital (Angradi et al. 2019). However, as a social-ecological system, the connections and feedbacks between remediation, restoration, and revitalization are not well-understood. In the Great Lakes region, remediation and restoration projects along urban waterfronts changed people's interactions with urban, aquatic ecosystems. For example, after projects were complete, increased use of trails and waterways, and changes in economic activity and land uses, improved people's quality of life (Krantzberg 2012; Hartig et al. 2019; Williams and Hoffman in review). However, a framework to identify the connection and feedbacks that led to the change in quality of life has not been available to researchers or managers. This impedes achieving a broader goal of maximizing ecological outcomes and social benefits to human well-being. Therefore, our goal was to develop this framework, recognizing that it should incorporate principles of Ecosystem-Based Management (EBM).

We developed R2R2R in the context of the Great Lakes Area of Concern (AOC) Program. The U.S.-Canada Great Lakes Water Quality Agreement (Annex 1, 2012) defines Areas of Concern as "geographic areas that fail to meet the general or specific objectives of the agreement where such failure has caused or is likely to cause impairment of beneficial use of the area's ability to support aquatic life." When the program was initiated in 1987, 43 AOCs were identified, 31 of which were entirely or partly in U.S. waters. Most AOCs are located near population centers, are within the Great Lakes coastal zone, and are degraded by legacy contaminants including heavy metals and persistent organic pollutants (Hartig et al. 2019). The program recognizes 14 distinct beneficial use impairments (BUIs). Nearly all AOCs have multiple impaired beneficial uses, which arise from multiple causes including sediment and water contamination, habitat loss, excess nutrients and sediment inputs, and improperly functioning storm or sewer systems (Hartig and Zarull 1992). It is common for Superfund sites to be located within an AOC, and for brownfields to be located nearby. The AOC program's goal is to remove BUIs through contaminated sediment remediation, aquatic habitat restoration, or both. In AOC communities, revitalization is primarily targeted towards urban waterfronts, and is inclusive of policies or actions on waterfronts or in adjacent aquatic areas.

Our objective is to describe the R2R2R framework and demonstrate how it builds upon EBM theory. We anticipate the R2R2R framework can be applied broadly to the problem of remediation, for example in brownfields or Superfund programs (Lipps et al. 2017) and recognize that important details of implementation will depend on each program's requirements and legal authorities. For example, under the Comprehensive Environmental Response, Compensation, and Liability Act (CERCLA, also known as Superfund), the definition of natural resources and the criterion for future use (e.g., residential) will influence remediation, restoration, and revitalization decisions (Burger 2008). For the purposes of this chapter, we will speak of R2R2R in general terms, and provide examples specific to the Great Lakes AOC program. We conclude with a case study which illustrates that integrating social and ecological knowledge is possible by integrating a formal decision-support system (Health Impact Assessment) designed to consider diverse knowledge and values into the R2R2R framework.

2 R2R2R Framework as a Decision-Support System

The R2R2R framework describes a process for achieving management goals and project objectives within a social-ecological system (Fig. 1). Generally, R2R2R sites involve substantial engineering to address ecological stressors causing identified ecological impairments. Community members, stakeholders, and agencies choose the acceptable ecological or human risk at the remediation site, the type and quantity of habitats to be restored (within biological constraints), and the future uses and activities that will occur at the site (Krantzberg 2003). Because it is an engineered project within a community, the ecosystem undergoing remediation and restoration is embedded within a social system which may be organized by neighborhoods, user groups, municipalities, and state, federal, and tribal agencies. It is a flexible, adaptive system that is built upon, and responsive to, diverse community members and organizations. As such, R2R2R can also be described as an adaptive co-management system (Folke et al. 2002). As with adaptive co-management, R2R2R focuses on specific places (AOCs) and environmental stressors (BUIs), and emphasizes learning through management actions, adaptively evolving management activities and governance (Dietz et al. 2003; Olsson et al. 2004).

A defining feature of R2R2R is its three feedback loops: a translational ecology (TE) feedback loop, an adaptive management (AM) feedback loop, and a project management (PM) feedback loop (Fig. 1). As illustrated, the R2R2R framework is cyclical and iterative, and thereby explicitly recognizes ecological and social

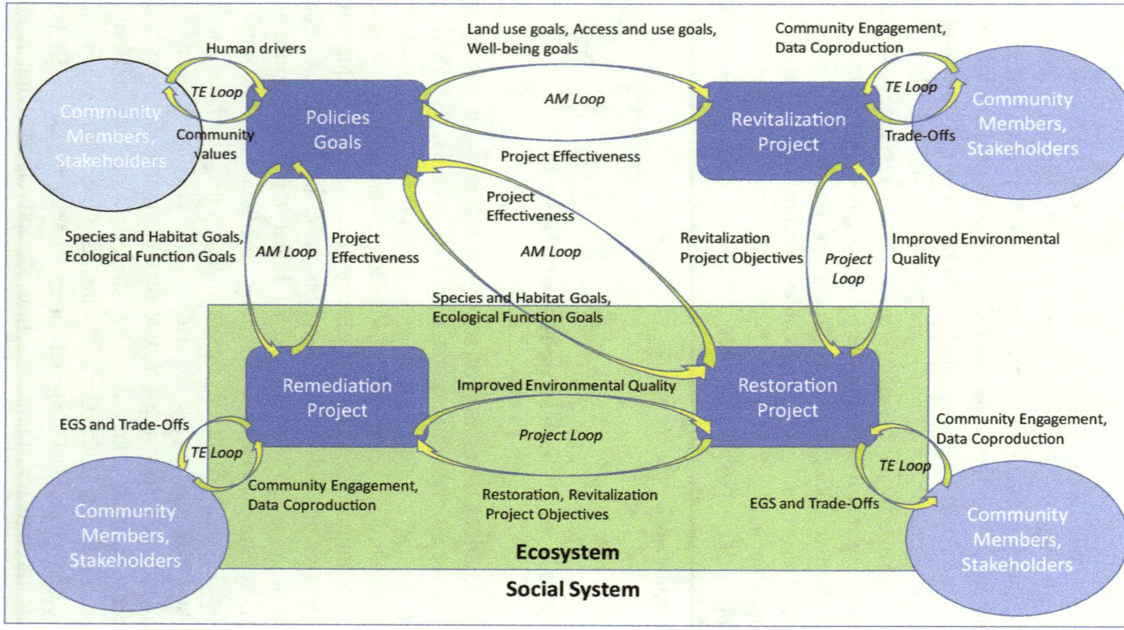

Fig. 1 Social-ecological system diagram of the Remediation to Restoration to Revitalization (R2R2R) framework. The framework includes four discrete stages (dark blue boxes): policy and goal setting, remediation projects, restoration projects, and revitalization projects. An adaptive management (AM) loop connects each of the project types to the policies and goals. The loop is based on the application of relevant goals to the projects and the measurement of project effectiveness to assess the project against the goals. The project loop connects each project, recognizing that the projects are dependent on each other such that the level of environmental quality or ecological integrity obtained is linked to the objectives of the other projects. A translational ecology (TE) loop connects each stage to the community and stakeholder through the application of local knowledge via data coproduction and community engagement to the project and the flow of ecosystem goods and services (EGS) from the project (and related EGS trade-offs when evaluating project alternatives) to benefit community members

complexity and interactions. For example, the cycle could begin with a plan for revitalization, which in turn may spark community, stakeholder, or decision-maker interest in remediation and restoration. These feedback loops are linked. For example, the TE and AM feedback loops intersect during community engagement as the community helps to define the project ecological and social goals.

In translational ecology (TE), scientists and community members collectively identify ecological and social goals, and address problems related to both environment and society (Enquist et al. 2017). Translation is built on knowledge exchange between scientists and stakeholders, and thereby promotes mutual learning. This approach is shared with EBM (Slocombe 1998) and is inherent to the design the Great Lakes AOC program (Krantzberg 2012). The R2R2R framework requires an iterative, participatory TE process that is dynamic and adaptive because the underlying ecological system is complex and the social system includes agencies, non-government organizations, and socio-economically diverse communities. For AOCs, R2R2R is inherently translational; when their original Remedial Action Plans (RAPs) were created in the late-1980s, state agencies were advised by scientists, local governments, and community members (Botts and Muldoon 2005; Hartig and Zarull 1992). The TE feedback loop ideally continues through four discrete stages of the R2R2R process (policies and goals, and remediation, restoration, and revitalization projects; Fig. 1). Scientists engage with community members and stakeholders to gather knowledge and co-produce scientific data, which are translated into potential ecosystem goods and services (EGS), with associated benefits or losses (trade-offs) identified based on project alternatives (e.g. Martin et al. 2018; DeWitt et al. 2020). This information is provided to decision-makers, and their decision in turn affects researchers and community members (Daily et al. 2009; Wall et al. 2017). In TE, a wide range of stakeholders and community members are included. To address related environmental justice concerns, extra effort may be required to engage under-represented and vulnerable populations that are directly or indirectly affected by the contaminated site (Geller et al. 2016).

In the adaptive management (AM) feedback loop, periodic evaluations based on project-effectiveness metrics determine whether the project is meeting identified targets relative to either goals (program or policy scale) or objectives (project scale; Slocombe 1998). If those goals are not met, decision-makers must modify the project to address identified shortcomings, and the cycle of evaluation and modification continues (e.g., McLain and Lee 1996). If decision-relevant endpoints are chosen that convey value to the public regarding the ecosystem services gained or lost through the remedy (e.g., water of sufficient quality to use, fish of sufficient quality to safely consume) or restoration (e.g., wetlands of sufficient quality to support ecologically sensitive, charismatic species), metrics can potentially be used for both public communication and in a translation ecological context (Allan et al. 2015; Angradi et al. 2016; Olander et al. 2017; Wall et al. 2017). Ideally, within the AOC program, AM for remediation and restoration projects begins with discrete management actions to support project implementation, followed by project implementation, monitoring, and evaluation, and ultimately (if impairments are successfully removed) AOC delisting. In between the time when all identified management actions are complete and delisting, there is opportunity for the AOC community (including state agencies, citizen advisory committee, and potentially stakeholders and community members) to address additional remediation and restoration needs as necessary.

Remediation goals are developed by evaluating the risk or impact to the ecosystem from chemical exposure, which can be accomplished formally though ecological risk assessment or through other evaluations to determine ecosystem impairments (Burger 2008). Remedy effectiveness is an approach to determine whether remediation goals are met, wherein physical, chemical, and biological lines of evidence (LOE) are measured prior to, potentially during, and after remediation (e.g., Meier et al. 2015). Multiple LOE such as sediment or porewater contaminant concentration, toxicity tests, contaminant bioaccumulation, indices of biotic integrity, or bioassays are measured to evaluate the ecosystem response. Ideally, at least some LOE have associated targets (i.e., a desired value post-remediation).

Restoration effectiveness is an emerging concept within the R2R2R framework. We assume here that the remediation project has at least some ecosystem restoration goals. The purpose of restoration is to address ecological stressors and in the context of the AOC program, which is focused on riverine, coastal, and riparian habitats, these stressors include aquatic habitat loss or degradation, sedimentation or erosion, and invasive species (Hartig and Zarull 1992). In R2R2R, identified ecological goals include multiple species and habitats, and may address supporting ecological functions (e.g., nutrient cycling or fish spawning habitat). The topic of developing habitat restoration targets has been addressed at-length (e.g., Ruiz-Jaen and Aide 2005; Palmer et al. 2005). Some targets, such as water quality criteria, must specifically address program needs. Other targets which are ecologically relevant (i.e., an index of biotic integrity or metric of stress) and devised for various levels of biological organization (species to ecological community) may support AM decisions.

Revitalization effectiveness is a relatively new concept in Great Lakes AOCs. Remediation has economic benefits, which have been measured in AOCs by hedonics and economic impact analysis (e.g., Isely et al. 2018). For R2R2R, revitalization is best understood as a positive change in community wellbeing (Krantzberg 2003). Community-level social and economic changes can occur in concert with remediation and restoration efforts (Krantzberg 2012). For example, AOC practitioners associate remediation with positive social changes, including increased recreational use of the waterfront and changes in land use or waterfront business (Williams and Hoffman in review). While environmental, economic, social, and governance metrics have been proposed (Angradi et al. 2019), few formal studies exist that demonstrate broad-based changes in community wellbeing or quality of life (Krantzberg 2012).

The PM feedback loop recognizes that the community and scientists may identify project-specific goals that require certain ecological targets to be met at the preceding step. Often, risk assessment, remediation, and restoration are conducted separately and independently; however, integration can achieve efficiencies in terms of time and energy (Burger 2008). In the R2R2R framework, we postulate that integration can yield a greater ecological and social impact than would be achieved by conducting remediation, restoration, and revitalization projects separately. However, we are not aware of studies evaluating the relative success of integrating these elements.

3 R2R2R and Ecosystem-Based Management

As a framework, R2R2R shares foundational elements with Ecosystem-Based Management (EBM; Grumbine 1994; Arkema et al. 2006; Ruckelshaus et al. 2008). These qualities include that it is normative (reflects specific values), principled (aims to improve public good), integrative of different interests by synthesizing a wide range of information and knowledge garnered through engagement and participation, accommodating of complexity and change through feedback loops, is explainable to a wide group of people through the translational ecology loop, and is adaptive (Box 1; Slocombe 1998). In R2R2R, project workflow is relational and directional, such that the outcomes of the last stage (revitalization) are linked to both the initial conditions and success of prior stages (remediation and restoration). As such, ideally, discussions regarding all three stages and the associated projects.

Box 1 Principles of the Remediation to Restoration to Revitalization Framework

Ecological integrity and sustainability

To restore ecological integrity, the program adheres to a sustainable relationship between humans and ecosystems recognizing that the capacity of ecosystems to support life has been substantially diminished because of human actions (i.e., ecological impairments). An important outcome is the creation or restoration of the cultural, spiritual, or experiential relationship between people and the river.

Spatial planning

The spatial distribution of people, resources, and ecosystems is critical to R2R2R. Remediation and restoration are spatially-specific activities that occur within a landscape mosaic of human and ecological communities. Sediment contamination and aquatic habitat loss or degradation are spatially discrete and heterogeneously distributed, though stressors may occur at larger spatial scales. The built environment is amenable to spatial planning.

Effectiveness metrics

Effectiveness metrics inform whether remediation, restoration, and revitalization project objectives are met. Adaptive management can occur in ecological and social domains, and therefore metrics and pre- and post-project monitoring should occur in both domains.

(continued)

Box 1 (continued)

Remediation, restoration, and revitalization adaptively linked

Objectives and outcomes of remediation, restoration, and revitalization projects flow from one project to another, and thus success is interlinked. While remediation and restoration objectives are often established by government agencies, because revitalization objectives are community-based, community values and use objectives can inform remediation and restoration objectives.

Agency of people

Where science and policy knowledge are produced together, the process undertaken to engage stakeholders and the public matters. In the context of decision-support (e.g., health impact assessment), collaborative group processes facilitate mutual learning. Trust and legitimacy, which are founded on facilitation and knowledge sharing, are important for project success.

Social system integral to the framework

The translational feedback loop integrates community values and knowledge with scientific knowledge to inform project objectives, which in turn aims to improve provisioning of ecosystem services for community wellbeing. Recommendations to achieve ecological and social goals arise from community members, stakeholders, and scientists, and are responsive to policy goals (e.g., removal of ecological impairments).

Participatory process that integrates different kinds of knowledge

Translational ecology is used to integrate scientific, local, and traditional ecological knowledge using participatory processes that adhere to principles of democracy, collaboration, communication, and equity. The participatory process is organized as a cooperation among individuals, nongovernment organizations, municipalities, and state, federal, and tribal agencies working at neighborhood, city, reservation, and state scales.

In R2R2R, as with EBM, ecosystem services provisioning is fundamental to sustainability over time (Levin and Lubchenco 2008). Ecosystem goods and services (EGS) are outputs of nature that contribute to human wellbeing when consumed, used, or enjoyed (Bruins et al. 2017). Coastal wetlands and river mouth estuaries (Larson et al. 2013), the most common habitat impacted in AOCs, provide a diverse array of ecosystem services (Sierszen et al. 2012). Beneficial use impairments in AOCs such as fish consumption advisories, beach closings, or dredging restrictions represent ecosystem services loss. Moreover, EGS respond directly to alteration of the biophysical state of the AOC (Yee et al. 2020), provide a consistent and comprehensive set of benefits for consideration, and can be mapped or quantified to illustrate trade-offs to the public (Angradi et al. 2016).

It is important to recognize in Fig. 1 the relationships between ecological integrity and ecosystem services on the ecosystem side, and wellbeing and equity on the social system side (Schoeman et al. 2014; Piet et al. 2020). An accounting of ecosystem services can provide measures of community benefits that may not be easily recognized (Olander et al. 2017). By embedding EGS as the connection to wellbeing and equity, we recognize trade-offs are central to decision-making (Angradi et al. 2016; Martin et al. 2018). As such, recommendations can be made to ensure that impacts do not disproportionately affect under-represented, disadvantaged, and vulnerable groups. Here again, we need decision support processes that integrate different kinds of knowledge, including traditional ecological knowledge (Berkes et al. 2000), recognize the importance of social inclusiveness, and provide recommendations to improve equity.

It is also important to recognize that the restored ecosystem and community are embedded within larger systems, such as a watershed and state, respectively. This external relationship may present concerns related to resiliency, reversibility, source control, long-term stewardship, or other factors that arise from outside the control of the project area or community and which need to be addressed as part of the solution (Adger et al. 2005; Levin and Lubchenco 2008; Palumbi et al. 2008). The Great Lakes AOC program can address AOC-specific structural goals (habitat amount, cultural features) and organizational goals (ecological productivity or connectivity, human use and development), but can only contribute to process goals (biodiversity and evolutionary complexity, quality of life; Slocombe 1998), which are driven by large-scale stressors (e.g., Allan et al. 2015).

One of the animating questions in EBM, and in R2R2R, is *who* will do this work? Substantial challenges exist with respect to our understanding of social-ecological systems for implementing integrated solutions. For example, social dimensions of social-ecological systems remain poorly defined (Brown 2009, 2014), and when considered are limited to studies of scale, governance, and institutions (Brown 2014; Turner 2014). Moreover, the focus on "the functionality of institutions and …normative issues as outcomes," instead of as integral parts of the system (Cote and Nightingale 2012) minimizes the attention given to relations of power, diverging interests, and social identities (Brown 2014; Turner 2014) that often challenge the creation of sustainable solutions. In R2R2R practice, we therefore ask, "what is the role of scientists, decision-makers and managers, public advisor groups, and citizens?" Are roles clarified, aligned, and sufficiently supported for R2R2R to be successful? How do we address ongoing conflict between government processes that traditionally rely on technological knowledge versus community processes that rely on integrated knowledge? Current environmental management practice often treats these different interests as competing, but an agonist approach (Mouffe 1999) suggests that finding ways to treat these different types of knowledge as equal can be productive (Cote and Nightingale 2012). To address the underlying drivers of environmental degradation, we need to engage community morals and values, as well as promote shared benefits of revitalization (Groenfeldt and Schmidt 2013; Daigneau 2015). This is where TE principles (collaboration, engagement, commitment, communication, process, and decision-framing) are critical to connect research and practice (Lawson et al. 2017; Wall et al. 2017).

R2R2R can address potential shortcomings with respect to community engagement that can occur in EBM (Krantzberg 2003). It does so by stipulating community engagement and translational science to occur at each part of the process, recognizing that different community groups may be engaged with and impacted by each part of the process differently. We postulate that for R2R2R to be successful, there must be a translational component that facilitates learning given that remediation and restoration projects generally rely on scientific or technological knowledge (Cote and Nightingale 2012; Partidario and Sheate 2013). To do so requires a decision-support process that follows TE principles and is based on constructivist learning theory, such as impact assessment (e.g., health impact assessment; Partidario and Sheate 2013) or structured decision-making (Sharpe et al. 2020). For implementation, R2R2R requires concepts and strategies from social science that aim to span the boundary between scientific knowledge and decision-making (Mollinga 2010; Williams 2015). Next, we demonstrate how using a constructivist approach can yield a broad range of potential political solutions and recommendations through community engagement (Cote and Nightingale 2012; Evans 2011). The following case study explores how this engagement may produce an integrated solution.

4 Implementing R2R2R Case Study

The U.S. Environmental Protection Agency designated the St. Louis River Area of Concern (AOC) in 1987 owing to historical degradation including discharge of untreated wastewater and debris from industrial and municipal facilities (MPCA and WDNR 1992). The AOC includes the lower 63 km of the St. Louis River. The port cities of Duluth, Minnesota and Superior, Wisconsin are situated at the river's mouth where it flows into Lake Superior. With respect to sediment contamination, chemicals of concern include polychlorinated biphenyls (PCBs), polychlorinated dibenzodioxins (PCDDs, or dioxins), polychlorinated dibenzofurans (PCDFs, or furans), polyaromatic hydrocarbons (PAHs), and heavy metals, all of which are present at multiple locations in the AOC (Crane et al. 2005). State (Minnesota, Wisconsin) and tribal (Fond du Lac Band of Lake Superior Chippewa) agencies coordinate sediment remediation and habitat restoration projects. In Duluth, the neighborhoods adjacent to the river were once home to numerous saw mills, coal tar processing facilities, and steel mills, but today are suffering from poverty and poor health outcomes (Williams et al. 2019).

With sediment remediation underway, the City of Duluth is revitalizing the "St. Louis River Corridor" through the development of active recreation opportunities including trails, regional parks, and improved access to the river in these same neighborhoods (City of Duluth 2018). The AOC and community revitalization processes intersect at the Kingsbury Bay and Grassy Point habitat restoration project. Kingsbury Bay is adjacent to a city park and campground that is also a historical Native American camp, along which a riverfront trail will be improved. Grassy Point is city-owned green space along the riverfront; it is a wetland complex and has an unimproved boat launch and a wetland boardwalk with a fishing pier that are in disrepair. Kingsbury Bay is impacted by long-term excess sedimentation that resulted in aquatic habitat loss. Grassy Point was the location of two former saw mills, resulting in sediment contamination from both wood wastes and dioxins. The two sites are about 2 km apart, and the combined project covers about 1 km^2 of river and involves dredging 268,000 m^3 of sediment. The restoration will beneficially use sediment dredged from Kingsbury Bay to build habitat at Grassy Point Park, while remediation work as Grassy Point includes wood waste removal and covering contaminated sediments. The project will improve river access in economically-disadvantaged neighborhoods with poor access (USEPA 2019). Project objectives include adding a river trail, a swimming beach, canoe and kayak landings, board-walks, fishing piers, and interpretative signage. The project area also has potential for supporting plants including wild rice that are culturally important to indigenous people.

Recognizing the complexity of design decisions, project implication for the health of adjoining neighborhoods, and a diversity of community members and stakeholders, we conducted a health impact assessment (HIA) to create evidence-based recommendations to inform the Minnesota Department of Natural Resources project design (i.e., sediment remediation and aquatic habitat restoration) and the City of Duluth park planning (i.e., site access and amenities). Health impact assessment (HIA) is a systematic process that uses a variety of data sources and community and stakeholder consultation to ascertain potential health impacts of a policy change or decision, and to make recommendations that mitigate negative health impacts or enhance positive health impacts (National Research Council 2011). Health impact assessment is built upon democratic participation, health equity, and a comprehensive approach to health (i.e., community wellbeing; Quigley et al. 2006). Further, HIA assumes the community is a stakeholder (Human Impact Partners 2011), which is important given a lack of consensus regarding long-term plans for redevelopment along the river (Johnston et al. 2017).

The inclusion of community input early in the process ensures community values are considered at a point when values can influence recommendations provided as part of the HIA (Iroz-Elardo 2014). In the Kingsbury Bay-Grassy Point habitat restoration project, we started the process with two introductory meetings for stakeholders, one for the general public and one for group representatives (i.e., city planners, non-government organizations, and natural resource managers). The meetings were nearly identical and included an introduction to HIA, an overview of the project, and opportunities for discussion and input. To ensure opportunities for meaningful inclusion, we captured different types of knowledge and experience. Participants in both meetings were invited to share their experiences with the two sites by placement of sticky notes on project maps, and to explain how the changes would impact their daily lives or experiences (Boschmann and Cubbon 2014; Johnston et al. 1995). The submitted comments reflected a variety of personal and professional experiences and conveyed individual perspectives regarding sense of place and identity. Comments included concern about park maintenance, safety, access, traffic, invasive species management, disturbance of an adjacent Superfund site, and identified opportunities for social gathering, birding, fishing, and boating.

Health impact assessment is designed to mediate power relations through an inclusionary and iterative process. In the Kingsbury Bay-Grassy Point habitat restoration project HIA, all submitted comments were analyzed to identify the social and environmental determinants of health and wellbeing valued by community members and stakeholders (Marmet et al. 2008). This community-based information was used to construct health pathways, which identify how project elements are likely to effect health outcomes. We identified impacted ecosystem services to relate the proposed biophysical changes at the site to public-friendly endpoints such as swimmable water or edible fish, and then to health benefits. Transparency was ensured throughout the process through ongoing communication and opportunities for input. Detailed meeting notes that include all comments, questions, and participants were distributed after each meeting, and community members and stakeholders could make suggestions or corrections to any set of notes produced. Community members and stakeholders were invited to participate in technical and community committees that advise the HIA through each step.

In total, the HIA yielded 77 unique recommendations to improve project health outcomes. Recommendations addressed trail safety, environmental quality, cultural resources, and social places for gathering. Recommendations also addressed the ecological quality of the project, including to protect existing high-quality wetlands and to implement invasive species controls. Importantly, the process was widely endorsed by community members, stakeholders and decision-makers as positive and constructive. We found that HIA is helpful in the AOC context because it provides a means to explore how sediment remediation, habitat restoration, and park management are interrelated, and how this will collectively impact community health as defined by community members and stakeholders. In this case study, the HIA informed both final landscape design for habitat restoration and will be included in future park master planning.

Broadly, HIA integrates with the three loops of R2R2R, and therefore may be widely useful. In this case study, HIA served as a formal approach to decision support that facilitated translational science. Other decision support approaches might also be used to support TE (e.g., Sharpe et al. 2020). The TE loop included the community through engagement and data co-production; community knowledge was foundational to the science-based pathways for impact analysis. The AM loop related HIA recommendations to restoration projects and city park plans, which changed in response to research and collective discussions about the value of Kingsbury Bay and Grassy Point Park. Finally, the PM loop linked revitalization outcomes such as safe swimming beaches and birdwatching opportunities to ecosystem services that were necessary to restore through remediation and restoration.

5 Conclusions

It has long been recognized that successful implementation of an ecosystem approach to integrate environmental and social decision-making requires attention to both the substance and process of the approach (Slocombe 1993; McLain and Lee

1996). The R2R2R framework builds on EBM theory by addressing the role of humans (e.g., natural resource agencies, community members) through multiple feedback loops, and recognizing their ability to learn and make choices that improve the environment through translational science. We demonstrate that integrating social and ecological knowledge is possible by utilizing the R2R2R framework with a specific process (HIA) designed to consider diverse knowledge and values. Ecosystem-Based Management is difficult because the benefits humans and society derive from ecosystem processes cannot be viewed as objectively existing "out there," but as entangled in social and political processes (Ernston 2013). The R2R2R framework provides interlinking loops of translational ecology, adaptive management, and project management as a system for integrating these diverse processes. As a community of practice, R2R2R is relatively new, and we anticipate that ongoing remediation and restoration efforts in the Great Lakes and elsewhere will contribute to our social capacity for inclusive, equitable decision-making in social-ecological systems.

Disclaimer This chapter has been subjected to Agency review and has been approved for publication. The views expressed in this paper are those of the author(s) and do not necessarily reflect the views or policies of the U.S. Environmental Protection Agency.

References

Adger, W. N., Hughes, T. P., Folke, C., Carpenter, S. R., & Rockström, J. (2005). Social-ecological resilience to coastal disasters. *Science, 309*, 1036–1039.

Allan, J. D., Smith, S. D., McIntyre, P. B., Joseph, C. A., Dickinson, C. E., Marino, A. L., Biel, R. G., Olson, J. C., Doran, P. J., Rutherford, E. S., & Adkins, J. E. (2015). Using cultural ecosystem services to inform restoration priorities in the Laurentian Great Lakes. *Frontiers in Ecology and the Environment, 13*, 418–424.

Angradi, T. R., Launspach, J. J., Bolgrien, D. W., Bellinger, B. J., Starry, M. A., Hoffman, J. C., Trebitz, A. S., Sierszen, M. E., & Hollenhorst, T. P. (2016). Mapping ecosystem service indicators in a Great Lakes Area of Concern. *Journal of Great Lakes Research, 42*, 717–727.

Angradi, T. R., Williams, K. C., Hoffman, J. C., & Bolgrien, D. W. (2019). Goals, beneficiaries, and indicators of waterfront revitalization in Great Lakes Areas of Concern: A natural capital perspective. *Journal of Great Lakes Research, 45*, 815–863.

Arkema, K. K., Abramson, S. C., & Dewsbury, B. M. (2006). Marine ecosystem-based management: From characterization to implementation. *Frontiers in Ecology and the Environment, 4*, 525–532.

Berkes, F., Colding, J., & Folke, C. (2000). Rediscovery of traditional ecological knowledge as adaptive management. *Ecological Applications, 10*, 1251–1262.

Boschmann, E. E., & Cubbon, E. (2014). Sketch maps and qualitative GIS: Using cartographies of individual spatial narratives in geographic research. *The Professional Geographer, 66*, 236–248.

Botts, L., & Muldoon, P. R. (2005). *Evolution of the Great Lakes water quality agreement.* Michigan State University Press.

Brown, K. (2009). Human development and environmental governance: A reality check. In W. N. Adger & A. Jordan (Eds.), *Governing sustainability* (pp. 32–51). London: Cambridge University Press.

Brown, K. (2014). Global environmental change I: A social turn for resilience? *Progress in Human Geography, 38*, 107–117.

Bruins, R. J., Canfield, T. J., Duke, C., Kapustka, L., Nahlik, A. M., & Schäfer, R. B. (2017). Using ecological production functions to link ecological processes to ecosystem services. *Integrated Environmental Assessment and Management, 13*, 52–61.

Burger, J. (2008). Environmental management: Integrating ecological evaluation, remediation, restoration, natural resource damage assessment and long-term stewardship on contaminated lands. *Science of the Total Environment, 400*, 6–19.

City of Duluth. (2018). *St. Louis river corridor: Connecting people to the river*. Retrieved from http://duluthmn.gov/media/543434/final-stlouis-river-corridor-brochure-2018.pdf.

Cote, M., & Nightingale, A. J. (2012). Resilience thinking meets social theory: Situating social change in socio-ecological systems (SES) research. *Progress in Human Geography, 36*, 475–489.

Crane, J. L., Richards, C., Breneman, D., Lozano, S., & Schuldt, J. A. (2005). Evaluating methods for assessing sediment quality in a Great Lakes embayment. *Aquatic Ecosystem Health and Management, 8*, 323–349.

Daigneau E. (2015). Just green enough. *Governing*. [online]. Retrieved from https://www.governing.com/topics/transportation-infrastructure/gov-green-gentrification-series.html.

Daily, G. C., Polsaky, S., Goldstein, J., Kareiva, P. M., Mooney, H. A., Pejchar, L., Ricketts, T. H., Salzman, J., & Shallenberg, R. (2009). Ecosystem services in decision making: Time to deliver. *Frontiers in Ecology and the Environment, 7*, 21–28.

DeWitt, T. H., Berry, W. J., Canfield, T. J., Fulford, R. S., Harwell, M. C., Hoffman, J. C., Johnston, J. M., Newcomer-Johnson, T. A., Ringold, P. L., Russel, M. J., Sharpe, L. A., & Yee, S. J. H. (2020). The final ecosystem goods and services (FEGS) approach: A beneficiary-centric method to support ecosystem-based management. In T. O'Higgins, M. Lago, & T. H. DeWitt (Eds.), *Ecosystem-based management, ecosystem services and aquatic biodiversity: Theory, tools and applications* (pp. 127–148). Amsterdam: Springer.

Dietz, T., Ostrom, E., & Stern, P. L. (2003). The struggle to govern the commons. *Science, 31*, 1907–1912.

Enquist, C. A. F., Jackson, S. T., Garfin, G. M., Davis, F. W., Gerber, L. R., Littell, J. A., Tank, J. L., Terando, A. D., Wall, T. U., Halpern, B., Hiers, J. K., Morelli, T. L., McNie, E., Stephenson, N. L., Williamson, M. A., Woodhouse, C. A., Yung, L., Brunson, M. W., Hall, K. R., Hallett, L. M., Lawson, D. M., Moritz, M. A., Nydick, K., Pairis, A., Ray, A. J., Regan, C., Safford, H. D., Schwartz, M. W., & Shaw, M. R. (2017). Foundations of translational ecology. *Frontiers in Ecology and the Environment, 15*, 541–550.

Ernston, H. (2013). The social production of ecosystem services: A framework for studying environmental justice and ecological complexity in urbanized landscapes. *Landscape and Urban Planning, 109*, 7–17.

Evans, J. P. (2011). Resilience, ecology and adaptation in the experimental city. *Transactions of the Institute of British Geographers, 36*, 223–237.

Folke, C., Carpenter, S., Elmqvist, T., Gunderson, L., Holling, C. S., & Walker, B. (2002). Resilience and sustainable development: Building adaptive capacity in a world of transformations. *Ambio A Journal of the Human Environment, 31*, 347–440.

Geller, A. M., Breville, M., Eisenhauer, E., Sykes, K., Fulk, F., Quackenboss, J., Zartarian, V., Jarabeck, A., Lee, C., Manibusan, M., Oxendine, S., Snyder, E., & Williams, K. (2016). *Environmental justice research roadmap, EPA/601/R-16/006*. Washington, DC: U.S. Environmental Protection Agency.

Groenfeldt, D., & Schmidt, J. J. (2013). Ethics and water governance. *Ecology and Society, 18*, 14.

Grumbine, R. E. (1994). What is ecosystem management? *Conservation Biology, 8*, 27–38.

Hartig, J. H., & Zarull, M. A. (1992). *Under RAPs: Toward grassroots ecological democracy in the Great Lakes basin*. Ann Arbor: University of Michigan Press.

Hartig, J. H., Krantzberg, G., Austin, J. C., & McIntyre, P. M. (2019). *Great lakes revival: How restoring polluted waters leads to rebirth of Great Lakes communities*. Ann Arbor: International Association of Great Lakes Research.

Human Impact Partners. (2011). *A health impact assessment toolkit: A handbook for conducting HIA* (3rd ed.). Oakland: Human Impact Partners.

Iroz-Elardo, N. (2014). Health impact assessment as community participation. *Community Development Journal, 50*, 280–295.

Isely, P., Isely, E. S., Hause, C., & Steinman, A. D. (2018). A socioeconomic analysis of habitat restoration in the Muskegon Lake area of concern. *Journal of Great Lakes Research, 44*, 330–339.

Johnston, R. J., Weaver, T. F., Smith, L. A., & Swallow, S. K. (1995). Contingent valuation focus groups: Insights from ethnographic interview techniques. *Agricultural and Resource Economics Review, 24*, 56–69.

Johnston, J. M., de Jesus Crespo, R., Harwell, M. C., Jackson, C., Myer, M., Seeteram, N., Williams, K., Yee, S., & Hoffman, J. (2017). *Valuing community benefits of final ecosystem goods and services: Human health and ethnographic approaches as complements to economic valuation, EPA/600/R-17/309*. Washington, DC: U.S. Environmental Protection Agency.

Krantzberg, G. (2003). Keeping remedial action plans on target: Lessons learned from Collingwood Harbor. *Journal of Great Lakes Research, 29*, 641–651.

Krantzberg, G. (2012). First off the list: The Collingwood Harbour study. In V. I. Grover & G. Krantzberg (Eds.), *Great Lakes: Lessons in participatory governance* (pp. 257–267). Boca Raton: CRC Press.

Larson, J., Trebitz, A. S., Steinman, A. D., Wiley, M., Carlson-Mazur, M., Pebbles, V., Braun, H., & Seelbach, P. (2013). Great Lakes rivermouth ecosystems: Scientific synthesis and management implications. *Journal of Great Lakes Research, 39*, 513–524.

Lawson, D. M., Hall, K. R., Yung, L., & Enquist, C. A. F. (2017). Building translational ecology communities of practice: Insights from the field. *Frontiers in Ecology and the Environment, 15*, 569–577.

Levin, S. A., & Lubchenco, J. (2008). Resilience, robustness, and marine ecosystem-based management. *Bioscience, 58*, 27–32.

Lipps, J., Harwell, M., Kravitz, M., Lynch, K., Mahoney, M., Pachon, C., & Pluta, B. (2017) *Ecosystem services at contaminated site cleanup*, Engineering Forum Issue Paper EPA 542-R-17-004. U.S. Environmental Protection Agency, Washington, DC.

Marmet, M., Friel, S., Bell, R., Houweling, T. A., & Taylor, S. (2008). Closing the gap in a generation: Health equity through action on the social determinants of health. *The Lancet, 372*, 1661–1669.

Martin, D. M., Mazzotta, M., & Bousquin, J. (2018). Combining ecosystem services assessment with structured decision making to support ecological restoration planning. *Environmental Management, 62*, 608–618.

McLain, R. J., & Lee, R. G. (1996). Adaptive management: Promises and pitfalls. *Journal of Environmental Management, 20*, 437–448.

Meier, J. R., Lazorchak, J. M., Mills, M., Wernsing, P., & Baumann, P. (2015). Monitoring exposure of brown bullheads and benthic macroinvertebrates to sediment contaminants in the Ashtabula River before, during, and after remediation. *Environmental Toxicology and Chemistry, 34*, 1267–1276.

Mollinga, P. P. (2010). Boundary work and the complexity of natural resources management. *Crop Science, 50*(Supplement 1), S1–S9.

Mouffe, C. (1999). Deliberative democracy or agonistic pluralism? *Social Responsibility, 66*, 745–758.

Minnesota Pollution Control Agency and Wisconsin Department of Natural Resources (1992) The St. Louis River System Remedial Action Plan: Stage One. Minnesota Pollution Control Agency, St. Paul.

National Research Council. (2011). *Improving health in the United States: The role of health impact assessment*. Washington, DC: National Academies Press.

Olander, L., Polasky, S., Kagan, J. S., Johnston, R. J., Wainger, L., Saah, D., Maguire, L., Boyd, J., & Yoskowitz, D. (2017). So you want your research to be relevant? Building the bridge between ecosystem services research and practice. *Ecosystem Services, 26*, 170–182.

Olsson, P., Folke, C., & Berkes, F. (2004). Adaptive comanagement for building resilience in social-ecological systems. *Environmental Management, 34*, 874–890.

Palmer, M. A., Bernhardt, E. S., Allan, J. D., Lake, P. S., Alexander, G., Brooks, S., Carr, J., Clayton, S., Dahm, C. N., Follstad Shah, J., Galat, D. L., Loss, S. G., Goodwin, P., Hart, D. D., Hassett, B., Jenskinson, R., Kondolf, G. M., Lave, R., Meyer, J. L., O'Donnell, T. K., Pagano, L., & Sudduth, E. (2005). Standards for ecologically successful river restoration. *Journal of Applied Ecology, 42*, 208–217.

Palumbi, S. R., McLeod, K. L., & Grünbaum, D. (2008). Ecosystems in action: Lessons from marine ecology about recovery, resistance, and reversibility. *Bioscience, 58*, 33–42.

Partidario, M. R., & Sheate, W. R. (2013). Knowledge brokerage-potential for increased capacities and shared power in impact assessment. *Environmental Impact Assessment Review, 39*, 26–36.

Piet, G., Delacámara, G., Kraan, M., Röckmann, G. C., & Lago, M. (2020). Advancing aquatic ecosystem-based management with full consideration of the social-ecological system. In T. O'Higgins, M. Lago, & T. H. DeWitt (Eds.), *Ecosystem-based management, ecosystem services and aquatic biodiversity: Theory, tools and applications* (pp. 17–38). Amsterdam: Springer.

Quigley, R., Furu, L. P., Bond, A., Cave, B., & Bos, R. (2006). *Health impact assessment international best practice principles*. Special publication series no 5. Retrieved March 24, 2017 from http://activelivingresearch.org/files/IAIA_HIABestPractice_0.pdf.

Ruckelshaus, M., Klinger, T., Knowlton, M., & DeMaster, D. P. (2008). Marine ecosystem-based management in practice: Scientific and governance challenges. *BioScience, 58*, 53–63.

Ruiz-Jaen, M. C., & Aide, T. M. (2005). Restoration success: How is it being measured? *Restoration Ecology, 13*, 569–577.

Schoeman, J., Allan, C., & Finlayson, M. (2014). A new paradigm for water? A comparative review of integrated, adaptive and ecosystem-based water management in the Anthropocene. *International Journal of Water Resources Development, 30*, 377–390.

Sharpe, L., Hernandez, C., & Jackson, C. (2020). Prioritizing stakeholders, beneficiaries and environmental attributes: A tool for ecosystem-based management. In T. O'Higgins, M. Lago, & T. H. DeWitt (Eds.), *Ecosystem-based management, ecosystem services and aquatic biodiversity: Theory, tools and applications* (pp. 189–212). Amsterdam: Springer.

Sierszen, M. E., Morrice, J. A., Trebitz, A. S., & Hoffman, J. C. (2012). A review of selected ecosystem services provided by coastal wetlands of the Laurentian Great Lakes. *Aquatic Ecosystem Health and Management, 15*, 92–106.

Slocombe, D. S. (1993). Implementing ecosystem-based management. *Bioscience, 43*, 612–621.

Slocombe, D. S. (1998). Defining goals and criteria for ecosystem-based management. *Environmental Management, 22*, 483–493.

Turner, M. D. (2014). Political ecology I: An alliance with resilience? *Progress in Human Geography, 38*, 616–623.

USEPA. (2019). Kingsbury bay and grassy point: A health impact assessment. Report (in preparation).

Wall, T. U., McNie, E., & Garfin, G. M. (2017). Use-inspired science: Making science usable by and useful to decision makers. *Frontiers in Ecology and the Environment, 15*, 551–559.

Williams, K. C. (2015). Building bridges in the Great Lakes: How objects and organization facilitate collaboration across boundaries. *Journal of Great Lakes Research, 41*, 180–187.

Williams, K. C., & Hoffman, J. C. (2020). Learning in Great Lakes Areas of Concern—connecting remediation, restoration, and revitalization. In: J. H. Hartig, & M. Munawar (Eds.), *Restoring Great Lakes Areas of Concern: A story of struggle and success, ecovision world monograph series* (in press). East Lansing, MI: Aquatic Ecosystem Health and Management Society.

Williams, K., Hoffman, J., & French, N. (2019). From remediation to restoration and revitalization: The St. Louis River story. In J. Hartig, G. Krantzberg, J. C. Austin, & P. McIntyre (Eds.), *Great Lakes revival: How restoring polluted waters leads to rebirth of Great Lakes communities* (pp. 61–66). Ann Arbor: International Association of Great Lakes Research.

Yee, S., Cicchetti, G., DeWitt, T. H., Harwell, M. C., Jackson, S. K., Pryor, M., Rocha, K., Santavy, D. L., Sharpe, L., & Shumchenia, E. (2020). The ecosystem services gradient: A descriptive model for identifying thresholds of meaningful change. In T. O'Higgins, M. Lago, & T. H. DeWitt (Eds.), *Ecosystem-based management, ecosystem services and aquatic biodiversity: Theory, tools and applications* (pp. 291–308). Amsterdam: Springer.

7

An Integrated Multi-Model Decision Support Framework for Evaluating Ecosystem-Based Management Options for Coupled Human-Natural Systems

Robert B. McKane, Allen F. Brookes, Kevin S. Djang, Jonathan J. Halama, Paul B. Pettus, Bradley L. Barnhart, Marc Russell, Kellie B. Vache, and John P. Bolte

Abstract Simulation models offer a way to achieve a comprehensive understanding of the consequences of alternative community planning scenarios. For example, a community might want to understand how a particular decision—such as expanding an urban growth boundary into lands zoned for agriculture—will result in ecological, economic, and social tradeoffs for various stakeholder groups. This chapter explores the utility of ENVISION, a spatially-explicit decision support framework that integrates various ecological and human systems model "plug-ins" for informing Ecosystem-Based Management (EBM) options. While ENVISION already has a reasonably large tool box of such plug-ins, its usefulness could be further extended to address a wider range of community and ecosystem types. We specifically examine how a suite of existing U.S. Environmental Protection Agency decision support tools (VELMA, HexSim, CORESET, Coral PF and others) could significantly extend ENVISION's plug-in toolbox for coastal ecosystem EBM, inclusive of terrestrial-marine interactions and restoration goals of coastal communities dependent on marine ecosystem services.

R. B. McKane (✉) · A. F. Brookes · J. J. Halama · P. B. Pettus
U.S. Environmental Protection Agency, Corvallis, OR, USA
e-mail: Mckane.Bob@epa.gov

K. S. Djang
Inoventures LLC, c/o U.S. Environmental Protection Agency, Corvallis, OR, USA

B. L. Barnhart
National Council for Air and Stream Improvement, Inc., Corvallis, OR, USA

M. Russell
US Environmental Protection Agency, Center for Computational Toxicology and Exposure, Gulf Breeze, FL, USA

K. B. Vache · J. P. Bolte
Oregon State University, Corvallis, OR, USA

Lessons Learned

- Advances in applications of ENVISION over the past decade—e.g., the Willamette Water 2100 application—represent a major step forward for identifying EBM solutions to intertwined and seemingly insoluble (aka, wicked) environmental-economic-social problems.
- Depending upon the complexity and scope of EBM objectives, applications of ENVISION or similar human-natural systems modeling frameworks can involve substantial effort and cost.
- Costs/benefits of applications of ENVISION or similar tools should be weighed against alternatives, such as applications of stand-alone EBM models, reliance on purely empirical studies, or some combination.

Needs to Advance EBM

- Develop and apply model plug-ins that extend ENVISION's applicability to coastal ecosystems, particularly for identifying policies and EBM best practices for reducing sources and runoff of terrestrial pollutants to estuarine ecosystems.

1 Introduction

Communities invest significant time and resources in planning and want to understand the consequences of decisions they are considering. Those consequences can be complex and far reaching, affecting a wide range of stakeholders having different priorities. One way to attempt to understand the consequences of change is by using models. For example, a community might want to understand how forest harvest will affect a watershed's capacity to provide clean drinking water. For this context they might model the hydrology and biogeochemistry of selected areas to estimate how stream flow and nutrient loading may change as a result of logging. But the consequences do not end with flow and nutrients. How, where, and when the logging is done will also affect costs, jobs, revenue, fish and wildlife populations, and recreation inside and outside the harvested areas.

To gain a more comprehensive understanding of the ecological, economic and social consequences of any given land management decision option, the community would ideally use a framework capable of modeling these consequences in an integrated way, since each can directly or indirectly influence the others. Integrated modeling that supports Ecosystem-Based Management (EBM) can be very difficult and time consuming. To simplify the task, communities would benefit from a modeling framework with a tool box of plug-in models that can be tailored to address their specific needs.

Such an EBM framework would need to allow integration of varied plug-in models with a way for the models to share data. Much of the data will be spatial, so the framework must provide a shared GIS platform. In addition, there should be a way to test multiple scenarios and to provide optimization. Finally, there should be a way to provide constraints on what decisions can be made corresponding to a location's laws and policies (e.g., land use laws).

One such framework is ENVISION, developed primarily at Oregon State University (Bolte and Vache 2010; Santelmann et al. 2012; Spies et al. 2014; Bradley et al. 2016; Villarreal et al. 2017). ENVISION provides all the features listed above and is a mature, publicly available product (http://envision.bioe.orst.edu/Downloads.aspx). Using an extensible array of integrated plug-in models, ENVISION has been used to analyze ecological, economic and social tradeoffs in response to alternative future scenarios for coupled human-natural systems in a variety of locations. Some examples include the Puget Sound region in Washington, the Big Woods drainage in Idaho, and the Willamette River Basin and Tillamook Bay in Oregon. (http://envision.bioe.orst.edu/CaseStudies.aspx).

Goals of this chapter are to provide (1) an overview of ENVISION and an example of a regional-scale application; (2) an example of the process by which an existing spatial model can be integrated with ENVISION as a plug-in; and (3) recommendations for developing additional model plug-ins to further extend ENVISION's capabilities for integrated EBM planning. Our intent is to illustrate a path for integrating various existing EBM models within a well-established decision support framework that community planners and stakeholders can use to explore ecological, economic, and social tradeoffs associated with different decision options.

2 ENVISION—A Decision Support Tool for Ecosystem-Based Management

2.1 ENVISION Overview

ENVISION is a framework for constructing alternative future scenario applications concerning ecological, economic and social outcomes of interest to communities and regional planners. The framework consists of four main components (http://envision.bioe.orst.edu/About.aspx):

1. A dynamic spatial (GIS) engine for representing polygonal, network, point, and grid-based landscape characterizations.
2. A multi-agent modeling framework for representing values and behaviors of different decision-makers (agents) on the landscape.
3. A rich representation of policies guiding and constraining agent decision-making, and scenarios describing alternative strategies for landscape management.
4. An extensible plug-in architecture for including:
    (a) Any number of conformant autonomous process models describing landscape change dynamics;
    (b) Any number of "evaluative models" reporting landscape production metrics, typically measured in terms of scarcity along biophysical, ecological, social or economic dimensions;

(c) Any number of "visualizers" for visual representation of spatial data and inputs.

Figure 1 illustrates how these ENVISION components are integrated and feed back upon each other over time and space.

Other EBM decision support frameworks—notably InVEST and ARIES—share many of these capabilities (see Lewis et al. 2020). ENVISION stands out for its emphasis on the integration of ecological, social and economic models (plug-ins) with a sophisticated agent-based modeling subsystem. It also enables a rich representation of local and regional policies guiding and constraining actions of human decision-makers (agents) in landscape-scale simulations. Agents can be individuals, such as landowners and other citizens; or organizations and institutions, such as governments and businesses.

A main purpose of ENVISION's human-natural systems modeling approach is to facilitate discourse among different decision-makers and enable them to interactively play out and compare consequences of alternative management and regulatory choices.

2.2 *Example Application—ENVISION Willamette*

The ENVISION Willamette Water 2100 project—aka, ENVISION Willamette—is a good example of this framework's capabilities. ENVISION Willamette is a regional human-natural systems application that closely engaged community and regional decision-makers concerned with current and future supplies, usage and management of water resources within the 30,000 km^2 Willamette River Basin in the state of Oregon, USA (e.g., Bolte et al. 2011; Jaeger et al. 2017). The Willamette River is the 13th largest river in the USA. The river and its basin support a mosaic of agricultural, timber, recreational resources and several rapidly growing urban centers dependent on surface water supplies.

The ENVISION Willamette project involved a collaborative effort of Oregon State University, the University of Oregon, Portland State University, and the University of California–Santa Barbara. The project team used a structured decision-making approach to set up this ENVISION application (Fig. 2). Through a series of stakeholder engagement workshops, local, state, and federal stakeholders identified pressures and ecosystems services deemed important to their immediate and long-term ecological, economic, and social goals. Models of landscape processes relevant to these goals were then identified and plugged in to ENVISION. Plug-in models included forest ecosystem dynamics, land use change, human population growth, watershed hydrology, and storage of carbon within ecosystems (see next section for plug-in details). Relevant spatial datasets, policies and decision alternatives were then assembled to inform these models and build project scenarios.

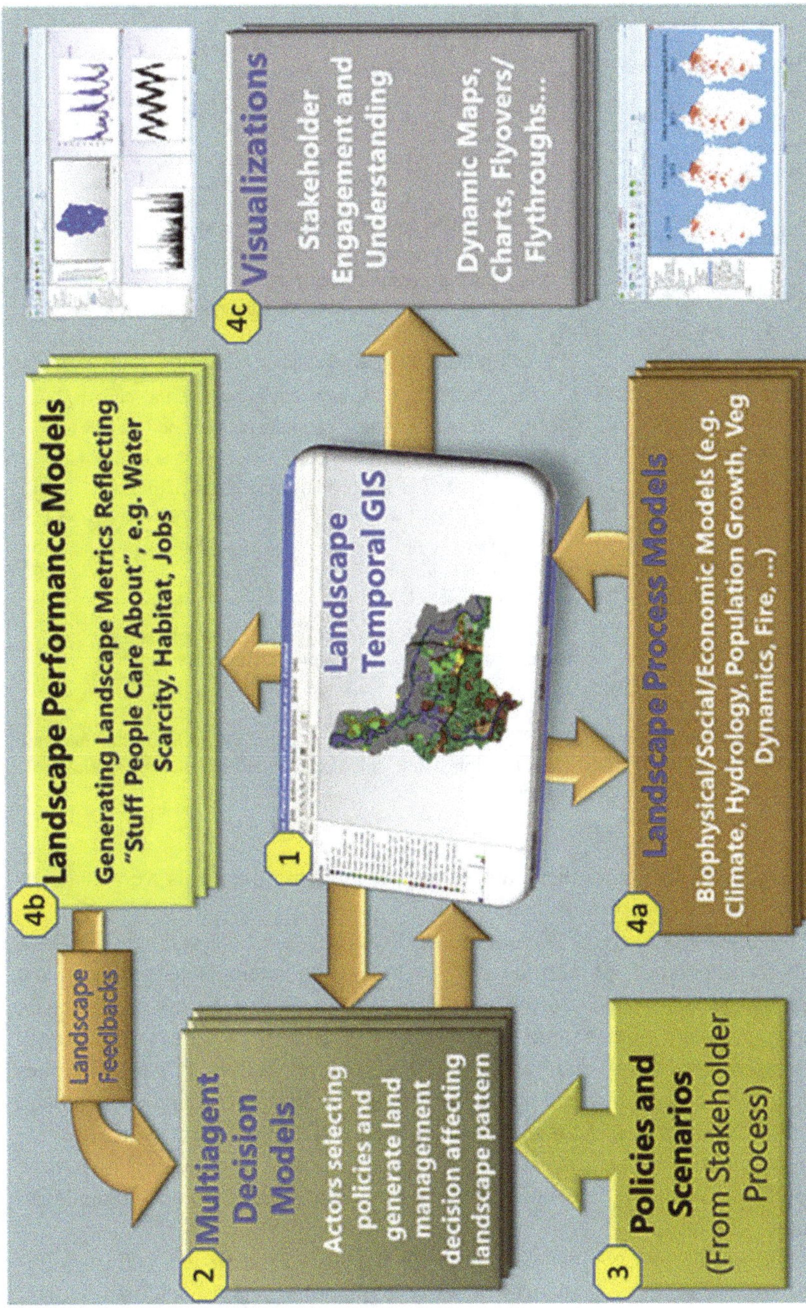

Fig. 1 ENVISION conceptual diagram. Numbers correspond to the list of four ENVISION components described in the text. (Reproduced from http://envision.bioe.orst.edu/About.aspx)

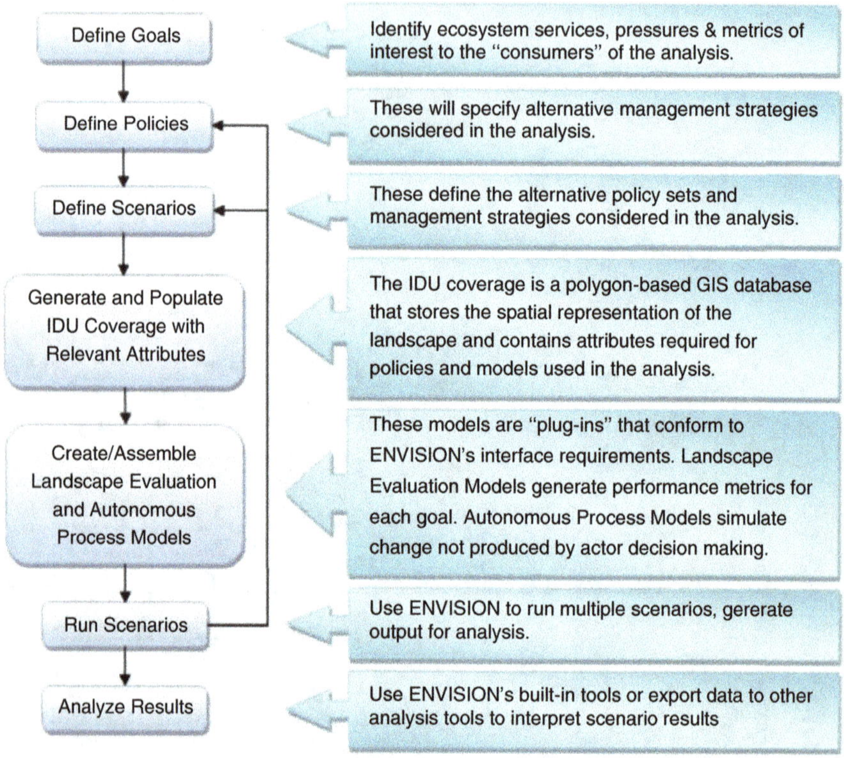

Fig. 2 Workflow for setting up ENVISION applications such as the ENVISION Willamette case study. "IDU" stands for Integrated Decision Unit, the user-defined landscape spatial unit. (Reproduced from http://envision.bioe.orst.edu/)

Modeled scenario results were evaluated, and iterative improvements were made, as necessary, to the workflow outlined in Fig. 2.

The resulting ENVISION Willamette application is a whole watershed model. Through its integration of various plug-in models, it attempts to represent the significant processes related to the supply and fate of water in the entire basin. These processes are both natural (e.g. precipitation, snow dynamics, infiltration, runoff, evapotranspiration) and human (e.g., reservoir operations, irrigation, municipal water use, crop choice). Application drivers include projections of climate, population, land use and income. The integrated ENVISION framework (Fig. 1) operates by simulating the processes across the entire basin per timestep—daily for some processes, annual for others (https://inr.oregonstate.edu/ww2100/analysis-topic/future-climate).

Key findings of the ENVISION Willamette Water 2100 case study address long-term (2010–2100) effects of an array of water regulations, land use alternatives, and climate and population scenarios on water scarcity, access to water, and associated ecological, economic and social tradeoffs.

A detailed summary of findings can be found on the ENVISION website (https://inr.oregonstate.edu/ww2100/key-findings) including several dozen peer-reviewed journal articles.

Briefly, ENVISION Willamette simulations broke out potential impacts for uplands (primarily forest) and lowlands (primarily urban and agricultural) during the twenty-first century. For example, upland snowpack is predicted to decline by 74–94% by 2100, limiting the region's main source of drinking water in summer, and contributing to as much as a ninefold areal increase in forest wildfires that, in turn, will reduce timber revenues and critical fish and wildlife habitat. In the lowlands, growing urban populations are predicted to increase water demand by up to 88%. ENVISION simulations also highlight water management implications and potential strategies for adapting to these complex and interconnected challenges.

3 ENVISION Plug-in Models

3.1 Existing ENVISION Plug-ins

This section reviews existing ENVISION plug-in models and their applications for the ENVISION Willamette project. In general terms, a plug-in is a software component that adds a specific feature to an existing software application, thereby increasing its functionality. Among the most familiar plug-ins are the innumerable "apps" that can be integrated into the operating systems of cellular phones to expand their functionality. Plug-ins can be very simple or quite complex depending on requirements.

ENVISION is distributed with a number of "standard" plug-ins (Table 1) that can be easily modified for specific applications. These are not required for ENVISION applications, but they have provided significant, commonly used functionality for many ENVISION applications.

Because of the complexity of the ENVISION Willamette application, a set of specialized plug-in models was also developed to achieve the project's hydrologic, ecological and socioeconomic goals. Examples include (1) *water system models* for simulating stream flow and temperature, and reservoir water management for the basin's stream/river network; (2) *ecosystem response models* for simulating effects of forestry, agriculture, fire and other disturbances on vegetation dynamics and stream habitat and fish populations; and (3) *socioeconomic models* for simulating changes in demand for drinking water and irrigation, and changes in land use, regulations, incentives and practices for managing rural and urban lands.

Additional details on ENVISION's standard and Willamette-specific plug-ins can be found in the ENVISION Developers Guide (http://envision.bioe.orst.edu/Downloads.aspx). The integration of these plug-in models within ENVISION provided a means for representing the complex feedbacks needed for estimating tradeoffs among ecological, economic and social processes and outcomes of concern to Willamette River Basin stakeholders. Characterization of such complex tradeoffs

Table 1 Standard ENVISION plug-ins

Plug-in name	Description
DynamicVeg	A sophisticated state transition model for vegetation in response to harvest, fire and other disturbances.
FLOW	Configurable hydrologic model representing water storage and movement within a landscape of user-defined polygons or a grid.
Reservoirs	Submodel of FLOW for defining individual reservoirs and their operations for holding and releasing water within a stream network.
WaterMaster	Submodel of FLOW that utilizes spatially explicit place-of-use, and point-of-diversion input data to simulate water right diversions as governed by a prior appropriated water-right system.
ModFlowAP	Implements the MODFLOW groundwater model.
Modeler	Allows users to define evaluative models for specified autonomous processes, including biophysical processes such as food and fiber production, water yield, etc.
ProgramEvaluator	Allows users to specify targets representing desired landscape condition. For instance, a community may want to preserve a certain percentage of the landscape in agricultural uses, or ensure some level of available buildable lands, or some other similar landscape statistic.
Trigger	Allows changes in one location to dynamically propagate changes into another location, once a specified threshold condition is met. For example, if a certain water quality regulatory threshold (TMDL) is exceeded at a given location, a remediation action such as implementation of riparian buffers can be triggered.
SpatialAllocator	Used for defining how processes (harvest, fire, fertilization...) are allocated spatially across the landscape to meet defined targets.
FlammapAP	Sophisticated fire behavior mapping and analysis program that computes potential fire behavior characteristics (spread rate, flame length, fire line intensity, etc.).
QuickFire	Simple fire model.
Target	Specifies a landscape variable whose total value across the landscape is represented as a trajectory.

within this human-natural system would not have been possible using the same or similar array of models applied in stand-alone mode.

3.2 *Development of New ENVISION Plug-ins—VELMA Example*

The list of ENVISION plug-ins is evolving as the framework is applied to new places and questions. In this section we provide an example of the general process by which an existing spatial model can be developed as a plug-in to extend ENVISION's capabilities for integrated EBM planning.

One of ENVISION's most pressing needs is a scalable, process-based biogeochemical model capable of addressing local and regional water quality and

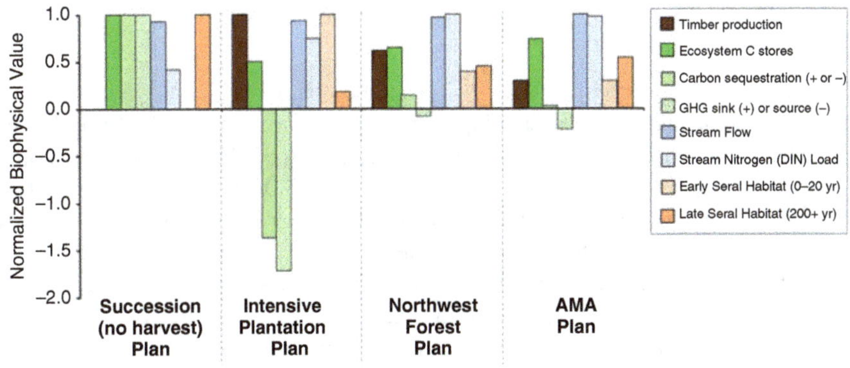

Fig. 3 VELMA results showing predicted ecosystem services and tradeoffs in the year 2200 when four alternative forest management scenarios (Cissel et al. 1999) were applied to present-day landscape conditions in the 230 km^2 Blue River watershed in Oregon. For each target ecosystem service—timber production, etc. (see legend)—simulated biophysical outputs (y-axis) across all four management scenarios are normalized with respect to the greatest simulated value for that ecosystem service. For example, timber production was at maximum for the Intensive Plantation Plan (y-axis value = 1.0), zero for the Succession plan (y-axis value = 0), and 62 and 30% of maximum, respectively, for the Northwest Forest Plan (NWFP) and the Adaptive Management Plan (AMA)

ecosystem services trade-off concerns. This gap could most easily be addressed by integrating a generalized landscape biogeochemical model plug-in with ENVISION's existing hydrologic FLOW plug-in (Table 1).

We chose the Visualizing Ecosystem Land Management Assessment (VELMA) model for this purpose. Developed by the EPA and Georgia Institute of Technology, VELMA is a spatially distributed, eco-hydrological model that links a land-surface hydrology model with a terrestrial biogeochemistry model for simulating the integrated responses of vegetation, soil, and water quality and quantity to interacting stressors. VELMA is applicable to essentially any terrestrial ecosystem (urban, agricultural, forest, grassland, prairie, wetland, etc.), as well as to mixed-use watersheds and regional basins (Abdelnour et al. 2011, 2013; McKane et al. 2014a, b; Hoghooghi et al. 2018; McKane et al. 2018a, b, c).

VELMA simulates how climate, land use, land cover and natural and engineered landscape features control the fate and transport of water, nutrients and toxics in watersheds, across scales ranging from small plots to large river basins, and from days to centuries. In addition to water quality assessments, VELMA has also been used to simulate climate and land use impacts on the capacity of ecosystems to provide a variety of ecosystem goods and services vital to human health and well-being—clean drinking water, flood prevention, food and fiber production, carbon sequestration, habitat for fish and wildlife, and others (e.g., Fig. 3).

3.3 Developing the ENVISION VELMA Plug-in

Using plug-in development procedures outlined on the ENVISION website (http://envision.bioe.orst.edu/Guides/CreatingPlugins.aspx), we developed a plug-in for VELMA's biogeochemical model, disconnected from VELMA's hydrologic model, and programmed to work interactively with ENVISION's existing hydrologic FLOW plug-in. FLOW provides many of the same functions as VELMA's hydrologic model, e.g., water infiltration, lateral surface and subsurface flow, stream routing, and evapotranspiration. Importantly, FLOW has the major advantage of being integrated with existing ENVISION plug-ins for informing management of the Willamette River basin's water, ecosystem and socioeconomic systems.

Figure 4 schematically illustrates the bidirectional exchange of information between the FLOW and VELMA biogeochemical plug-ins and, potentially, with ENVISION's standard plug-ins. All plug-ins are housed in a shared ENVISION library on disk. This library also contains plug-in input parameters (XML), environmental drivers (csv), spatial data (asc), and a project file (envx) that ENVISION uses to initialize a particular application.

The Flow XML file (HBV.xml in Fig. 4) describes a set of methods in the shared library that will be executed as part of the FLOW plug-in for each time step in the ENVISION application. The methods include (1) a hydrologic method that works on a specified landscape grid; (2) a built-in evapotranspiration method; and (3) the VELMA plug-in that is also placed in the shared library.

The VELMA plug-in is initialized using another XML file containing code that signals when the plug-in will be called, in this case at the beginning of each daily time step.

For testing purposes, a "BlueRiver_VelmaPlugin.xml" file was created and applied to demonstrate that ENVISION calls the VELMA plug-in, successfully engaging VELMA's biogeochemical model for the Blue River project application described above.

4 Potential Additional ENVISION Plug-ins for Coastal Ecosystem Applications

To further explore how ENVISION could be extended for EBM assessments, we compiled a list of existing and potential model plugins (Table 2) that are in current use by EPA and other researchers engaged in coastal ecosystem recovery planning.

Table 2 reflects our interest in the integration of models that have been or could be integrated for informing ongoing restoration activities across the EPA's National Estuary Program (NEP) (https://www.epa.gov/nep/local-estuary-programs). Many of these models have already been developed as ENVISION plug-ins for EBM projects at a number of NEP sites—Guánica Bay, Puerto Rico; Tampa Bay, Florida; and Puget Sound, Washington. Applications at these sites aim to assist community

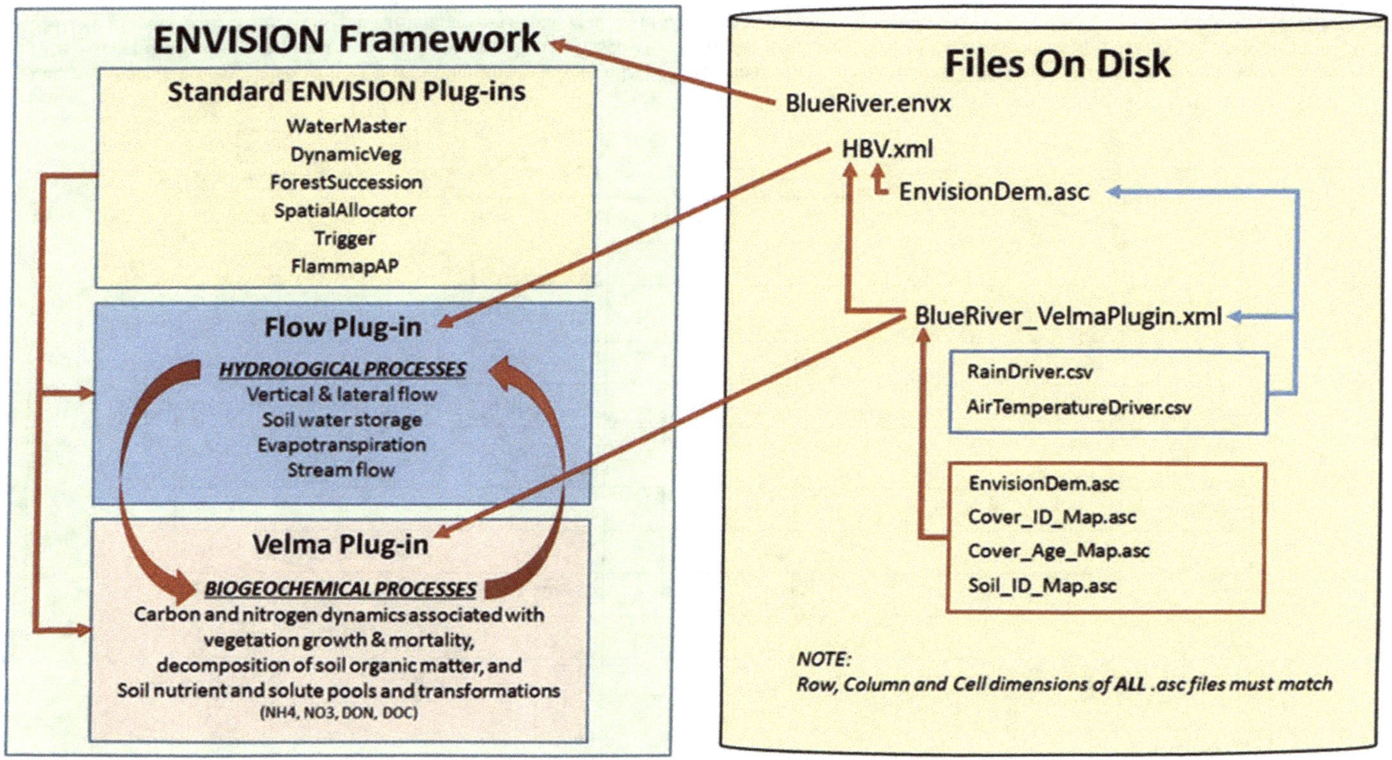

Fig. 4 Schematic showing data requirements and exchange of information among VELMA, FLOW and other ENVISION plug-ins

Table 2 Models currently in use by EPA for various community-based EBM projects designed to simulate ecological, economic, and/or human health outcomes for alternative decision scenarios

Model Name	Model Type	Landscape Unit, Scale	Timestep	Climate Regulation	Hazard Mitigation	Clean Air	Clean Water	Flood Prevention	Food, Fiber, Fuel	Carbon Sequestra-tion	Aesthetics, Recreation	Biodiversity	Human Health	Valuation
VELMA	Hydrology, Biogeochemistry	Grid, Flexible	Daily	√	√		√	√	√	√	√			
SMURF	Fish Habitat, Populations	Reaches, NHD	Seasons						√		√	√		
CORESET	Biodiversity Indicators	Grid, 500m	Yearly									√		
Coral PF	Coral Ecosystems	Flexible	Static		√							√		√
HexSim	Wildlife Habitat, Populations	Hexagons, Flexible	Flexible								√	√		
EPA H2O	Hydrology	NHD, Flexible	Static		√		√	√			√			√
HWBI	Human Well-Being Index	Flexible	Static										√	
Salish Sea Model	Estuarine Hydrodynamics, Biogeochemistry	Flexible	Variable		√		√		√		√	√	√	
CMAQ	Air Quality	Region, Continent	Sub-daily		√	√					√			
BenMAP-CE	Air Quality Health Impacts	Flexible	Daily			√							√	√

Legend
Existing ENVISION plug-in
Planned plug-in
No plug-in currently planned

Legend colors indicate whether an ENVISION plug-in has been or may be developed for a particular model. Listed models were developed in whole or in part by EPA, except the Pacific Northwest National Laboratory's Salish Sea Model. Model references: VELMA (Abdelnour et al. 2011, 2013; McKane et al. 2014b); SMURF (Snyder et al. 2019); CORESET (Melbourne-Thomas et al. 2011a, b); HexSim (Schumaker et al. 2014); EPA H2O (Russell et al. 2015); HWBI (Orlando et al. 2017); Salish Sea Model (Khangaonkar et al. 2018, 2019); CMAQ (Luecken et al. 2019); BenMAP-CE (Davidson et al. 2007; Berman et al. 2012)

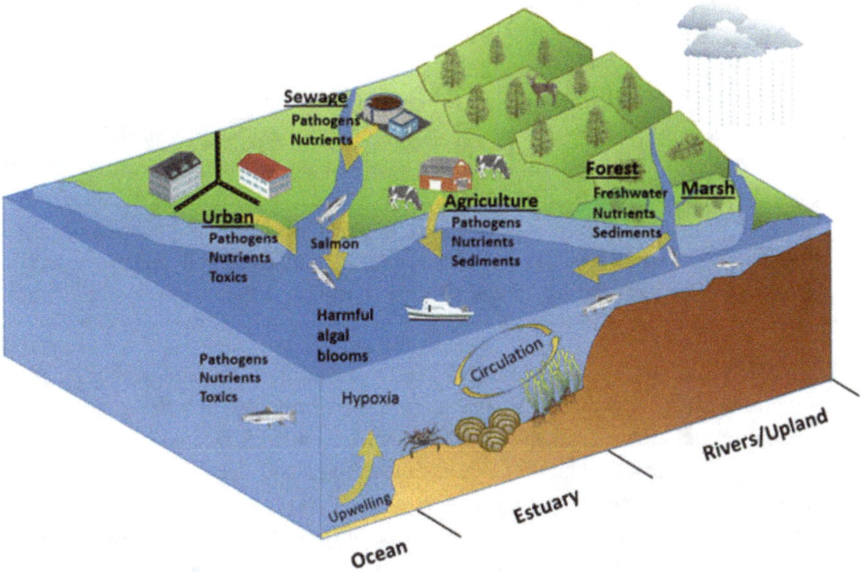

Fig. 5 Generalized representation of major terrestrial pollutant sources and transfers to marine waters in populated coastal ecosystems. Image: Darryl Marois

decision makers in balancing ecological, economic and social criteria over timescales relevant to immediate needs and long-term planning goals.

A major challenge in modeling coastal ecosystems is the establishment of a coupled human-natural modeling framework capable of addressing transfers of terrestrial nutrients and toxic chemicals to marine waters, and consequent impacts on marine life and ecosystem goods and services. Figure 5 schematically summarizes typical coastal terrestrial-to-marine pollutant transfers, highlighting some major challenges for modeling EBM of NEP sites, such as Puget Sound's 31,000 km^2 basin and its mosaic of terrestrial, marine and human environments.

Characterizing such transfers and impacts is critical to estuarine restoration efforts in densely populated coastal areas. For example, in the Puget Sound National Estuary, killer whales (orcas) and their prey, Chinook salmon, have accumulated dangerously high levels of organic chemicals such as PCBs, PBDEs and PAHs (https://ecology.wa.gov/Water-Shorelines/Puget-Sound/Orca-task-force). This well-publicized situation has made these endangered species iconic indicators of the rapidly declining condition of Puget Sound's estuarine food web, perhaps foreshadowing a system-wide collapse such as those observed in Chesapeake Bay and other estuaries globally (Gelfenbaum et al. 2006).

We developed the ENVISION VELMA plug-in to better simulate EBM strategies for reducing transfers of terrestrial nutrients and contaminants to estuarine ecosystems in general, and to Puget Sound ongoing case studies in particular. Both ENVISION and VELMA have been extensively but separately applied to the terrestrial portion of the Puget Sound basin (e.g., Bolte and Vache 2010 for ENVISION; McKane et al. 2018a, b, c for VELMA). With VELMA's capabilities for estimating effects of land use and other disturbances on water quality, we anticipate that the ENVISION VELMA plug-in, when integrated with existing and proposed ENVISION plug-ins (Tables 1 and 2), can effectively extend ENVISION's functionality for addressing terrestrial and marine water quality and ecosystem service objectives pertinent to coastal ecosystem EBM goals. The State of Washington's Puget Sound Partnership has formalized such goals in terms of ~25 terrestrial-marine Vital Signs (ecosystem services), each monitored and managed under a specific Implementation Strategy (https://www.eopugetsound.org/articles/puget-sound-vital-signs-0).

Figure 6 conceptually illustrates ENVISION's flexibility for integrating existing and proposed plug-ins for coastal ecosystem EBM projects such as Puget Sound. This example includes a structured decision-making workflow involving development and application of stakeholder-relevant policies and decision scenarios, ecological production functions, ecosystem goods and services production functions, and benefit functions. The intent of such integration is to enable analyses of how alternative EBM options impact tradeoffs among ecological, socioeconomic and human health endpoints of concern to stakeholders.

In collaboration with Puget Sound community, tribal, state and federal partners, we have initiated development of such a human-natural systems framework, inclusive of coastal terrestrial-marine-human system interactions. This effort would merge the following:

- Existing Oregon State University ENVISION Puget Sound applications (Bolte and Vache 2010; http://envision.bioe.orst.edu/StudyAreas/PugetSound/).
- Ongoing EPA Puget Sound applications of VELMA (McKane et al. 2018a, b, c)
- Pacific Northwest National Laboratory applications of the Salish Sea Model (SSM) (Khangaonkar et al. 2018, 2019), coupled with National Oceanic and Atmospheric Administration applications of the Atlantis model (Levin et al. 2009), for simulating the circulation and fate of terrestrial pollutant inputs within the Puget Sound marine ecosystem and consequent impacts on water quality, fisheries and threatened food web species.
- Existing and new ENVISION plug-ins developed by EPA and others for extending ENVISION's applicability to coastal ecosystems (Tables 1 and 2).

Besides specific case study objectives, the overarching intent of these activities is to establish a generally applicable, coupled human-natural modeling framework that coastal ecosystem community stakeholders and restoration planners can use to anticipate and visualize how effects of EBM options in any particular location can propagate downstream and downcurrent with far-reaching benefits and tradeoffs for terrestrial and marine ecosystem services.

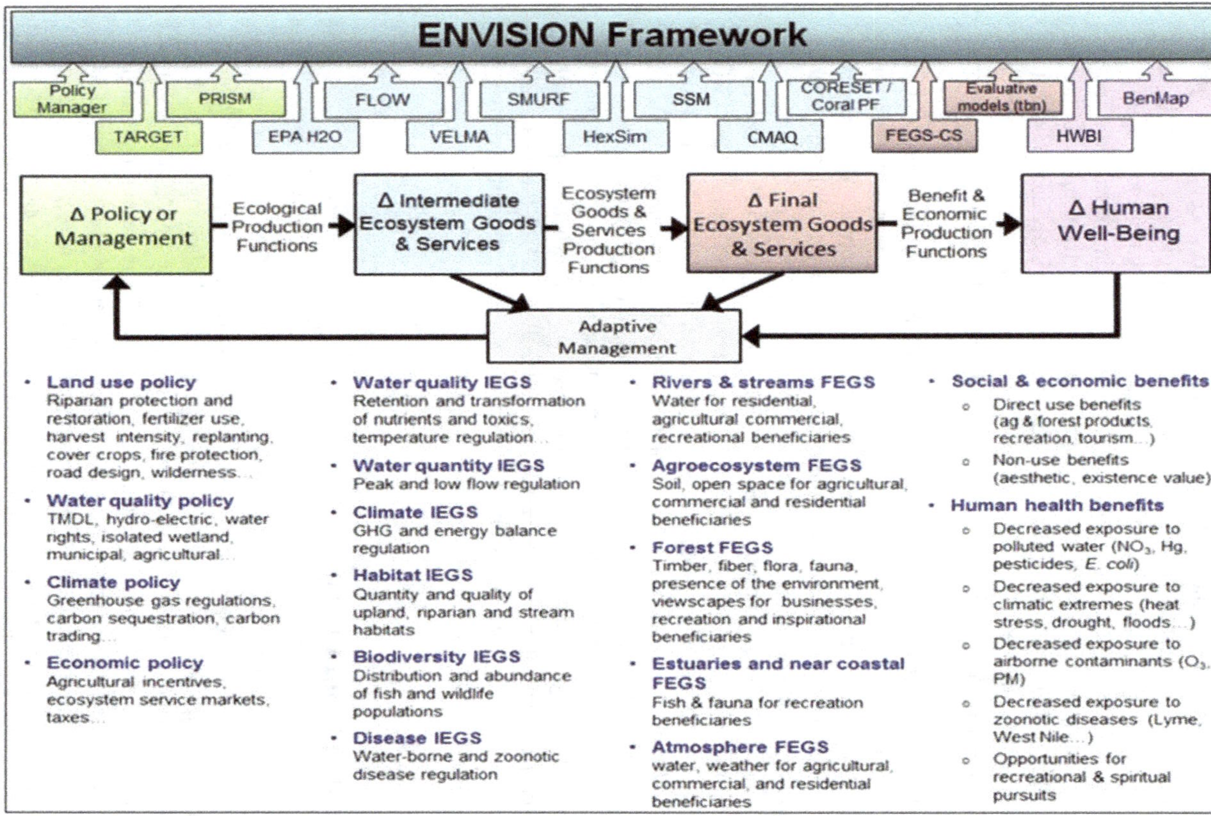

Fig. 6 Hypothetical ENVISION configuration for exploring the effects of alternative policies and land use decisions on intermediate ecosystem goods and services, final ecosystem goods and services, and human well-being for human-natural coastal ecosystems. This configuration includes existing and proposed plug-ins described in Tables 1 and 2, respectively. Integration of these plug-ins within ENVISION can potentially provide comprehensive tradeoff assessments for ecological, economic, social, and human well-being tradeoffs for alternative policies and EBM decision choices

5 Addressing Uncertainty Propagation within Multi-Model Frameworks

ENVISION and other multi-model decision support frameworks (Lewis et al. 2020) have emerged as important tools for informing EBM. A common concern about such frameworks is that uncertainties in one model can propagate and magnify as information gets passed from one model to the next, putting into question a framework's final outputs. Uncertainties in outputs can originate from uncertainties in input data that include measurement and processing errors; in individual models that include conceptual, mathematical, and computational errors; and in model-to-model incompatibilities such as differences in spatial and temporal scales that can propagate errors.

To reduce these sources of uncertainty, ENVISION employs a Monte-Carlo simulation approach that uses statistical descriptors of decision processes and, optionally, scenario-specific statistical descriptors of model inputs. This approach allows a given scenario to be run multiple times to produce a probabilistic distribution of possible outcomes (Bolte et al. 2012).

This uncertainty-based approach has additional advantages. For example, it is well-suited to the explore-then-test paradigm routinely emphasized in ENVISION applications to more effectively engage stakeholders in the design and analysis of alternative future scenarios (http://envision.bioe.orst.edu/CaseStudies.aspx). This approach also facilitates assessments of landscape vulnerability—the predisposition of a system to be adversely affected by stressors and lacking in capacity to adapt to environmental change (Bolte et al. 2012; IPCC 2014). And, when coupled with a multi-criteria assessment procedure (Hajkowicz and Collins 2007; Kiker et al. 2005), this approach can be used to estimate stakeholder weighting-of-importance for various ecosystem services.

6 Summary and Conclusions

This chapter has summarized key elements of the ENVISION decision support framework and how it has been used to help a wide range of communities develop and apply alternative decision scenarios in support of EBM planning. We used the ENVISION Willamette case study in Oregon to highlight key ENVISION features for this purpose. These include a dynamic GIS engine for representing changing landscape characteristics; a multi-agent framework for representing different decision makers; a rich representation of policies limiting decision maker's options; and an extensible array of model plug-ins for simulating ecological, economic and social tradeoffs in response to alternative future scenarios.

Using EPA's VELMA biogeochemical model, we demonstrated the relative ease of developing ENVISION plug-ins for enhancing water quality and ecosystem service assessments. We also demonstrated a hypothetical example and recommendations for developing additional model plug-ins to extend ENVISION's capabilities for integrated EBM planning in coastal ecosystems. This demonstration focused on our ongoing case studies in Puget Sound and other National Estuaries experiencing increasing terrestrial pollutant loads and consequent impacts on marine ecosystem goods and services.

Many of the recommended plug-ins already exist for extending ENVISION to simulate coastal terrestrial-marine interactions. The integration and application of these plug-ins via ENVISION would represent an important step in helping restoration planners and managers identify terrestrial pollutant sources and best practices for reducing terrestrial loadings to the marine ecosystem.

In conclusion, depending upon the scope of EBM questions and objectives, ENVISION and other coupled human-natural modeling frameworks (chapter "Projecting Changes to Coastal and Estuarine Ecosystem Goods and Services: Models and Tools") can require a substantial and expensive effort, potentially involving local, state, federal and academic organizations working in concert to assess how alternative decision scenarios may impact the capacity of local and regional ecosystems to sustainably provide vital ecosystem services.

That said, an argument can be made that the development and application of coupled natural-human modeling frameworks are a cost-effective means for discovering and guiding the implementation of EBM solutions to "wicked" ecological-economic-social problems. That is, problems that are "difficult or impossible to solve for as many as four reasons: incomplete or contradictory knowledge, the number of people and opinions involved, the large economic burden, and the interconnected nature of these problems with other problems" (Kolko 2012).

Obviously, no model can completely address these difficulties, but the integrative and analytical advantages of EBM frameworks such as ENVISION can at least provide informed and useful approximations based on best available data and science.

Acknowledgements The information in this document has been funded in part by the U.S. Environmental Protection Agency (U.S. EPA). It has been subjected to the Agency's peer and administrative review, and it has been approved for publication as an EPA document. We thank Rich Fulford, Tim O'Higgins, Ted DeWitt, Matt Harwell and Chloe Jackson for review comments on earlier manuscript versions.

References

Abdelnour, A., Stieglitz, M., Pan, F., & McKane, R. (2011). Catchment hydrological responses to forest harvest amount and spatial pattern. *Water Resources Research, 47*, W09521. https://doi.org/10.1029/2010WR010165

Abdelnour, A., McKane, R., Stieglitz, M., Pan, F., & Cheng, Y. (2013). Effects of harvest on carbon and nitrogen dynamics in a Pacific Northwest forest catchment. *Water Resources Research, 49* (3), 1292–1313.

Berman, J. D., Fann, N., Hollingsworth, J. W., Pinkerton, K. E., Rom, W. N., Szema, A. M., Breysse, P. N., White, R. H., & Curriero, F. C. (2012). Health benefits from large-scale ozone reduction in the United States. *Environmental Health Perspectives, 120*(10), 1404–1410.

Bolte, J., & Vache, K. (2010). *Envisioning Puget Sound alternative futures: PSNERP final report.* Washington State Department of Fish and Wildlife.

Bolte, J., McKane, R., Phillips, D., Schumaker, N., White, D., Brookes, A., & Olszyk, D. (2011). In Oregon, the EPA calculates nature's worth now and in the future. *Solutions, 2*(6), 35–41.

Bolte, J., McKane, R., Phillips, D., Schumaker, N., White, D., Brookes, A., Olszyk, D., Burdick, C., & Papenfus, M. (2012). An extensible decision support system for evaluating ecosystem services under alternative future scenarios—A Willamette River Basin case study. *Clearance Number ORD-002136.* Washington, DC: U.S. Environmental Protection Agency.

Bradley, M. P., Fisher, W., Dyson, B., Yee, S., Carriger, J., Gambirazzio, G., Bousquin, J., & Huertas, E. (2016). *Application of a structured decision process for informing watershed management options in Guánica Bay, Puerto Rico.* National Health and Environmental Effects Research Laboratory, Office of Research and Development, US Environmental Protection Agency.

Cissel, J., Swanson, F., District, B. R. R., & Leader, E. T. (1999). Blue River Landscape Study: Testing an alternative approach. *Unpublished paper.* Blue River Ranger District, Blue River, Oregon, 97413.

Davidson, K., Hallberg, A., McCubbin, D., & Hubbell, B. (2007). Analysis of PM2. 5 using the environmental Benefits Mapping and Analysis Program (BenMAP). *Journal of Toxicology and Environmental Health, Part A, 70*(3–4), 332–346.

Gelfenbaum, G., Mumford, T., Brennan, J., Case, H., Dethier, M., Fresh, K., Goetz, F., van Heeswijk, M., Leschine, T. M., & Logsdon, M. (2006). *Coastal habitats in Puget Sound: A research plan in support of the Puget Sound Nearshore Partnership* (No. TR-2006-1). Seattle, WA: Corps of Engineers.

Hajkowicz, S., & Collins, K. (2007). A review of multiple criteria analysis for water resource planning and management. *Water Resources Management*, 21(9), 1553–1566.

Hoghooghi, N., Golden, H., Bledsoe, B., Barnhart, B., Brookes, A., Djang, K., Halama, J., McKane, R., Nietch, C., & Pettus, P. (2018). Cumulative effects of low impact development on watershed hydrology in a mixed land-cover system. *Water, 10*(8), 991.

IPCC. (2014). Climate Change 2014: Impacts, adaptation and vulnerability. In *Contribution of Working Group II to the Fifth Assessment Report of the Intergovernmental Panel on Climate Change* (Vol. 1, p. 32). New York: Cambridge University Press.

Jaeger, W. K., Amos, A., Bigelow, D. P., Chang, H., Conklin, D. R., Haggerty, R., Langpap, C., Moore, K., Mote, P. W., Nolin, A. W., & Plantinga, A. J. (2017). Finding water scarcity amid abundance using human–natural system models. *Proceedings of the National Academy of Sciences, 114*(45), 11884–11889.

Khangaonkar, T., Nugraha, A., Xu, W., Long, W., Bianucci, L., Ahmed, A., Mohamedali, T., & Pelletier, G. (2018). Analysis of hypoxia and sensitivity to nutrient pollution in Salish Sea. *Journal of Geophysical Research—Oceans, 123*(7), 4735–4761.

Khangaonkar, T., Nugraha, A., Xu, W., & Balaguru, K. (2019). Salish Sea response to global climate change, sea level rise, and future nutrient loads. *Journal of Geophysical Research—Oceans, 124*(6), 3876–3904.

Kiker, G. A., Bridges, T. S., Varghese, A., Seager, T. P., & Linkov, I. (2005). Application of multicriteria decision analysis in environmental decision making. *Integrated Environmental Assessment and Management: An International Journal, 1*(2), 95–108.

Kolko, J. (2012). *Wicked problems: Problems worth solving: A handbook & a call to action.* Austin, TX: AC4D.

Levin, P. S., Fogarty, M. J., Murawski, S. A., & Fluharty, D. (2009). Integrated ecosystem assessments: Developing the scientific basis for ecosystem-based management of the ocean. *PLoS Biology, 7*(1).

Lewis, N. S., Marois, D. E., Littles, C. J., & Fulford, R. S. (2020). Projecting changes to coastal and estuarine ecosystem goods and services - models and tools. In T. O'Higgins, M. Lago, & T. H. DeWitt (Eds.), *Ecosystem-based management, ecosystem services and aquatic biodiversity: Theory, tools and applications* (pp. 235–254). Amsterdam: Springer.

Luecken, D. J., Yarwood, G., & Hutzell, W. T. (2019). Multipollutant modeling of ozone, reactive nitrogen and HAPs across the continental US with CMAQ-CB6. *Atmospheric Environment, 201*, 62–72.

McKane, R. B., Brookes, A., Djang, K., Papenfus, M., Ebersole, J., Phillips, D., Halama, J., Pettus, P., Burdick, C., & Russell, M. (2014a). Sustainable and healthy communities Pacific Northwest demonstration study. *Report No. ORD-007386 of the U.S. Environmental Protection Agency*. Washington, DC.

McKane, R. B., Brookes, A., Djang, K., Stieglitz, M., Abdelnour, A., & Pan, F. (2014b). Enhanced version of VELMA ecohydrological modeling and decision support framework to address engineered and natural applications of GI for reducing nonpoint inputs of nutrients, contaminants, and sediments. *Report No. ORD-010080 of the U.S. Environmental Protection Agency*. Washington, DC.

McKane, R. B., Barnhart, B., Pettus, P., Halama, J., Brookes, A., Djang, K., Khangaonkar, T., Harvey, C., Kaplan, I., Luna, H., Schmidt, M., Howe, E., & Levin, P. (2018a). An integrated environmental and human systems modeling framework for Puget Sound restoration planning. *Salish Sea Ecosystem Conference*, Seattle, Washington.

McKane, R. B., Barnhart, B., Pettus, P., Halama, J., Brookes, A., Djang, K., Khangoankar, T., Kaplan, I., Harvey, C., Morzaria Luna, H., Schmidt, M., Howe, E., Levin, P., Francis, T., Baker, J., Stanley, S., & Hume, C. (2018b). A science-governance partnership for integrating ecosystem services into Puget Sound restoration planning. *ACES 2018 Conference Proceedings*, Washington, DC.

McKane, R. B., Halama, J., Pettus, P., Barnhart, B., Brookes, A., Djang, K., Blair, G., Hall, J., Kane, J., Swedeen, P., & Benson, L. (2018c). How Visualizing Ecosystem Land Management Assessments (VELMA) modeling quantifies co-benefits and tradeoffs in Community Forest management. *Keynote presentation to the Northwest Community Forest Forum*, Astoria, OR, May 10–11, 2018.

Melbourne-Thomas, J., Johnson, C. R., Fung, T., Seymour, R. M., Chérubin, L. M., Arias-González, J. E., et al. (2011a). Regional-scale scenario modeling for coral reefs: A decision support tool to inform management of a complex system. *Ecological Applications, 21*, 1380–1398. https://doi.org/10.1890/09-1564.

Melbourne-Thomas, J., Johnson, C. R., Perez, P., Eustache, J., Fulton, E. A., & Cleland, D. (2011b). Coupling biophysical and socioeconomic models for coral reef systems in Quintana Roo, Mexican Caribbean. *Ecology and Society, 16*, 23.

Orlando, J., Yee, S., Harwell, L., & Smith, L. (2017). Technical guidance for constructing a Human Well-Being Index (HWBI): A Puerto Rico example. *U.S. Environmental Protection Agency Report No. EPA/600/R-16/363*.

Russell, M., Harvey, J., Ranade, P., & Murphy, K. (2015). EPA H2O user manual. *EPA/600/R-15/090*. Washington, DC: US EPA Office of Research and Development.

Santelmann, M., McDonnell, J., Bolte, J., Chan, S., Morzillo, A. T., & Hulse, D. (2012). Willamette water 2100: River basins as complex social-ecological systems. *WIT Transactions on Ecology and the Environment, 155*, 575–586.

Schumaker, N. H., Brookes, A., Dunk, J. R., Woodbridge, B., Heinrichs, J. A., Lawler, J. J., Carroll, C., & LaPlante, D. (2014). Mapping sources, sinks, and connectivity using a simulation model of northern spotted owls. *Landscape Ecology, 29*(4), 579–592.

Snyder, M. N., Schumaker, N. H., Ebersole, J. L., Dunham, J. B., Comeleo, R. L., Keefer, M. L., Leinenbach, P., Brookes, A., Cope, B., Wu, J., & Palmer, J. (2019). Individual based modeling of fish migration in a 2-D river system: Model description and case study. *Landscape Ecology, 34*(4), 737–754.

Spies, T. A., White, E. M., Kline, J. D., Fischer, A. P., Ager, A., Bailey, J., Bolte, J., Koch, J., Platt, E., Olsen, C. S., & Jacobs, D. (2014). Examining fire-prone forest landscapes as coupled human and natural systems. *Ecology and Society, 19*(3), 9.

Villarreal, M. L., Labiosa, B., & Aiello, D. (2017). *Evaluating land-use change scenarios for the Puget Sound Basin, Washington, within the ecosystem recovery target model-based framework* (No. 2017-1057). US Geological Survey.

8

Ecosystem Services in U.S. Environmental Law and Governance for the Ecosystem-Based Management Practitioner

Donna R. Harwell

Abstract This chapter provides an overview of ecosystem services issues in United States (U.S.) law and governance for the Ecosystem-Based Management (EBM) practitioner. A brief overview summary of a suite of U.S. federal environmental laws where ecosystems services are relevant is presented along with a high-level overview of ecosystem services in federal and state agency regulations as it helps inform ecosystem-based management. As with the published science-based literature on ecosystem services, there is also a sizeable law-based literature available on ecosystem services. A HeinOnline law journal library focused search identified 1903 legal articles that contained reference to ecosystem services. Focusing on a snapshot of key literature, this chapter presents an overview of those articles that contained "ecosystem services" or "ecosystem based management" just in the article's title. From this survey across the breadth of law journals, a suite of ecosystem services topics related to EBM in environmental law are identified and summarized. Overall, the goal of this chapter is to present a high-level overview and direct the reader to resources to find more in-depth legal analyses of select ecosystem services topics.

Lessons Learned

- EBM practitioners need to have the large suite of federal environmental laws that impact EBM as a frame of reference.
- There is active legal scholar literature on the intersection between ecosystem services and environmental laws.
- The majority of the active legal scholar literature is focused on the core environmental laws, however, the summary table informs the reader of the potential applications of other legal and governance frameworks to ecosystem services and EBM.

D. R. Harwell (✉)
Palm Beach State College, Palm Beach Gardens, FL, USA
e-mail: harwelld@palmbeachstate.edu

- The community of EBM practitioners should take advantage of the legal scholar literature; the chapter demonstrates to the reader the value of adding the intersection between ecosystem services and environmental laws into the information space for EBM practitioners.
- In the recent past, a few U.S. states have started to add ecosystem services language into their statutory and regulatory materials.

Needs to Advance EBM

- EBM practitioners need to add information from the environmental law literature to their background information as part of efforts to frame ecosystem services information in their EBM activities.
- Practitioners can look to the environmental law literature to identify examples where relevant information might be transferable to their specific scenarios, such as the examples of watershed-based services (e.g., Funk et al. 2020).
- The suite of traditional EBM practitioners needs to expand to include law and governance practitioners in order to merge and create a large overlap and cross-information exchange between the disciplines.

1 Introduction to Ecosystem Services and EBM in Law and Governance

There are multiple ways to present the intersection between ecosystem services and environmental law and policy as it informs Ecosystem-Based Management (EBM). One perspective involves considering how the authority of individual law or regulations may influence the access, condition, protection, and/or utilization of nature. Examples of high-level overviews of key environmental laws include Ruhl and Salzman (2007), Thompson (2008), Davis (2010), Ruhl et al. (2013), and Farber and Findley (2014). Another perspective might be from an ecosystem-type lens. For example, Ruhl et al. (2013) explores three examples of United States (U.S.) environmental laws and regulations from wetland, coastal, and forest resources protection perspectives. The full suite of U.S. statutes for natural resources are highly domain-specific in character (Scarlett and Boyd 2015); Ruhl (2005b) successfully argues that, "(U.S.) ecosystem management law is a cobbled-together body of law, if it can even be called that much."

There is a large breadth of U.S. federal environmental laws that can be broadly organized around overarching purposes of protection/conservation, restoration/remediation, and regulations focused on socio-ecological interactions (Fig. 1; note that acronyms are captured in Table 1's compilation of these laws). As an individual law is inherently complex (e.g., containing multiple goals and objectives), its placement on a Venn diagram showing the relationships between protection and conservation, restoration and remediation, and socio-ecological perspectives shown in Fig. 1 is overly simplistic.

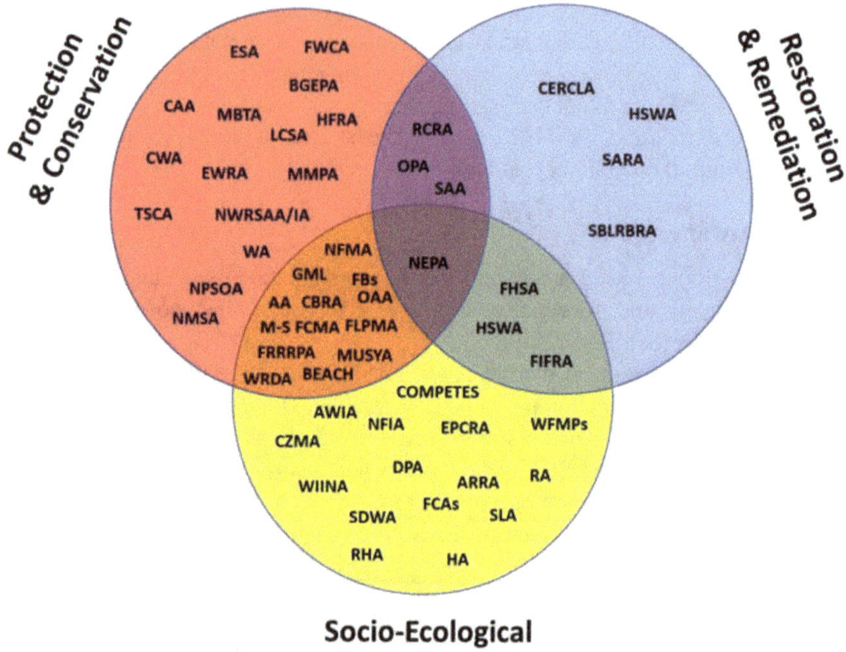

Fig. 1 Venn diagram identifying U.S. Federal environmental laws organized by three main perspectives. See Table 1 for acronyms used

Ruhl et al. (2013) highlights how the breadth of the U.S. environmental regulatory infrastructure from the 1970s–1990s became considered "top heavy." Hirokawa and Porter (2013) argue that, "the effort to integrate ecosystem services valuation into law has yielded complicated and unsatisfactory results." Further, legal scholars have called for the evolution of the application of environmental laws and regulatory tools to not only increase protection of ecosystems, but also the services they provide to people (e.g., Markell 2007; Ruhl et al. 2013). Examples of this perspective of characterizing how to protect ecosystem services through the use of regulations include Markell (2007), Davis (2010), and Pardy (2014).

The objectives of this chapter are two-fold: (1) to present an overview of the suite of U.S. Federal environmental laws and regulations with connections to ecosystem services; and (2) to present a survey of the legal scholar literature for a synopsis of ecosystem services issues in U.S. environmental law. These reviews are not intended to be fully exhaustive, but rather capture the broad suite of ecosystem services topics in U.S. environmental law and translated for EBM practitioners. For a recent overview of case law on ecosystem services, including U.S. examples, the reader is directed to Sharon et al. (2018).

Table 1 U.S. Federal environmental laws

Law & citation	Summary
Foundational laws over 100 years old	
Swamp Land Act 43 U.S.C. § 23 *et seq.*	The 1850 **SLA** provided legislation for giving Federal lands to the States in order to convert swamp lands into land for provision of agricultural and flood protection services.
Homestead Act 12 Stat. 392	The 1862 **HA** provided 160 acres of public land to homesteaders who paid a filing fee and lived on the land for five years before receiving the deed in order to promote westward expansion and the productive use (as a service) of the frontier. This was mostly repealed in 1976.
General Mining Law 30 U.S.C. § 22 *et seq.* (as amended)	The 1872 **GML** established that all valuable mineral deposits, and the lands where found, belonging to the United States were to be free and open to exploration and purchase for extractive services.
Organic Administration Act 16 U.S.C. § 551 *et seq.*	The 1897 **OAA** provided the authorizing legislation for the National Forest Service. The National Forest Service focuses on supporting forestry-based services.
Rivers and Harbors Act 33 U.S.C. § 407	The **RHA** of 1899 prohibited the construction over or in navigable waterways of the U.S. without Congressional approval and provided initial protection from water quality pollution.
Reclamation Act 43 U.S.C. § 391 *et seq.*	The **RA** was put into place in 1902 to set up water development (irrigation) projects in the U.S. west to support increasing westward settlement, including the productive use lands as a service, at the beginning of the twentieth century.
National Park Service Organic Act 16 U.S.C. § 1b *et seq.*	The 1916 **NPSOA** authorized the establishment of the U.S. National Park Service. The broad ecogeographic spectrum of National Parks encompasses a range of services associated with coastal, aquatic, and terrestrial services, along with public use and recreational-focused services.
Federal Migratory Bird Treaty Act 16 U.S.C. §§ 703–712	The 1918 **MBTA** is the federal enactment of the Migratory Bird Treaty (United States and Great Britain, acting on behalf of Canada; Mexico, Japan, and Russia subsequently signed onto this treaty) providing protections making it illegal to take, possess, sell or purchase any migratory bird (or parts) without a federal permit. The MBTA supports ecosystem services protection for recreational experiences and use for spiritual and ceremonial purposes.

(continued)

Table 1 (continued)

Law & citation	Summary
"Granddaddy" of U.S. Environmental Law	
National Environmental Policy Act 42 U.S.C. § 4321 *et seq.*	The 1969 **NEPA** law establishes the U.S.' goal to live harmoniously with nature by identifying where there is a federal nexus for the consideration of actions on the environment. It created the Environmental Impact Statement process that requires all federal agencies to review all potential actions for their impact on the environment. As a foundational piece of environmental legislation, there is a broad range of potentially relevant ecosystem services. (see Sect. 2.3)
Air Resources	
Clean Air Act 42 U.S.C. § 7401 *et seq.*	The **CAA** (including the 1970 Amendments) created a regulatory system to control most of the commonly produced and significant air pollutants. It set up air quality control regions and established National Ambient Air Quality Standards. Relevant ecosystem services may include air pollution removal and breathable air for human health and well-being. (see Sect. 2.3)
Water Resources	
Flood Control Acts 33 U.S.C. § 15 *et seq.*	A suite of legislation starting 1917, the **FCA**s provided authorizations for federal water control and flood protection services-based projects. The U.S. Army Corps of Engineers was established by the 1941 FCA with the authority to implement flood control policies such as the Flood Control and Coastal Emergency Act (Pub. L. No. 84–99).
Federal Water Pollution Control Act (**Clean Water Act**) 33 U.S.C. § 1251 *et seq.*	The 1972 **CWA** established that dumping in U.S. waters was not a right, that any person or corporation that wanted to dump anything into U.S. waters must have a permit, and that all wastewater must be treated, no matter the condition of the receiving waters. The CWA may involve many types of relevant ecosystem services, including pollution removal and supporting habitat condition for commercial and recreationally valuable species. (see Sect. 2.3)
Coastal Zone Management Act 16 U.S.C. § 1451 *et seq.*	The 1972 **CZMA** promotes development in the coastal zone area using a national perspective. but attempts to limit pollution arising from such development.
Safe Drinking Water Act 42 U.S.C. § 201 *et seq.*	The 1974 **SDWA** protects public drinking water supplies across the nation. It requires the U.S. Environmental Protection Agency to establish national primary and secondary drinking water standards to limit contaminants in drinking water, supporting the service of drinkable water.

(continued)

Table 1 (continued)

Law & citation	Summary
Coastal Barrier Resources Act 16 U.S.C. § 3501 *et seq.*	The 1982 **CBRA** was passed to protect and conserve coastal barriers, habitats that provide important flood protection and storm mitigation services. It prevents individuals who build in these zones from receiving any federal assistance, including federal flood insurance policies.
Emergency Wetlands Resources Act 16 U.S.C. §§ 3901–3932	The 1986 **EWRA** instructs the U.S. Fish & Wildlife Service to map the status and conditions of wetlands (and resulting services) across the U.S. and create a National Wetlands Priority Conservation Plan.
Water Resources Development Acts e.g., WRDA 2000 Pub. L. No. 106–541	**WRDA**s are a suite of laws focusing on a range of water resource management, protection, and utilization activities (and services) involving a federal nexus. For example, WRDA 2000 authorized most projects for the Comprehensive Everglades Restoration Plan in addition to other water resources protection plans across the country.
Beaches Environmental Assessment and Coastal Health (BEACH) Act Pub. L. No. 106–284	The Beaches Environmental Assessment and Coastal Health (**BEACH**) Act (2000) amends part of the Clean Water Act and focuses on monitoring and notifying the public about possible human health problems related to the ecosystem service of use for coastal recreation.
Water Infrastructure Improvements for the Nation Act Pub. L. No. 114–322	**WIINA** was passed in 2016 to focus on aspects of the U.S. drinking water infrastructure involving public health, supporting the service of drinkable water.
National Flood Insurance Act 42 U.S.C. § 4001 *et seq.*	The 1968 **NFIA** encourages states to create floodplain management programs that place restrictions on the location and type of construction, supporting the ecosystem service of flood protection. It also has a buy-out program component to take people out of floodplains to reduce rebuilding costs.
Deepwater Port Act 33 U.S.C. § 1501 *et seq.*	The 1974 **DPA** focuses on construction, operation, and decommissioning of deepwater ports (located beyond the U.S. territorial sea boundaries) and minimization of adverse impacts on the marine environment and the services they provide.
America COMPETES Reauthorization Act of 2010 Pub. L. No. 111–358	The 2010 **COMPETES** law instructs the Administrator of the National Oceanic and Atmospheric Administration to "identify emerging and innovative research and development priorities to enhance United States competitiveness, support development of new economic opportunities based on NOAA research, observations, monitoring modeling, and predictions that sustain ecosystem services." 33 U.S.C. § 893(b)(1)

(continued)

Table 1 (continued)

Law & citation	Summary
America's Water Infrastructure Act of 2018 Pub. L. No. 115–270	The 2019 **AWIA** law includes asking the National Academy of Science to examine how the U.S. Army Corps of Engineers approaches formulation, evaluation, and budget determination for water resources development projects, including "an analysis of whether such principles and methodologies fully account for all of the costs of project alternatives, including potential societal costs, such as lost ecosystem services, and full lifecycle costs for such alternatives." (Sec. 1103)
Land, Fish & Wildlife Resources	
Farm Bills 7 U.S.C. covers Agriculture	A suite of legislation starting in 1933, the **FB**s provided authorizations for, among other things, efforts on development and sale of agricultural products and other agroservices, research, and conservation. The Agricultural Act of 2014 (**AA**) called for an update to the strategic plan for forest inventory and analysis, with the implementation of an, "annualized inventory of trees in urban settings, including the status and trends of trees and forests, and assessments of their ecosystem services, values, health, and risk to pests and diseases." (Pub. L. No. 113–179; Sec. 8301)
Fish and Wildlife Coordination Act 16 U.S.C. § 661 *et seq.*	The 1934 **FWCA** law created the U.S. Fish and Wildlife Service, established the National Wildlife Refuge System, and requires the Secretary of the Interior to protect and conserve wildlife resources and authorize the public-use service of hunting of overabundant species.
Federal Wildland Fire Management Policies	The **WFMP**s are a suite of cross-Federal Agency wildland fire policies dating back to 1935 and have been revised following large-scale fire seasons (e.g., post-1998 fire season and the Yellowstone National Park fires). These policies have focused on supporting resource objectives in federal wildlands, including balancing the use of prescribed natural fires and ecosystem services (e.g., recreational access to nature) and potential negative ecosystem services effects such as smoke and stream sedimentation.
Bald and Golden Eagle Protection Act 16 U.S.C. § 668 *et seq.*	The 1962 **BGEPA** focuses on a suite of protections for two eagle species, including addressing issues of interference and abuse related to different aspects of shelter, breeding, nest abandonment, and feeding. BGEPA supports ecosystem services protection for recreational experiences and use for spiritual and ceremonial purposes.

(continued)

Table 1 (continued)

Law & citation	Summary
Multiple Use—Sustained Yield Act 16 U.S.C. § 528 *et seq.* (amended 1996)	The 1960 **MUSYA** established the system of National Forests for multiple outdoor recreation, range, timber, watershed, and fish and wildlife purposes (and services).
Wilderness Act 11 U.S.C. § 1131 *et seq.*	The **WA** of 1964 was established to preserve and protect, for current and future generations, certain wilderness lands in their "natural condition" with a number of relevant ecosystem services related to recreation and existence services.
National Wildlife Refuge System Administration Act 16 U.S.C. § 668dd	The 1966 **NWRSAA** is the U.S. Fish and Wildlife Service's "Organic Act" analog of the Park Service for management of the National Wildlife Refuge System for the purpose of protecting lands for the conservation of fish and wildlife, including threatened and endangered species, and puts boundaries on the ecosystem services of public access/use of refuge lands and waters.
Marine Mammal Protection Act 16 U.S.C. §§ 1361–1407	The 1972 **MMPA** represents the first legislation for ecosystem-based management for marine resources and was established to manage marine mammal species and population stocks as components of the ecosystems of which they are a part.
National Marine Sanctuaries Act 16 U.S.C. § 1431 *et seq.* (amended 2000)	The 1972 **NMSA** allowed for the designation and protection of special areas of the marine environment supporting a broad suite of coastal, recreational, and fisheries-related services.
Endangered Species Act 16 U.S.C. § 1531 *et seq.*	The 1973 **ESA** established protections for species, and their habitats, that have been listed as endangered or threatened. This law closed down the U.S. market in endangered wildlife, established heavy criminal penalties and fines for "taking" a member of an endangered or threatened species, and prohibits any federal actions that may impact the life or habitat of a listed endangered or threatened species. Habitats and supporting biodiversity preservation are often cited as the most relevant ecosystem services. (see Sect. 2.3)
Forest and Rangeland Renewable Resources Planning Act 16 U.S.C. § 1600 *et seq.*	The 1974 **FRRRPA** law gives authority to the U.S. Forest Service and U.S. Department of Agriculture to assess the Nation's renewable resources and develop and prepare a national renewable resources program for forestry and agriculture-based services.
Magnuson-Stevens Fishery Conservation and Management Act 16 U.S.C. § 1801 *et seq.* (amended 2007)	The 1976 **M-S FCMA** focuses on the valuable and renewable natural resources of U.S. fisheries across a range of geographic boundaries, ranging from anadromous species which spawn in rivers or estuaries, to migratory species of the high seas, and species in U.S. federal waters of the continental shelf. Ecosystem services related to fishery resources include harvesting for food supply and recreational opportunities.

(continued)

Table 1 (continued)

Law & citation	Summary
Federal Land Policy and Management Act 43 U.S.C. § 1701 *et seq.*	The 1976 **FLPMA** established the Bureau of Land Management, including inventories for present and future resource use.
National Forest Management Act 16 U.S.C. § 1600 *et seq.*	The 1976 **NFMA** Amends the Forest and Rangeland Renewable Resources Planning Act of 1974.
National Wildlife Refuge System Improvement Act Pub. L. No. 105–57	The 1997 **NWRSIA** updated the guidance for overall management of the National Wildlife Refuge System, including focus on maintaining the environmental health of the Refuge System while providing for determination of "compatible uses" of refuges for ecosystem services such as hunting and fishing, wildlife observation and photography, and environmental education and interpretation.
Healthy Forests Restoration Act Pub. L. No. 108–148	The 2003 **HRFA** was designed to minimize impacts of destructive wildfires on federal lands, including by allowing timber harvests on protected National Forests, and creating the ability for expedited NEPA review for projects under HFRA.
Chemicals	
Federal Hazardous Substances Act 15 U.S.C. § 1261 *et seq.*	The 1960 **FHSA** required labeling of hazardous household products to help consumers safely store and use products and allow for the ban of certain products that are dangerous or hazardous to human health. Ecosystem services are potential endpoints for a risk assessment process to examine the potential adverse effects of chemicals on the environment.
Federal Insecticide, Fungicide and Rodenticide Act 7 U.S.C. § 135 *et seq.*	The 1964 **FIFRA** law covers all chemicals manufactured to deal with pests in relation to agriculture and requires all of these chemicals to be registered with the U.S. Environmental Protection Agency prior to their use. Ecosystem services are potential endpoints for a risk assessment process to examine the potential adverse effects of chemicals on the environment.
Toxic Substances Control Act 15 U.S.C. § 2601 *et seq.*	The 1976 **TSCA** law requires the U.S. Environmental Protection Agency to keep a registry of chemicals used and introduced into everyday life. Ecosystem services are potential endpoints for a risk assessment process to examine the potential adverse effects of chemicals on the environment.
Emergency Planning and Community Right to Know Act 42 U.S.C. § 11001 *et seq.*	The 1986 **EPCRA** was created to help communities plan for chemical emergencies. It also requires industry to report on the storage, use and releases of hazardous substances to federal, state, and local governments. EPCRA requires state and local governments, and Indian tribes to use this information to prepare their community for potential risks. Ecosystem services are potential endpoints for a risk assessment process to examine the potential adverse effects of chemicals on the environment.

(continued)

Table 1 (continued)

Law & citation	Summary
Frank R. Lautenberg Chemical Safety for the 21st Century Act Pub. L. No 114–182, 130 Stat. 448	The 2016 **LCSA** is an update of the TSCA requiring the U.S. Environmental Protection Agency to have ongoing evaluations of chemicals registered under the TSCA using a risk-based standard. It also requires the EPA to impose fees on chemical manufacturers to pay for these evaluations. Ecosystem services are potential endpoints for a risk assessment process to examine the potential adverse effects of chemicals on the environment.
Environmental Remediation	
Resource Conservation and Recovery Act 42 U.S.C. § 6901 *et seq.*	The 1976 **RCRA** law regulates the waste cycle by defining wastes and dictating how they are to be treated. It aims to prevent the release of hazardous wastes into the environment. RCRA can support the protection or restoration of a suite of ecosystem services that are location specific.
Comprehensive Environmental Response, Compensation and Liability Act 42 U.S.C. § 9601 *et seq.*	The 1980 **CERCLA**, also referred to as Superfund, is an administrative system for removing hazardous materials from sites where they had been improperly dumped into the environment in years past. Sites are identified and placed on a National Priorities List, then assessed through the hazard ranking system. CERCLA can support the protection or restoration of a suite of ecosystem services that are location specific. (see Sect. 2.3)
Hazardous and Solid Waste Amendments Pub. L. No. 98–616, 98 Stat. 3221	The 1984 **HSWA** law requires the U.S. Environmental Protection Agency to develop criteria for identifying hazardous waste: ignitability; corrosivity; reactivity; and toxicity. It added stringent regulation of land disposal of hazard wastes to RCRA. HSWA can support the protection or restoration of a suite of ecosystem services that are location specific.
Superfund Amendments and Reauthorization Act Pub. L. No. 96–510, 94 Stat. 2767	**SARA** (1986) is the reauthorization of CERCLA and required that the hazard ranking system be updated and all identified sites were reviewed for possible water contamination due to run off. SARA supports the protection or restoration of a suite of ecosystem services that are location specific.
Oil Pollution Act 33 U.S.C. § 2701 *et seq.*	The 1990 **OPA** established a trust fund to clean up spills when the responsible party is incapable or unwilling to do so and outlines requirements for facilities (e.g., aboveground storage facilities) and vessels (e.g., oil tankers) to detail how they will respond to large discharges. OPA clean-up activities can support the protection or restoration of a suite of ecosystem services that are location specific.

(continued)

Table 1 (continued)

Law & citation	Summary
Small Business Liability Relief and Brownfields Revitalization Act Pub. L. No. 107–118, 115 Stat. 2356	The 2002 **SBLRBRA**, referred to as Brownfields, amended CERCLA to increase funding for cleanup at urban and suburban CERCLA sites. It focuses on cleanup of sites with petroleum or other hazardous waste contamination. Brownfields revitalization can support the protection or restoration of a suite of ecosystem services that are location specific.
American Recovery and Reinvestment Act Pub. L. No. 111–5, 123 Stat. 115	The 2009 **ARRA** updated CERCLA to add a large amount of stimulus monies to the Superfund in order to accelerate ongoing clean-up activities. As a result, ARRA can support the protection or restoration of a suite of ecosystem services that are location specific.
Supplemental Appropriations Act Pub. L. No. 111–212	The 2010 **SAA** included the call for an "ecosystem services impact study" by the National Academy of Sciences to, "conduct a study of the long-term ecosystem service impacts of the Deepwater Horizon oil discharge. Such study shall assess long-term costs to the public of lost water filtration, hunting, and fishing (commercial and recreational), and other ecosystem services associated with the Gulf of Mexico." (Sec. 2004)

The laws are grouped by thematic areas and year, and where the reader can find more information on an individual law via the full title, year, and legal citation information. U.S.C. = U.S. Code; § = Section; §§ = Sections; *et seq.* (*et sequentes*) = "and what follows"; Pub. L. No. = Public Law Number; Stat. = Statutes at Large

A high-level summary provides initial information about each law, its commonly used acronym (used for Fig. 1), and relevant examples to the field of ecosystem services

2 Ecosystem Services in U.S. Federal Environmental Laws

This Section presents a brief overview of a suite of U.S. federal environmental laws where ecosystem services and ecosystem-based management topics may be relevant. Here, this chapter expands beyond the list of what are considered "key" environmental laws to highlight the larger breadth of U.S. laws, and where the reader can turn to find more information, that may have relevance to ecosystem services and ecosystem-based management. Additionally, this chapter includes information on early U.S. federal legislation (over 100 years old), ancillary legislation not considered part of the "traditional suite" of environmental laws, and select Executive Orders that speak to environmental law related to ecosystem services.

2.1 Foundational Legislation

The foundational elements of U.S. environmental law predate the flurry of activities in the 1970s–1990s (Ruhl et al. 2013) and ultimately can be anchored in Roman law's "recognition that the general public had inalienable rights to access and use certain resources, namely the sea and seashore, rivers, and the air" (Connolly 2009). This is referred to as the "Public Trust Doctrine" (cf., Sax 1970; Ruhl 2005a; Ruhl and Salzman 2006); the first case addressing this in the U.S. occurred in 1842 (Smith and Sweeney 2006). Examples of 100+ year old U.S. Federal legislation that set the stage for identifying the importance of (protecting and valuing) ecosystems include: the Swamp Land Act (1850), the Homestead Act (1862); the General Mining Act (1872), the Organic Administrations Act (1897), the Rivers and Harbors Act (1899), the Reclamation Act (1902), the National Park Service Organic Act (1916), and the Migratory Bird Treaty Act (1918) (Donahue 2007; Tarlock 2007; Hirokawa 2011c; Cosens and Fremier 2014; Robbins 2018b).

2.2 U.S. Federal Environmental Laws—Overview

An overview of approximately 50 U.S. Federal environmental laws is presented in Table 1, including a description of each law's goals and purpose and an initial identification of which parts of a given law that may have relevance to ecosystem services and ecosystem-based management. Examples of U.S. Federal regulations that directly speak to ecosystem services are presented in Table 2. For a broader overview of the suite of U.S. environmental laws, the reader is referred to Farber and Findley (2014) and Salzman and Thompson (2003). At, or near, the "top" of the key list of U.S. Federal environmental laws are the National Environmental Policy Act, the Clean Water Act, the Clean Air Act, the Endangered Species Act, and the Comprehensive Environmental Response, Compensation, and Liability Act. The rest of this section briefly introduces these key laws and their intersection with ecosystem services.

2.3 "Key" U.S. Federal Environmental Laws

One of the primary U.S. Federal environmental laws with relevance to ecosystem services is the National Environmental Policy Act (NEPA), with the requirement that federal agencies evaluate a suite of alternatives (including a "no action" scenario) for developing pros/cons lists before a decision is made (Anderson 2011). Fischman (2001) argues for the direct utility of ecosystem services assessments as they may be "exactly the kind of assessment NEPA envisions, providing a means to inform the public and decision-makers about what we stand to gain or lose in several alternative

Table 2 Current U.S. federal regulations capturing "ecosystem services"

Agency	Code section	Summary	Citation
U.S. Department of Agriculture, Natural Resources Conservation Service	Healthy Forests Reserve Program Compensation for Easements and 30-year Contracts	This is part of the Healthy Forests Reserve Program to assist landowners to restore, enhance, and protect forestland resources on private land.	7 CFR § 625.8
	Grasslands Reserve Program Definitions	These definitions include conservation values which covers sustaining and enhancing ecosystem functions of grasslands.	7 CFR § 1415.3
	Wetlands Reserve Program Market Based Conservation Issues	Establishes the use of environmental credits for entities that implement conservation practices and activities.	7 CFR § 1467.20
	Agricultural Conservation Easement Program Environmental Markets	This section gives ecosystem service credits to landowners for conservation improvements to wetland reserve easements.	7 CFR § 1468.10
	National Forest System Land Management Planning Assessment; Sustainability; Multiple Use	This part deals with land management plans and the assessments of plan developments to include collaborative and science-based input so the lands involved are ecologically sustainable and have the capacity to provide ecosystem services to people and the community.	36 CFR § 219.XX (0.1; 0.6; 0.8; 0.10; 0.19)
U.S. Environmental Protection Agency	National Ambient Air Quality Standards Revisions to the Guideline on Air Quality Models	This Appendix provides guidelines for air quality modeling related to the derivation of Ozone National Ambient Air Quality Standards, noting that "emissions of NO_X, sulfur oxides, NH_3, mercury, and secondary pollutants such as ozone and particulate matter" can affect ecosystem services provided by forests and natural areas. (40 CFR Part 51, Appendix W)	40 CFR Part 51 Appendix W

CFR, Code of Federal Regulations; §, Section

scenarios." Fischman (2001) gives an example list of five types of NEPA-relevant decision activities:

1. Community-scale development activities with a federal nexus (e.g., highways; flood protection);
2. Development and use of renewable resource on public lands (e.g., logging and grazing);

3. Use (e.g., development, generation, and transmission) of renewable energy production, including coal, petroleum, and natural gas;
4. Use (e.g., development, processing, and transport) of non-energy mineral resources; and
5. Implementation of water projects, including permitting (e.g., wetland modification).

Cross-walking these examples with the Millennium Ecosystem Assessment's approach (Millennium Ecosystem Assessment 2005; Carpenter et al. 2009; da Silva and de Carvalho 2018) to classifying ecosystem services into four main categories:

- Provisioning (e.g., food/fiber; fuel);
- Regulating (e.g., water, disease);
- Cultural (spiritual; recreational; aesthetic); and
- Supporting (e.g., primary production; nutrient cycling)

it becomes clear that there is extensive relevance of applying NEPA to a range of ecosystem services that may be considered as part of NEPA consultations. Some example topics within NEPA's umbrella that are relevant include: property (Sect. 1.1); valuation (including cost/benefits analysis and markets; Sect. 1.2); development of alternative scenarios (including mitigation; Sect. 2); environmental impact assessments (not discussed here); and habitat evaluations (not discussed here). Recent legal scholar publications on NEPA and ecosystem services include Fischman (2001), Hirokawa and Porter (2013), Ruhl (2015). As a side note, the Millennium Ecosystem Assessment framework has been applied to environmental law issues around a number of topics (Thompson 2008; Ruhl 2015), including agrosystems (Ruhl 2008), public lands (Ruhl 2010a), aquatic resources (Ruhl 2010b), as well as the evolution of the ecosystem approach in international environmental law (Enright and Boteler 2020; Le Tissier 2020; O'Hagan 2020).

The Federal Water Pollution Control Act (commonly referred to as the Clean Water Act) includes provisions to protect aquatic ecosystems from human activities in order to protect a range of ecosystem services, including pollution removal (dilution and breakdown), providing habitat for wildlife (including those harvested commercially and recreationally), and assimilation and sequestration of nutrients (e.g., removal of excess nitrogen) (Salzman et al. 2001; Craig 2008; Ruhl 2010b; Smith et al. 2010). The Clean Water Act also includes provisions for mitigation banking, a mechanism of preservation, enhancement, or restoration of a specific natural resource area in order to provide compensation for the loss or degradation of another natural resource (see Sect. 2.1; Davis 2010). Additionally, there are several current topics of legal discussion with the Clean Water Act, including issues of jurisdiction (e.g., Craig 2008), setting Total Maximum Daily Loads (Ruhl 2010b), and filling wetlands (Ruhl et al. 2009) that are outside the scope of this chapter. As a side note, the Beaches Environmental Assessment and Coastal Health (BEACH) Act (2010), as an amendment to the CWA, is an example of additional legislation focused on the intersection between identification of human health issues and the ecosystem service of coastal beach use for recreation.

The Clean Air Act focuses on air quality protection and establishment of standards and intersects with ecosystem services in a number of areas, including, nutrient pollution removal (e.g., nitrogen, sulfur), and regulation of greenhouse gas emissions (e.g., Lazarus 2008; McGuire 2015).

The Endangered Species Act focuses on single-species management of threatened and endangered species, but with capacity to give attention to related habitats and for programmatic and multi-species consultations. Consideration of the intersection with ecosystem services is established in the literature for issues related to critical habitat (Salzman 1997; 2006), (indirect) protection of biodiversity (Thompson 2008; McGuire 2015), the use of Habitat Conservation Plans (Davis 2010), and the use of credits (Davis 2010).

The Comprehensive Environmental Response, Compensation, and Liability Act ("Superfund") has ecosystem services related connections to damage assessment (Wilson 2004; Desjardins 2014) as well as approaches to enhance cleanups (e.g., Green Remediation; Lipps et al. 2017) and redevelopment (Thompson 2008). See Sect. 4 on ecosystem services and remediation.

2.4 Non-Traditional Suite of Laws Related to Environmental Law and Ecosystem Services

There are other regulations not considered part of the suite of traditional environmental laws that are related to how ecosystem services are considered. For a land-use example, the National Flood Insurance Program, authorized by the National Flood Insurance Act, include the influences on, and distortion of, land prices that influence coastal and flood-plain development decoupled from other ecosystem valuation efforts for these important ecosystem landscapes (McGuire 2015). In contrast, the 1990 Conservation Reserve Program, established by the "Farm Bill," assesses (ranks) land parcels with the highest environmental benefits based on multiple criteria (Boyd et al. 2001; Davis 2010).

2.5 Executive Orders

Another suite of U.S. federal tools that can be used to examine related issues are Executive Orders (EOs), directives from the U.S. President to the Executive Branch of the government, including covering rulemaking for federal agencies such as the U.S. Environmental Protection Agency, U.S. Department of the Interior, and the U.S. Army Corps of Engineers. It is important to acknowledge that EOs represent policies, which are the operational applications of laws. That is, they capture different approaches to governing and interpretations of the execution of laws by the Executive Branch. This introduction to EOs is not intended to present an exhaustive survey of EOs relevant to ecosystem services, rather introduce this type of mechanism to the reader. For example, the National System of Marine Protected Areas was established in 2000 through EO 13158 ("Marine Protected Areas"). While EOs have extensive authority in that they are implemented at the same level as a regulation, they do not overrule an individual law, they are not legislatively approved, and they can be rescinded with the stroke of a pen by subsequent administrations.

Examples of EOs that explore further development of environmental-related cost-benefit analyses (Thompson 2008) include a suite of EOs on "Regulatory Planning and Review": EO 12866 (1993; 58 FR 51735), EO 13258 (2002; 67 FR 9385), EO 13422 (2007; 72 FR 2763), EO 13563 (2011; 76 FR 3821), and EO 13777 (2017; 82 FR 12285). This example suite of Executive Orders spans across multiple Presidential administrations.

In another example, EO 13547 ("Stewardship of the Ocean, Our Coasts, and the Great Lakes"; 2010; 75 FR 43023) explicitly referred to ecosystem services in providing guidance for coastal and marine spatial planning, specifically identifying those areas, "most suitable for various types or classes of activities in order to reduce conflicts among uses, reduce environmental impacts, facilitate compatible uses, and preserve critical ecosystem services to meet economic, environmental, security, and social objectives." This EO was revoked in 2018 and replaced by EO 13840 ("Ocean Policy to Advance the Economic, Security, and Environmental Interests of the United States"; 2018; 83 FR 29431) that did not reference the "ecosystem services" that nature provides to people but does reference the "benefits" the ocean provides the U.S. economy. As a side note, the reader is directed to Craig (2007) to learn more about coastal ecosystem services and environmental law and policy.

3 Themes in Ecosystem Services, EBM and Environmental Law

This Section presents a high-level literature review analysis of the existing legal scholarly literature on several current ecosystem services topics within U.S. federal environmental law. A literature search of abstracts, titles, and keywords published in the legal scholar literature was conducted using the HeinOnline law journal library search engine to identify potential peer-reviewed sources. The period of record for HeinOnline searches ranged from the date of inception for each legal journal in their database through March 2019. As a frame of reference, the HeinOnline search identified a total of 1903 legal articles that contained reference to ecosystem services. Focusing in on a snapshot of key literature, this chapter presents an overview of those articles that contained "ecosystem services" or "ecosystem based management" in the article's title. Although this search was not exhaustive, it provides a high-level snapshot of the current state of emphasis within the legal scholarly literature. The analysis presents a suite of ecosystem services themes in this literature, including fundamental elements (property and ownership; valuation, accounting and markets), conservation and protection (conservation and mitigation banking, public lands), and remediation on the "back end," including an introduction on natural resources damages.

3.1 Property and Ownership

Some aspects of ecosystem services may be subject to property rights law, that is, whether the value of an ecosystem service can be reduced to ownership (Hirokawa 2011c; Ruhl 2015). Pardy (2014) outlines one property premise related to developing approaches to protecting ecosystem services, namely that, "although some ES have no market value because they are not the subject of property rights and/or are not easily exchanged, all ES have an economic value that can be calculated by measuring their actual or potential importance to human well-being." Robbins (2018b) provides a general characterization that the case law for ecosystem services-based regulatory takings generally does not expressly treat ecosystem services as a property interest. And Hirokawa (2011c) argues that because ecosystem services may not have discrete boundaries, they could be considered property interests within another's property boundaries. One area of intersection between environmental law, property law, and ecosystem services is in "ecosystem energy services" (Hodas 2013). Ruhl (2005b) argues that the nuisance aspects of "common law" may be applicable for ecosystem services because the structure of this vehicle is flexible to handle changes, such as those encountered in the evolution of both the science of ecosystem services and its consideration in society (Hirokawa 2011c). The reader is directed towards Abrams (2007) for an overview on nuisance law and ecosystem services.

In the case of conservation easements, areas established to maintain essential habitat for species that can also provide ecosystem services, Cooley and Olander (2012) and Robbins (2018b) argue that because human value for ecosystem services can be extrapolated from easements, ecosystem-services related easements are considered property. Additional areas of development in environmental law and policy include the potential applicability of easements, and the services they may provide, for use in markets (see Sect. 1.2), such as for carbon credits, which require establishing a permanence of the market for credits (Ristino 2010). Easements, however, may not be permanent property instruments, and thus the ownership of the benefits (including delivery of ecosystem services) from a given easement, is an area of active development in the law (McLaughlin 2015).

From a technical perspective, there is a difference between an ecosystem good and an ecosystem service, namely that a good represents a market product (e.g., harvestable timber), while a service represents an ecosystem process or function (e.g., wetland filtering out water pollution) (Brown et al. 2007). Furthermore, the delineation of those ecosystem goods and services into *intermediate* (supporting products and processes not directly used by humans) and *final* (those used directly by humans) services to advance classification systems and environmental accounting (DeWitt et al. 2020; Russell et al. 2020) may also inform future discussions on ecosystem services and property law as it relates to matters of ownership. As a side note, there is continuing debate about biodiversity as an ecosystem service (Goble 2007), including whether it represents a final ecosystem service (DeWitt et al. 2020) directly benefiting people.

3.2 Ecosystem Services Accounting, Markets

The field of ecosystem services accounting and valuation is an ongoing area of scientific development. Pardy (2014) describes the three primary approaches for protection of ecosystem services as: (1) a regulatory approach (e.g., da Silva and de Carvalho 2018); (2) payments to protect ecosystem services (e.g., Hirsch 2007; Ruhl 2008; Benjamin 2013; Salzman et al. 2018), including investments in green infrastructure (Cosens and Fremier 2014; Salzman et al. 2014; da Silva and de Carvalho 2018); and (3) market-based approaches (e.g., Salzman 2005; Hirsch 2007; Glicksman and Kaime 2013; Kaime 2013). From an environmental law perspective, the authority for using valuation and accounting, and the range of potential approaches and methodologies themselves are all areas of ongoing development, case law, and legal debate. The primary legal spaces include natural resource damages, the consideration of compensation and mitigation, and the establishment of markets.

One primary approach for valuation includes "(focus) on a traditional, tort-like derivation of damages through per-unit calculations of past, present, and future damages" (Desjardins 2014). In one example, the Habitat Equivalency Analysis approach, a CERCLA provision using an accounting approach for habitat status/condition that is used to look at lost and restored services from a one-to-one comparison perspective (Ray 2009; Shaw and Wlodarz 2013), has received attention in a number of areas of environmental policy management, including natural resource damage assessments (NOAA 2000), NEPA projects (e.g., Ray 2009), and restoration decision making (Snyder and Desvousges 2013). Equivalence assessment approaches, including Habitat Equivalency Analysis and the related Resource Equivalency Analysis approach, are used for measuring losses and gains in habitat and biodiversity have been developed for a range of purposes (Desjardins 2014; Bezombes et al. 2017). Another suite of ecosystem services valuation focuses on the "willingness-to-pay" approaches, such as Contingency Valuation, the application of methodologies for natural resources that have no established market (Carson et al. 2001).

Ruhl et al. (2009) and Womble and Doyle (2012) explore mitigation banking in wetland and stream ecosystems resulting from the Clean Water Act and the 2009 Compensatory Mitigation Rule and its focus on market-based assessments of these ecosystems from a compensatory mitigation perspective. Further discussion on issues associated with geographic boundaries in environmental law and policy are outside the scope of this chapter, but the reader is directed to Womble and Doyle (2012) and Ruhl et al. (2009) for more information. Ruhl and Salzman (2007) and Salzman et al. (2018) present an overview of payments for ecosystem services, including both positive and negative incentives ("carrots vs. sticks"; Salzman et al. 2018) from a mitigation context.

A decade ago, there were more than 700 ecosystem services markets in the U.S. (Ristino 2010), with more than 2400 markets by 2016 (Bennett et al. 2016). Ruhl and Salzman (2007) provides examples of markets for forests. For carbon-based markets, one area of policy development is in carbon offsets, an accounting approach whereby the reduction in carbon emissions by one source could be used to offset the need for reduction in carbon emissions by another source. Carbon credits is one area of property law that is still in development (Ristino 2010; Glicksman and Kaime 2013; Ruhl et al. 2013). While this chapter does not explore current cap-and-trade issues, the reader is directed towards Glicksman and Kaime (2013) and Ruhl et al. (2013) to learn more. Likewise, the reader is pointed towards Brown et al. (2007) to learn more about technical and policy issues associated with measuring "carbon dioxide-equivalents" or the "social cost of carbon."

There are a number of environmental law and policy issues related to the development and implementation of markets, including property law, credits, banking, and accountability and oversight (e.g., Ristino 2010; Glicksman and Kaime 2013). One area of on-going property-based efforts is focused on real property instruments, the legal vehicles used to assign ownership of property (Ristino 2010). Another relates to the potential use of conservation easements (see Sect. 2.1), and whether easements can create permanency of both the credits themselves, and ownership of those credits, as it relates to how those credits are considered from a market or governance perspective.

3.3 Conservation, Protection & Mitigation Banking Tools

The Wilderness Act (1964) was primarily focused on the protection of public lands, including non-extractive services (Kammer 2013). The Federal Land Policy and Management Act (1976), focusing on the management of public lands, has a "no degradation" requirement that including taking necessary actions to prevent unnecessary or undue degradation of public lands (Donahue 2007). The 1978 Public Rangeland Improvement Act, implemented by the Bureau of Land Management, characterizes "less-than-potential production of ecosystem services, namely, 'wildlife habitat, recreation, forage, and water and soil conservation benefits,' is evidence of rangelands' 'unsatisfactory condition.'" (Donahue 2007). There is a large breadth of rangeland improvements called for by the Public Rangelands Improvement Act (and subsequent regulations), including soil resources, water resources, fish and wildlife habitat resources, and improvements for livestock and wild horse management Penderly (1997). Ecosystem services elements of forests are broad (Neuman

2007); Federal forest lands are "administered for outdoor recreation, range, timber, watershed, and wildlife and fish purposes." (16 U.S.C. § 528). Three U.S. Federal laws overseeing forestry services include the National Forest Management Act (1976), the Forest and Rangeland Renewable Resources Planning Act (1974), and the Multiple-Use Sustained-Yield Act (1960). From an environmental law perspective, attention in forestry ecosystem services includes advancing concepts of "payment for services" (Ruhl and Salzman 2007), balancing vegetation management plans, provisioning of forestry goods, and potential impacts on watershed function (Hirokawa and Porter 2013).

There are a number of federal "incentive programs," whereby the government pays private landowners to protect ecosystems and their services, including the Conservation, Wetlands, and Grasslands Reserve programs (Table 2), Environmental Quality Incentives Program, the Farm and Ranch Lands Protection Program, the Conservation Security Program, and the Forestland Enhancement Program (Brown et al. 2007; Ruhl 2008). In a related tool, the U.S. Department of Interior oversees the Land and Water Conservation Fund (1965), designed, in part, to "preserve ecosystem benefits for local communities" for both public and private lands (Land and Water Conservation Fund 2017). Areas of current attention in agricultural ecosystem services is in markets for carbon offsets (Davis 2010; see Sect. 3.2 for more on markets), and the U.S. Department of Agriculture's use of, "ecosystem service values as a basis for payments under traditional conservation program payments" (Ruhl 2015).

The U.S. government's Council of Environmental Quality's regulation on mitigation includes a section on, "compensating for the impact by replacing or providing substitute resources or environments" (40 CFR 1508.20(e)). Mitigation banking, an approach to protect, enhance, or create a habitat (particularly wetlands) as compensation for the impacts at other locations, is one tool used for providing compensation for ecosystem impacts. Examples of mitigation banking include those established through the authority of Sect. 404 of the Clean Water Act. U.S. federal agencies involved in mitigation banking include the U.S. Environmental Protection Agency, U.S. Department of Agriculture, the U.S. Fish and Wildlife Service, and the U.S. Army Corps of Engineers. Conservation easements are established to maintain essential habitat for species, where Cooley and Olander (2012) and Robbins (2018b) argue that human value for the ecosystem service can be extrapolated. Here, ecosystem services related easements are considered property. Discussions on environmental law and policy perspectives on ecosystem services and mitigation banking include Salzman and Ruhl (2000), Boyd et al. (2001), Hirsch (2007), and Robbins (2018b). A spin on wetland mitigation banking for use as a market for biodiversity offsets is presented in Spurgeon (2008).

3.4 Remediation on the "Back End"—Natural Resources Damages

There are a number of examples of natural resource damage provisions in U.S. Federal law that mandate valuation in response to a loss, or deprivation of ecosystem function and services, including the Deepwater Port Act (1974), the Oil Pollution Act (1990), the National Marine Sanctuaries Act (1972), and the Comprehensive Environmental Response, Compensation, and Liability Act (1980) (commonly referred to as Superfund) (Boyd et al. 2001; Wilson 2004; Smith et al. 2010). Salzman (1997) provides an early environmental law analysis of the need for information on ecosystem services information markets to feed the design of remediation strategies in Superfund. Another example of another ecosystem services related regulation that focuses on violation/penalties is the Natural Resources Damage Assessment that focuses on assessing compensation for injuries to natural resources (see Boyd et al. 2001 and Davis 2010 for overviews). In characterizing the U.S. Environmental Protection Agency's three-prong approach to enforcement (deterrence; fairness; swift resolution of environmental problems), Markell (2007) provides an overview of three tools: penalties for violations; injunctive relief (i.e., a court-driven order to address a problem); and Supplemental Environmental Projects (SEPs) as a form of relief in case settlements. A number of U.S. Federal environmental laws focus on prevention of ecosystem contamination (Table 1) using ecological risk assessment characterizations as an important tool. For an overview of efforts to advance ecosystem services as assessment endpoints in the ecological risk assessment process, the reader is directed to Munns et al. (2016).

4 Ecosystem Services and Environmental Law at Different Scales

One guiding principle of EBM involves the interaction across different scales (federal, state, and local) to address geographic-based management issues (Nugent and Cantral 2006; Green et al. 2014). This section presents an overview of where ecosystem services are captured at different scales, including state agency laws, and several examples at regional and local scales. Examples of current State environmental laws are introduced in Table 3. At present, just the three U.S. west coast states (Washington State, Oregon, and California) and Rhode Island have laws that explicitly refer to ecosystem services.

Examples of ecosystem services captured within U.S. regional-scale environmental law issues include:

- Ecosystem-Based Management of the western U.S. (e.g., Smith 1999);
- The use of the Endangered Species Act as an overarching framework for northwest Montana (Guercio and Duane 2009);

Table 3 Current U.S. state laws capturing "ecosystem services"

State	Code Section	Summary	Citation
Oregon	Public Health and Safety	Chapter 468 of the Oregon statutes deals with environmental quality. "It is the policy of this state to support the maintenance, enhancement and restoration of ecosystem services throughout Oregon, focusing on the protection of land, water, air, soil and native flora and fauna." Or. Rev. Stat. 468.583 (2018)	468.581; 468.583; 468.585; 468.587 (2018)
	Forestry and Forest Products	These sections of the Oregon statutes advance the continuation of the Forest Resource Trust to promote establishment and management of nonindustrial state forestland through the use of paying landowners for preserving ecosystem services.	526.695; 526.703; 526.705 (2018)
	Water Resources: Irrigation, Drainage, Flood Control, Reclamation	This Oregon statute establishes the Water Resources Department to develop and implement a holistic water usage plan for the entire state of Oregon. It includes ecosystem services as a point to consider when developing such plan.	536.220 (2018)
Washington	Forest and Forest Products	This Washington statute establishes a forest maintenance plan to wisely use timber resources and replenish such, including payments to forest landowners for ecosystem services provided to the public in preserving timber resources.	76.09.010; 76.09.020 (2019)
	Forest and Forest Products	This Washington statute discusses the Forestry Riparian Easement Program and reimbursement to small forest landowners for preservation of timber resources and ecosystem services supported by the program.	76.13.120 (2019)
	Public Lands	This Washington statute discusses the community forest trust program and that preservation of "ecosystem services such as clean water protection or carbon storage." Wash. Rev. Code s 79.155.030(2)(c)	79.155.030 (2019)
California	Fish and Game	This California statute defines "Ecosystem-based management" as "an environmental management approach relying on credible science, as defined in Section 33, that recognizes the full array of interactions within an ecosystem, including humans, rather	Cal. Fish & Game Code § 43 (West 2019)

(continued)

Table 3 (continued)

State	Code Section	Summary	Citation
		than considering single issues, species, or ecosystem services in isolation." Cal. Fish & Game Code § 43 (West 2019)	
	Fish and Game	This California statute gives definitions of words used in the state's advance mitigation and regional conservation investment strategies and includes incorporating the benefits of ecosystem services as part of the "regional conservation assessment" definition.	Cal. Fish & Game Code § 1851 (West 2019)
	Public Resources	This California statute instructs the Ocean Protection Council to support sharing of information between state agencies and making that information publicly available with respect to "social, economic, and cultural values, including the value of coastal and ocean ecosystems for providing ecosystem services." Cal. Pub. Res. Code § 35620(a)(2)(E) (West 2019)	Cal. Fish & Game Code § 35620 (West 2019)
Rhode Island	Health and Safety	This statute is the legislative findings for the Rhode Island Climate Risk Reduction Act of 2010 where the legislature states that "natural ecosystems and habitats, both coastal and upland, provide critical ecosystem services including, fisheries habitat, drinking water, and flood protection." 23 R.I. Gen. Laws § 23-84-2(6)	§ 23-84-2 (2019)

- The Northwest Forest Plan's framework (e.g., Neuman 2007); and
- Legal and regulatory authorities for managing the coastal resources of the Gulf of Mexico (Nugent and Cantral 2006), such as the use of the Manguson-Stevens Fishery Conservation and Management Act (1976) for red snapper and other fishery species (Pace 2009).

Importantly, Federal laws have spurned state and local laws, often looking at the scale of a watershed, the boundary of which may not necessarily align with political or governance boundaries. Looking at a broader suite of examples related to water protection, Greenwalt and McGrath (2009) explore the tenets of a pay-for-ecosystem-services (PES) model at a watershed scale. For a specific example, the Safe Drinking Water Act spurred New York City to implement local regulations on protecting the Catskills and Delaware watersheds providing the primary source of clean drinking water for its citizens (Thompson 2008; Salzman et al. 2001; Salzman

2011; Robbins 2018a, b). Other watershed-scale examples include water-based natural and engineered services in the Columbia River Basin (Cosens and Fremier 2014), and the production (Greenwalt and McGrath 2009), purification (Salzman et al. 2001), and apportionment of water for municipal and other uses (Ruhl 2003). Green et al. (2014) examine EBM issues at different legal scales for coral reefs, an example where upstream land-use decisions may not align with different scales and domains of existing environmental regulations on the downstream resource of interest.

Local-scale forestry examples in the legal literature includes the ecosystem management of Tillamook State Forest, involving a range of stakeholders, different scales of regulatory hierarchy in forest and adjacent lands, and a suite of forest-related ecosystem services (Neuman 2007 and citations therein). Other local-scale examples include urban forest planning (Hirokawa 2011a, b), land-use policies for agrosystems (Ruhl 2008), and salmon fisheries (Hirokawa and Gottlieb 2011). Other urban ecosystem services issues, also considered local scale, are outlined in Salzman et al. (2014).

5 Conclusions

EBM practitioners work in an interdisciplinary universe, spanning a range of science, engineering, and management/policy backgrounds and expertise. This chapter presents an overview of the large spectrum of U.S. Federal environmental laws, with particular relevance to the field of ecosystem services. Anchored by a review of the extant legal scholarly literature, this chapter presents a review of a broad suite of ecosystem services topics in U.S. Federal environmental law specifically translated for EBM practitioners as the primary audience, pointing the reader towards resources to learn more about individual elements presented throughout the chapter. This chapter provides EMB practitioners information from the environmental law literature to inform how they frame the legal context of ecosystem services information in their EBM activities. Finally, this chapter helps the reader identify examples where relevant information might be transferable to their specific scenarios, such as how policy and legal directives are framed in the watershed-based EBM example of the Danube Basin (Funk et al. (2020)).

Acknowledgements Ben Boteler and Ahjond Garmestani are thanked for valuable reviews of this chapter. The views expressed in this chapter are those of the author, and do not necessarily reflect the views or policies of Palm Beach State College. Mention of trade names or commercial products does not constitute endorsement or recommendation for use.

References

Abrams, R. H. (2007). Broadening narrow perspectives and nuisance law: Protecting ecosystem services in the ACF Basin. *Journal of Land Use & Environmental Law, 22*(2), 243–298.

Anderson, F. R. (2011). *NEPA in the courts: A legal analysis of the National Environmental Policy Act*. Washington DC: Resources for the Future Press.

Benjamin, A. H. (2013). Payment for ecosystem services. *Potchefstroom Electronic Law Journal, 16*(2), 1–5.

Bennett, G., Carroll, N., Sever, K., Neale, A., & Hartley, C. (2016). *An atlas of ecosystem markets in the United States*. Washington, DC: Forest Trends' Ecosystem Marketplace. Retrieved August 9, 2019, from https://www.forest-trends.org/wp-content/uploads/2017/03/doc_5440.pdf.

Bezombes, L., Gaucherand, S., Kerbiriou, C., Reinert, M. E., & Spiegelberger, T. (2017). Ecological equivalence assessment methods: What trade-offs between operationality, scientific basis and comprehensiveness? *Environmental Management, 60*(2), 216–230.

Boyd, J., King, D., & Wainger, L. A. (2001). Compensation for lost ecosystem services: The need for benefit-based transfer ratios and restoration criteria. *Stanford Environmental Law Journal, 20*, 393–412.

Brown, T. C., Bergstrom, J. C., & Loomis, L. B. (2007). Defining, valuing, and providing ecosystem goods and services. *Natural Resources Journal, 47*(2), 329–376.

Carpenter, S. R., Mooney, H. A., Agard, J., Capistrano, D., DeFries, R. S., Díaz, S., Dietz, T., et al. (2009). Science for managing ecosystem services: Beyond the millennium ecosystem assessment. *Proceedings of the National Academy of Sciences, 106*(5), 1305–1312.

Carson, R. T., Flores, N. E., & Meade, N. F. (2001). Contingent valuation: Controversies and evidence. *Environmental and Resource Economics, 19*(2), 173–210.

Connolly, P. J. (2009). Saving fish to save the bay: Public trust doctrine protection for Menhaden's foundational ecosystem services in the Chesapeake Bay. *Boston College Environmental Affairs Law Review, 36*(1), 135–170.

Cooley, D., & Olander, L. (2012). Stacking ecosystem services payments: Risks and solutions. *Environmental Law Review News and Analysis, 42*(2), 10150–10165.

Cosens, B., & Fremier, A. (2014). Assessing system resilience and ecosystem services in large river basins: A case study of the Columbia River basin. *Idaho Law Review, 51*(1), 91–125.

Craig, R. K. (2007). Valuing coastal and ocean ecosystem services: The paradox of scarcity for marine resources commodities and the potential role of lifestyle value competition. *Journal of Land Use & Environmental Law, 22*(2), 355–410.

Craig, R. K. (2008). Justice Kennedy and ecosystem services: A functional approach to Clean Water Act jurisdiction after Rapanos. *Environmental Law, 38*(3), 635–666.

da Silva, K. R., & de Carvalho, D. W. (2018). Initial contributions to a legal protection of ecosystem services. *Veredas do Direito, 15*(32), 87–115.

Davis, A. I. (2010). Ecosystem services and the value of land. *Duke Environmental Law & Policy Forum, 20*(2), 339–384.

Desjardins, M. (2014). Ecosystem services: Unifying economic efficiency and ecological stewardship via natural resource damage assessments under CERCLA. *George Mason Law Review, 21* (3), 717–751.

DeWitt, T. H., Berry, W. J., Canfield, T. J., Fulford, R. S., Harwell, M. C., Hoffman, J. C., Johnston, J. M., Newcomer-Johnson, T. A., Ringold, P. L., Russel, M. J., Sharpe, L. A., & Yee, S. J. H. (2020). The final ecosystem goods and services (FEGS) approach: A beneficiary-centric method to support ecosystem-based management. In T. O'Higgins, M. Lago, & T. H. DeWitt (Eds.), *Ecosystem-based management, ecosystem services and aquatic biodiversity: Theory, tools and applications* (pp. 127–148). Amsterdam: Springer.

Donahue, D. L. (2007). Federal Rangeland policy: Perverting law and jeopardizing ecosystem services. *Journal of Land Use & Environmental Law, 22*(2), 299–352.

Enright, S.R., & Boetler, B. (2020). The ecosystem approach in international law. In T. O'Higgins, M. Lago, & T. H. DeWitt (Eds.), *Ecosystem-based management, ecosystem services and aquatic biodiversity: Theory, tools and applications* (pp. 333–352). Amsterdam: Springer.

Farber, D. A., & Findley, R. W. (2014). *Environmental law in a nutshell*. Saint Paul, MN: West Academic Publishing.

Fischman, R. L. (2001). The EPA's NEPA duties and ecosystem services. *Stanford Environmental Law Journal, 20*, 497–536.

Funk, A., O'Higgins, T. G., Borgwardt, F., Trauner, D., & Hein, T. (2020). Ecosystem-based management to support conservation and restoration efforts in the Danube Basin. In

T. O'Higgins, M. Lago, & T. H. DeWitt (Eds.), *Ecosystem-based management, ecosystem services and aquatic biodiversity: Theory, tools and applications* (pp. 431–444). Amsterdam: Springer.

Glicksman, R. L., & Kaime, T. (2013). A comparative analysis of accountability mechanisms for ecosystem services markets in the United States and the European Union. *Transnational Environmental Law, 2*(2), 259–283.

Goble, D. D. (2007). What are slugs good for—Ecosystem services and the conservation of biodiversity. *Journal of Land Use & Environmental Law, 22*(2), 411–418.

Green, O. O., Garmestani, A. S., Hopton, M. E., & Heberling, M. T. (2014). A multi-scalar examination of law for sustainable ecosystems. *Sustainability, 6*(6), 3534–3551.

Greenwalt, T., & McGrath, D. (2009). Protecting the city's water: Designing a payment for ecosystem services program. *Natural Resources & Environment, 24*(1), 9–13.

Guercio, L. D., & Duane, T. P. (2009). Grizzly bears, gray wolves, and federalism, oh my—The role of the endangered species act in de facto ecosystem-based management in the Greater Glacier Region of Northwest Montana. *Journal of Environmental Law & Litigation, 24*(2), 285–366.

Hirokawa, K. H. (2011a). Sustainability and the urban forest: An ecosystem services perspective. *Natural Resources Journal, 51*(2), 233–259.

Hirokawa, K. H. (2011b). Sustaining ecosystem services through local environmental law. *Pace Environmental Law Review, 28*(3), 760–826.

Hirokawa, K. H. (2011c). Three stories about nature: Property, the environment, and ecosystem services. *Mercer Law Review, 62*(2), 541–604.

Hirokawa, K. H., & Gottlieb, C. (2011). Sustainable habitat restoration: Fish, farms, and ecosystem services. *Fordham Environmental Law Review, 23*(1), 1–54.

Hirokawa, K. H., & Porter, E. J. (2013). Aligning regulation with the informational need: Ecosystem services and the next generation of environmental law. *Akron Law Review, 46*(4), 963–991.

Hirsch, D. D. (2007). Trading in ecosystem services: Carbon sinks and the clean development mechanism. *Journal of Land Use & Environmental Law, 22*(2), 623–639.

Hodas, D. R. (2013). Law, the laws of nature and ecosystem energy services: A case of wilful blindness. *Potchefstroom Electronic Law Journal, 16*(2), 66–121.

Kaime, T. (2013). Framing the law and policy for ecosystem services. *Transnational Environmental Law, 2*(2), 211–216.

Kammer, S. (2013). Coming to terms with wilderness: The wilderness act and the problem of wildlife restoration. *Environmental Law, 43*(1), 83–124.

Land and Water Conservation Fund. (2017). *U.S. Department of the Interior*. Retrieved August 9, 2019, from www.doi.gov/lwcf.

Lazarus, R. J. (2008). Super wicked problems and climate change: Restraining the present to liberate the future. *Cornell Law Review, 94*(5), 1153–1234.

Le Tissier, M. (2020). Unravelling the relationship between ecosystem-based management, integrated coastal zone management and marine spatial planning. In T. O'Higgins, M. Lago, & T. H. DeWitt (Eds.), *Ecosystem-based management, ecosystem services and aquatic biodiversity: Theory, tools and applications* (pp. 403–416). Amsterdam: Springer.

Lipps, J. M., Harwell, M. C., Kravitz, M., Lynch, K., Mahoney, M., Pachon, C., & Pluta, B. (2017). *Ecosystem services at contaminated site cleanups*. U.S. Environmental Protection Agency, EPA/542/R-17/004.

Markell, D. (2007). Is there a possible role for regulatory enforcement in the effort to value, protect, and restore ecosystem services. *Journal of Land Use & Environmental Law, 22*(2), 549–598.

McGuire, C. J. (2015). Valuing ecosystem services in coastal management policy: Looking beyond the here and now. *Natural Resources & Environment, 30*(2), 42–45.

McLaughlin, N. A. (2015). Interpreting conservation easements. *Probate & Property, 29*(2), 30–35.

Millennium Ecosystem Assessment. (2005). *Ecosystems and human well-being: Synthesis*. Washington, DC: Island Press.

Munns, W. R., Jr., Rea, A. W., Suter, G. W., Martin, L., Blake-Hedges, L., Crk, T., Davis, C., Ferreira, G., Jordan, S., Mahoney, M., & Barron, M. G. (2016). Ecosystem services as assessment endpoints for ecological risk assessment. *Integrated Environmental Assessment and Management, 12*(3), 522–528.

National Oceanic and Atmospheric Administration (NOAA). (2000). *Habitat equivalency analysis: An overview. NOAA damage assessment and restoration program.* Washington, DC: NOAA. Retrieved August 9, 2019, from https://casedocuments.darrp.noaa.gov/northwest/cbay/pdf/cbhy-a.pdf.

Neuman, J. (2007). Thinking inside the box: Looking for ecosystem services within a forested watershed. *Journal of Land Use & Environmental Law, 22*(2), 173–205.

Nugent, I., & Cantral, L. (2006). Charting a course toward ecosystem-based management in the Gulf of Mexico. *Duke Environmental Law & Policy Forum, 16*(2), 267–292.

O'Hagan, A. M. (2020). Ecosystem-based management (EBM) and ecosystem services in EU law, policy and governance. In T. O'Higgins, M. Lago, & T. H. DeWitt (Eds.), *Ecosystem-based management, ecosystem services and aquatic biodiversity: Theory, tools and applications* (pp. 353–372). Amsterdam: Springer.

Pace, N. L. (2009). Ecosystem-based management under the Magnuson-Stevens Act: Managing the competing interests of the Gulf of Mexico red snapper and shrimp fisheries. *Sea Grant Law & Policy Journal, 2*(2), 1–28.

Pardy, B. (2014). The logic of ecosystems: Capitalism, rights and the law of ecosystem services. *Journal of Human Rights & Environment, 5*(2), 136–152.

Penderly, B. M. (1997). Reforming livestock grazing on the public domain: Ecosystem management-based standards and guidelines blaze a new path for range management. *Environmental Law, 27*(2), 513–602.

Ray, G. L. (2009). *Application of habitat equivalency analysis to USACE projects.* EMRRP Technical Notes Collection. ERDC TN-EMRRP-EI-04. U.S. Army Engineer Research and Development Center, Vicksburg, MS. Retrieved August 9, 2019, from https://apps.dtic.mil/dtic/tr/fulltext/u2/a501248.pdf.

Ristino, R. (2010). Conservation easements in an ecosystem services age. *Natural Resources & Environment, 24*(3), 56–58.

Robbins, K. (2018a). Complementary authority and the one-way ratchet: Ecosystem services property, regulation, and wildlife conservation. *Environmental Law, 48*(2), 291–310.

Robbins, K. (2018b). Allocating property interests in ecosystem services: From chaos to flowing rivers. *Harvard Environmental Law Review, 42*(1), 197–229.

Ruhl, J. B. (2003). Equitable apportionment of ecosystem services: New water law for a new water age. *Journal of Land Use & Environmental Law, 19*(1), 47–57.

Ruhl, J. B. (2005a). Ecosystem services and the common law of the fragile land system. *Natural Resources & Environment, 20*(2), 3–69.

Ruhl, J. B. (2005b). Toward a common law of ecosystem services. *St. Thomas Law Review, 18*(1), 1–19.

Ruhl, J. B. (2008). Agriculture and ecosystem services: Strategies for state and local governments. *NYU Environmental Law Journal, 17*, 424–459.

Ruhl, J. B. (2010a). Ecosystem services and federal public lands: Start-up policy questions and research needs. *Duke Environmental Law & Policy Forum, 20*(2), 275–290.

Ruhl, J. B. (2010b). Ecosystem services and the Clean Water Act: Strategies for fitting new science into old law. *Environmental Law, 40*(4), 1381–1399.

Ruhl, J. B. (2015). In defense of ecosystem services. *Pace Environmental Law Review, 32*(1), 306–335.

Ruhl, J. B., & Salzman, J. (2006). Ecosystem services and the public trust doctrine: Working change from within. *South Eastern Environmental Law Journal, 15*(1), 223–239.

Ruhl, J. B., & Salzman, J. (2007). The law and policy beginnings of ecosystem services. *Journal of Land Use & Environmental Law, 22*(2), 157–172.

Ruhl, J. B., Salzman, J., & Goodman, I. (2009). Implementing the new ecosystem services mandate of the section 404 compensatory mitigation program—A catalyst for advancing science and policy. *Stetson Law Review, 38*(1), 251–272.

Ruhl, J. B., Kraft, S. E., & Lant, C. L. (2013). *The law and policy of ecosystem services.* Washington, DC: Island Press.

Russell, M. J., Rhodes, C., Sinha, R. K., Van Houtven, G., Warnell, G., & Harwell, M. C. (2020). Ecosystem-based management and natural capital accounting. In T. O'Higgins, M. Lago, & T. H. DeWitt (Eds.), *Ecosystem-based management, ecosystem services and aquatic biodiversity: Theory, tools and applications* (pp. 149–164). Amsterdam: Springer.

Salzman, J. (1997). Valuing ecosystem services. *Ecology Law Quarterly, 24*(4), 887–904.

Salzman, J. (2005). Creating markets for ecosystem services: Notes from the field. *NYU Law Review, 80*(3), 870–961.

Salzman, J. (2006). A field of Green—The past and future of ecosystem services. *Journal of Land Use & Environmental Law, 21*(2), 133–151.

Salzman, J. (2011). What is the emperor wearing—The secret lives of ecosystem services. *Pace Environmental Law Review, 28*(2), 591–613.

Salzman, J., & Ruhl, J. B. (2000). Currencies and commodification of environmental law. *Stanford Law Review, 53*(3), 607–694.

Salzman, J., & Thompson, B. H. (2003). *Environmental law and policy*. New York: Foundation Press.

Salzman, J., Thompson, B. H., Jr., & Daily, G. C. (2001). Protecting ecosystem services science, economics, and law. *Stanford Environmental Law Journal, 20*, 309–332.

Salzman, J., Arnold, C. A., Garcia, R., Hirokawa, K., Jowers, K., LeJava, J., Peloso, M., & Olander, L. (2014). The most important current research questions in urban ecosystem services. *Duke Environmental Law & Policy Forum, 25*(1), 1–47.

Salzman, J., Bennett, G., Carroll, N., Goldstein, A., & Jenkins, M. (2018). Payments for ecosystem services: Past, present and future. *Texas A&M Law Review, 6*(1), 199–227.

Sax, J. L. (1970). The public trust doctrine in natural resource law: Effective judicial intervention. *Michigan Law Review, 68*(3), 471–566.

Scarlett, L., & Boyd, J. (2015). Ecosystem services and resource management: Institutional issues, challenges, and opportunities in the public sector. *Ecological Economics, 115*(C), 3–10.

Sharon, O., Fishman, S. N., Ruhl, J. B., Olander, L., & Roady, S. E. (2018). Ecosystem services and judge-made law: A review of legal cases in common law countries. *Ecosystem Services, 32*(A), 9–21.

Shaw, W. D., & Wlodarz, M. (2013). Ecosystems, ecological restoration, and economics: Does habitat or resource equivalency analysis mean other economic valuation methods are not needed? *Ambio, 42*(5), 628–643.

Smith, R. E. (1999). The canyon country partnership and ecosystem-based management on the East-Central Colorado plateau. *Journal of Land Resources & Environmental Law, 19*(1), 19–38.

Smith, G. P., & Sweeney, M. W. (2006). The public trust doctrine and natural law: Emanations within a penumbra. *Boston College Environmental Affairs Law Review, 33*(2), 307–344.

Smith, L. C., Jr., Smith, L. M., & Ashcroft, P. A. (2010). Analysis and ecosystem services deprivation: From Cuyahoga to the Deepwater Horizon. *Albany Law Review, 74*(1), 563–585.

Snyder, J. P., & Desvousges, W. H. (2013). Habitat and resource equivalency analyses in resource compensation and restoration in decision making. *Natural Resources & Environment, 28*(1), 3–6.

Spurgeon, J. (2008). Economic incentives for biodiversity and ecosystem services. *Business Law Review, 29*(4), 84–94.

Tarlock, A. D. (2007). Ecosystem services in the Klamath Basin: Battlefield casualties or the future. *Journal of Land Use & Environmental Law, 22*(2), 207–242.

Thompson, B., Jr. (2008). Ecosystem services & natural capital: Reconceiving environmental management. *NYU Environmental Law Journal, 17*(1), 460–489.

Wilson, M. A. (2004). *Ecosystem services at superfund redevelopment sites revealing the value of revitalization landscapes through the integration of ecological economics spatial information group subcontract to systems research and applications corporation*. Prepared for the U.S. EPA, OSWER (Contract 68-W-01-058).

Womble, P., & Doyle, M. (2012). The geography of trading ecosystem services: A case study of wetland and stream compensatory mitigation markets. *Harvard Environmental Law Review, 36* (1), 229–296.

9

The Promise and Pitfalls of Ecosystem Services Classification and Valuation

Stephen Flood, Timothy G. O'Higgins, and Manuel Lago

Abstract We are currently facing the triple interlocked threats of climate change, unsustainable land use change, and the sixth mass species extinction. This chapter firstly outlines these interlinked threats making the case for urgent action. It then documents the efforts to assign economic and non-economic value to our biodiversity (Ecosystem Services) through environmental and ecological economics, highlighting the fundamental philosophical principles underlining both approaches. This sets up a discussion on the development and potential of Ecosystem Services (ESS) as a discipline in its own right, the challenges of application, and the awareness of, and priority assigned to, ESS by policymakers and the private sector. The chapter closes by outlining specific methodological challenges and recommendations for substantially increasing the level of attention and action needed to protect and enhance our invaluable ecosystems in our age of potential ecological collapse.

Elements of this chapter are based on the PhD thesis of Flood, S. (2012), which is available online at: http://mural.maynoothuniversity.ie/4760/

S. Flood (✉)
MaREI Centre for Energy, Climate and Marine and Renewable Energy, Environmental Research Institute, University College Cork, Cork, Ireland

School of Geography, Environment & Earth Sciences, Victoria University of Wellington, Wellington, New Zealand
e-mail: stephen.flood@ucc.ie; stephen.flood@vuw.ac.nz

T. G. O'Higgins
MaREI Centre for Energy, Climate and Marine and Renewable Energy, Environmental Research Institute, University College Cork, Cork, Ireland
e-mail: tim.ohiggins@ucc.ie

M. Lago
Ecologic Institute, Berlin, Germany
e-mail: manuel.lago@ecologic.eu

Lessons Learned

- We have documented the development of ESS, from its foundations in environmental and ecological economics to its evolution into FEGS and CICES classifications and valuation frameworks.
- In doing so we have highlighted the challenges of valuation (including issues of scale, biodiversity awareness or literacy, and polycentric governance), and provided lessons and recommendations going forward for the successful application of ESS.
- Combining a number of different tools and methods can help strengthen assessment
- A persistent gap persists between ESS applications and their ability to provide easily usable information for decision-makers
- ESS can help shine a light on what we will lose if we fail to protect valuable (and indeed invaluable) ecosystems and earth's flora and fauna in general.

Needs to Advance EBM

- Keep it simple – Decision-makers are interested in simple, easy-to-use decision support tools that are understandable and can be easily incorporated into science-policy processes (Ruckelshaus et al. 2015; IPBES 2019; Dunford et al. 2018). Scientists should note that even basic tools are ample for parameterizing and interpreting data at the early stages of applying Biodiversity and Ecosystem Services (BES) information (Ruckelshaus et al. 2015);
- It's not always about the money – Attributing economic values to biophysical ecosystem service estimates is an important conceptual advance. This ability to follow biophysical estimates though to economic value has allowed decision makers to begin having conversations they did not previously engage with, and lead to new policy outcomes (Dunford et al. 2018; Ruckelshaus et al. 2015; Barton et al. 2018);
- Relate BES change to livelihoods and other wellbeing metrics.

1 Introduction

"We are the first generation that has a clear picture of the value of nature and its integral link with human well-being. We are also the last generation that has the opportunity to prevent the collapse of our planet's biodiversity in the face of habitat destruction and climate change." (WWF 2018, p. 10).

The scientific evidence indicates that the Earth's climate is changing (IPCC 2014) and, without taking appropriate and early action, climate change will have severe impacts on many of the planet's species and habitats (Scheffers et al. 2016). The 2006 Stern Review emphasises that the benefits of strong early action on climate change is likely to outweigh the costs, and values the cost of inaction at 5% of global GDP each year indefinitely (Stern 2006). It is important to note the value of global biodiversity is not fully captured in this percentage and the fact that crossing critical tipping points in our ecosystems, that would lead to extensive and run-away species and habitat loss, is also not captured. Yet research into the value of ecosystem services reveals that eco-services contribute more than twice as much to human well-being as global GDP (Costanza et al. 2014) and greater investment into the restoration and protection of the ecosystems and habitats that make those services possible can increase resilience to climate change.

1.1 Climate and Biodiversity Crises and the Need for Change

The secretariat of the Convention on Biodiversity considered the interlinkages between climate change adaptation and mitigation and biodiversity in a technical report published in 2009. The report established that biodiversity and climate change are interconnected because climate change effects biodiversity and because changes in biodiversity affects our ability reduce our atmospheric greenhouse gas levels (e.g. our natural carbon sinks) and to adapt to and mitigate against the impacts of climate change (CBD Secretariat 2009). It also highlighted the potential of ecosystem-based adaptation to create co-benefits for climate action and biodiversity conservation.

Furthermore, a recent special report by the Intergovernmental Panel on Climate Change (IPCC) indicates significant impacts to biodiversity and other sectors are set to occur even if we keep climate change to 1.5 °C over preindustrial levels, which are below business-as-usual global average temperature increases by mid-century (IPCC 2018). Biodiversity is at the forefront of climate change impacts globally. Headline results from the 2018 Living Planet Report, published by the World Wildlife Fund (WWF 2018), reveal that Earth is losing biodiversity at a rate seen only during mass extinctions. The report finds that global losses in populations of vertebrate species—mammals, fish, birds, amphibians and reptiles—have averaged 60% between 1970 and 2014. Overexploitation of species, agriculture, land conversion, and climate change are the main drivers of biodiversity decline, with climate change becoming a growing threat (Ibid.).

The 2018 Conference of the Parties to the Convention on Biodiversity and the 2019 report from the Intergovernmental Science-Policy Platform on Biodiversity and Ecosystem Services (IPBES) also echo the findings of the WWF Living Planet report in highlighting the critical role of biodiversity and ecosystems functions and services for human well-being with the IPBES reporting that the health of ecosystems, on which we all and all other species depend, is deteriorating at a rapid rate, and only through 'transformative change' can nature be conserved, restored and used sustainably (CBD Secretariat 2018; IPBES 2019). Transformative change should be understood as a fundamental, system-wide reorganisation across technological, economic and social factors, including paradigms, goals and values (IPBES 2019). It was also recognised that climate change is a major and growing driver of biodiversity loss, and that biodiversity and ecosystem functions and services, significantly contribute to climate change adaptation, mitigation and disaster risk reduction.

The IPCC's (2019) Special Report on Climate Change and Land states with high confidence that increasing impacts on land, ecosystems and biodiversity are projected under all greenhouse gas emission scenarios with cascading risks occurring across systems and sectors (IPCC 2019). It also states with high confidence that near-term actions to promote sustainable land management will help reduce land and food-related vulnerabilities, provide both short-term positive economic returns and longer-term benefits for climate change adaptation and mitigation, biodiversity and enhance ecosystem functions and services.

1.2 Ecological Damage as an Externality

Traditional neo-classical economic approaches neglect to account for market failures of ecological damage due to absence of markets for many environmental public goods.

The field of environmental economics was established to address these types of market failures or 'externalities' and aims to internalise market externalities through considering and capturing social and environmental costs relating to economic activities (Flood 2012; Tietenberg and Lewis 2007).

There are two requirements for decision-making when it comes to quantifying environmental damages.

The first is to determine one's fundamental philosophical position and, contingent on one's worldview, the second is the need to know the extent to which people are willing-to-pay to prevent damages or the willingness-to-accept compensation for damages suffered (Spash 1997).

The philosophical position assumed by environmental economists is that the net utility from the consequences of an action determines whether the action is right or wrong.

Cost-benefit analysis and its tools, such as the contingent valuation method, assume that individuals are able and willing to consider trade-offs in relation to public goods, i.e. that individuals follow a utilitarian philosophy (Ibid.).

The contingent valuation method involves directly questioning people, in a survey or interview, how much they would be willing to pay for specific environmental services.

It is called "contingent" valuation, because people are asked to state their willingness to pay, contingent on a specific hypothetical scenario and description of the environmental service.

This utilitarian standpoint is the approach from which the majority of socioeconomic impacts associated with ecological damage are approached in the literature (Flood 2012).

This tendency towards the single metric of monetary valuation and the reluctance of the mainstream to consider other numéraires finds its roots in the epistemology of the Enlightenment or Age of Reason (Flood 2012). Enlightenment thinking originates with 17th and eighteenth century European thinkers such as Voltaire, Rousseau, Kant and Hegel, with foundations built upon the theories of Descartes (Van Asselt and Rotmans 2002). In unpacking the field of economic evalution of ecosystem services, the question of substitutability is one that burns at the very core of the

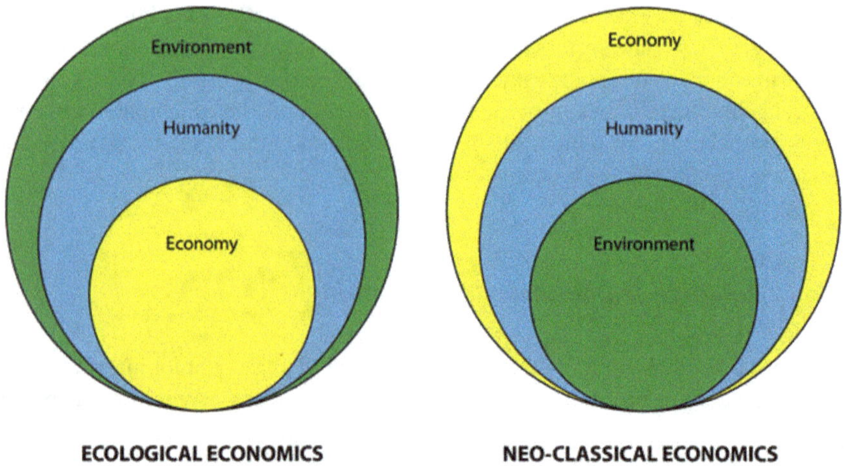

Fig. 1 Foundations of ecological economics and environmental economics

debate and is intertwined with questions of ethics (Flood 2012). The implicit utilitarian viewpoint of environmental economics, and in particular cost-benefit analysis, precludes the preservationist perspective which focuses on non-human intrinsic values associated with environmental systems (Spash 1997). Most environmental policy is couched in terms of calculating the usefulness to humans of preserving specific environmental goods and services provided by environmental systems. This contrasts with the foundations of ecological economics.

Ecological economics is holistic in its approach and much less anthropocentric than environmental economics. It also tends towards rights-based thinking. Figure 1 displays the fundamental differences between ecological economics and traditional neo-classical economics approaches, in terms of their view of the environment, economy and humanity (Flood 2012). Neo-classical economics tends to view the environment and humanity as embedded within the economy. Ecological economics takes a more holistic approach and considers the economy as a part of humanity living within its environment. Making decisions on a utilitarian basis is considered the most sensible approach by the majority of economists (Spash 1997).

However, we must be careful not obviate the fact that it is the analysis of public policy choices (housing, transport, etc) that also marks the way analytical approaches are designed for environmental protection. If we seek economic efficiency in public policy choices, to ascertain if the investment is worth it or money would be better employed somewhere else, the rationalist would suggest the logic of employing the same approach to inform policy choices for the environment in the absence of alternative decision-making techniques or approaches. Moreover, the challenge is that if we take decisions about the environment outside conventional decision making approaches, we may well end up marginalising environmental decisions and, as a society, be unable to assess where money is best employed. In other words, the risk of the ecological economics approach is that environmental decision making may suffer from isolation from other public policy areas, as it is very difficult to consider or mix monetary budget lines with other metrics. It can be argued that looking at environmental choices in isolation will not lead to real-world solutions, and rather they should be considered in the context of a wider public policy debate.

1.3 Traction: ESS as a Discipline in Its Own Right

Ecosystem Services (ESS) are benefits humans recognise as obtained from an ecosystem and that support, directly or indirectly, their survival and quality of life (Millennium Ecosystem Assessment 2005). This recognition of the idea of 'natural capital' was first coined in the book Small is Beautiful by E.F Schumacher in the 1970s (Schumacher 1973). The term 'environmental services' was introduced in a 1970 report: The Study of Critical Problems (MIT Press 1970). The services listed in the report included flood control, climate regulation, insect pollination, and fisheries. This concept ESS has continued to develop and expand to include both conservation and socio-economic objectives. Ecosystem goods and ecosystem services were combined by Robert Costanza and his colleagues in the Millennium Ecosystem Assessment (MEA 2005). The Assessment conceptualised the interactions between biodiversity, ecosystem services, human well-being, and drivers of change (Fig. 2).

Changes in drivers that indirectly affect biodiversity, such as population, technology, and lifestyle (upper right corner of Figure), can lead to changes in drivers directly affecting biodiversity, such as the catch of fish or the application of fertilizers (lower right corner). These result in changes to ecosystems and the services they provide (lower left corner), thereby affecting human well-being. These interactions can take place at more than one scale and can cross scales. For example, an international demand for timber may lead to a regional loss of forest cover, which increases flood magnitude along a local stretch of a river. Similarly, the interactions can take place across different time scales. Different strategies and interventions can be applied at many points in this framework to enhance human well-being and conserve ecosystems.

2 State of the Art ESS Concepts Complexity and Simplicity

The growing focus on the application of ESS to real world problems has led to a continuous refinement of Ecosystem Services concepts which has reflected the multidisciplinary nature of the research area, at the interface between ecology and society. While the classification system of the Millennium Ecosystem Assessment (MEA), which includes the supporting services, incorporates the full complexity of interactions between ecosystems and human beings, in practice inclusion of

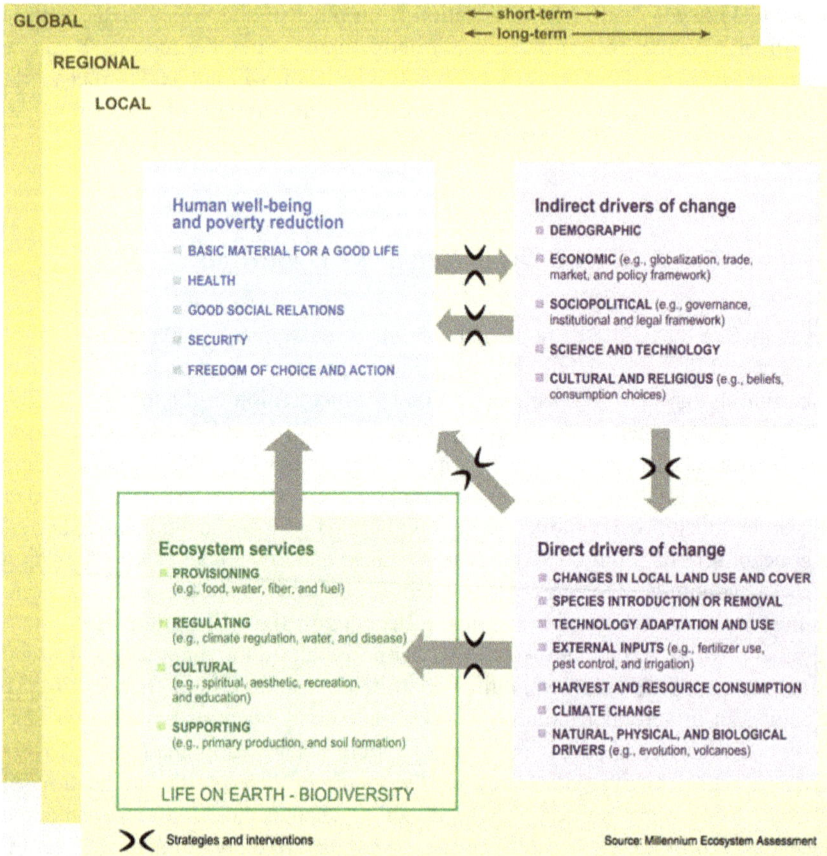

Fig. 2 Millennium ecosystem assessment conceptual framework of interactions between biodiversity, ecosystem services, human well-being, and drivers of change

supporting services in economic analysis can results in double counting. For example, a salmon may result in benefits to human beings, either through commercial harvest (as a provisioning service) or through recreational harvest (as a cultural service). The resulting economic benefits are, the market value (for commercial fisheries) or recreational enjoyment value of the caught salmon (for the recreationally caught fish), less the cost incurred in catching it. If the supporting services such as, habitat provision by freshwater, insect production as a food source for the fish, were also to be valued economically, the *human* benefits from nature would be double counted. By analogy in valuing a car, we consider the value of the final product but do not separately account for the value of the car manufacturing plant (built capital) and the raw materials (inputs). This practical challenge has been met by development of the Final Ecosystem Goods and Services (FEGS) concept. *Final ecosystem services* are "components of nature, directly enjoyed, consumed, or used to yield human well-being" (Boyd and Banzhaf 2007). It should also be noted

Table 1 CICES classification system, with examples of class type

Section	Division	Group	Class	Code	Class type
Provisioning (biotic)	Biomass	Reared aquatic animals for nutrition, materials or energy	Animals reared by in-situ aquaculture for nutritional purposes	1.1.4.3	Salmon
Regulating and maintenance (biotic)	Transformation of biochemical or physical inputs to ecosystems	Mediation of wastere or toxic substances of anthropogenic origin by living processes	Bio-remediation by micro-organisms, algae, plants and animals	2.1.1.1	In coastal marine systems
Cultural (abiotic)	Direct, in-situ and outdoor interactions with natural physical systems that requires presence in the environmental setting	Intellectual and representative interactions with abiotic components of the natural environment	Natural, abiotic characteristics of nature that enable spiritual symbolic and other interactions	6.1.2.1	Enjoyment of a land or sea scape

that in making just environmental decisions, the consideration of beneficiaries (who benefits) as well as benefits (what type of benefits) is all vital. Applications of the FEGS approaches are described in more detail in this volume by DeWitt et al. (2020) and Yee et al. (2020) and the FEGS concept is essential in the application for the economic valuation of Ecosystem Services and these concepts are commonly employed by ecological economists.

However, the FEGS are only a subset of all the ecosystem services which contribute to human well-being. Internationally, in order to standardise efforts in environmental accounting, considerable efforts have been expended in developing standardised classification systems. Within Europe, the Common International Classification of Ecosystem Services (CICES) is widely adopted, while at the national level with the U.S. the National Ecosystem Services Classification System (NESCS) is emerging (USEPA 2015). These systems attempt to standardise and codify the analysis of ecosystem services, with the aim of informing efforts such as the United Nations System of Environmental-Economic Accounting. The CICES classification system is analogous to the Linnaean biological classification system (phylum, class, genus species) and is comprised of four discrete categories (Table 1).

The CICES classification is the most widely used in Europe, and while developed for the purposes of environmental accounting, it has also been used in efforts to map the supply of ecosystem services (Maes et al. 2015). As with other classification systems, the supporting services of the millennium ecosystem assessment have been dropped and are now incorporated as regulating and maintenance services. CICES also recognises the benefits from non-living components of the environment, as abiotic services. The inclusion of some services which may be seen as intermediate rather than final services is one critique of this classification system, and may reflect a disciplinary bias toward ecology in the development of the system.

The focus of ES research, whether encompassing demand-side, human used of FEGS or covering the whole suite of supporting services and final services often depends on the disciplinary background of the researcher. Economists tend to focus only on FEGS and beneficiaries while ecological researchers with an interest in ecosystem services tend to focus on the full range of services, reflecting their interests in the functioning of ecosystems as a whole and the objective of the analysis, whether for simple accounting or to justify actions to maximize specific services.

2.1 Challenge of Valuation

While all these systems are designed for the purposes of accounting, the challenges of constructing of an agreed international standard are considerable, and will no doubt continue for many years, in many cases the process of (both monetary and non-monetary) valuation itself presents major challenges. For non-market goods and services there are two major categories of valuation methodologies. Revealed preference methodologies are an indirect methods of estimating the monetary value of an ecosystem service based on how much people spend to access or travel to a site (Silvertown 2015).

Revealed preferences are based on indirect calculations, deriving monetary values from the effects of behavioural change associated with the service (or the lack of it) in real markets (Spangenberg and Settele 2010). They are made up of non-use values (existence values) such as knowing about the existence of a deer population in a region; non-consumptive use values (watching them) and consumptive use values (hunting them) (Ibid.). The two main assessment methods for revealed preference are hedonic pricing and travel cost estimates.

Travel cost is mainly applicable to leisure and holiday activities where travelling is voluntary. In these cases, as the *homo economicus* is always maximising their utility, and will only be travelling to a particular location if the time spent there provides more utility than saving the cost and abstaining from the visit. (Ibid.). The travel cost is therefore stand-in for the value of what has been enjoyed at the destination. The method gives higher amenity value to a visitor who travels by car than someone who travels on foot or by bicycle even though the former involves the least effort and is the most environmentally damaging (Silvertown 2015).

Knoche and Lupi (2007) calculate the value of the white-tailed deer by assessing the demand for deer hunting via the hunters' travel costs. As a result, the value of 10,000 deer more per county is the result of additional travel expenditures of US $3.94 per hunting trip for firearm hunters, and of US$1.75 per trip for archery hunters.

Hedonic pricing valuations use relationships between land property prices and property characteristics to value changes in the characteristics (Swinton et al. 2007,

p. 248).They start with the assumption that services/disservices like improved or diminished environmental quality change the willingness to pay for a good associated with them, and this is reflected in the market price (with an implicit assumption of full knowledge and perfect markets), in particular in the housing market. The price change is then a measure of the value of the ecosystem services enjoyed, like a price increase due to the establishment of a nature reserve in the neighbourhood. However, empirical work comparing the changes in individual well-being caused by pollution to housing prices have shown that they do not necessarily reflect the local environmental quality changes (Rehdanz and Maddison 2008; Spangenberg and Settele 2010).

2.2 *Stated Preference- Methodological Advances and Subtleties*

Stated preference approaches simulate a market for ecosystem services through the generation of surveys on hypothetical (policy-driven) changes in the provision of ecosystem services. (TEEB 2012, p. 20). Stated preference methods can be used to estimate non-use and use values of ecosystems. The main types of stated preference techniques are:

1. Contingent valuation method (CV): Uses questionnaires to ask people how much they would be willing to pay to increase or enhance the provision of an ecosystem service, or alternatively, how much they would be willing to accept for its loss or degradation;
2. Choice modeling (CM): Attempts to model the decision process of an individual in a given context (Hanley et al. 1998; Philip and MacMillan 2005). Individuals are faced with two or more alternatives with shared attributes of the services to be valued, but with different levels of attribute (one of the attributes being the money people would have to pay for the service); and
3. Group valuation: Combines stated preference techniques with elements of deliberative processes from political science (Spash 2001; Wilson and Howarth 2002), and are being increasingly used as a way to capture value types that may escape individual based surveys, such as value pluralism, incommensurability, non-human values, or social justice (Spash 2008).

There is a vibrant economic literature on the refinement of stated preference techniques, and these techniques are constantly being defined and developed.

One major drawback of non-market valuation studies, is that (unlike market values) these figures require significant time, effort and expertise to establish, and while real non-market values will vary based on any given unique situation, it is practically unfeasible for reasons of expense and expertise to carry out such studies in every case. As a result, the use of benefits transfer (the practice of taking values from existing studies and applying them at another site) is commonplace. However this practice results in uncertainty which can undermine their usefulness in decision making contexts. The potential for big data to contribute to the volume of information on human recreational use patterns is beginning to improve the capacity to tailor economic valuation studies to specific locations (Adamowicz et al. 2011).

2.3 Splitters and Lumpers: Real World Complexity, Bundling and Its Implications

In reality ecosystems services generally do not exist in isolation but emerge as a bundle, jointly produced from a range of ecosystem processes and components (Fisher et al. 2009). For example, a healthy river ecosystem provides clean drinking water and salmon and waste remediation services as well as opportunities for active and passive recreational use and the interactions between the ecosystem processes which contribute to these bundled services operate together as a system. Developing quantitative understandings of the ecosystem processes themselves and understanding how these systems respond to human activities may highly complex ecological modelling approaches (see Fulford et al. 2020), yet the joint bundled of benefits to humans is something relatively easily understood intuitively and without deep ecological knowledge. Recognition and communication of the characteristics of joint ecosystem services supply may in some cases be more useful than, providing an exhaustive list of services provided by a given system or developing a mechanistic understanding of how the different ecosystem components interact to produce services.

Carefully developed frameworks and classifications such as the Millennium Ecosystem Assessment (MEA), Final Ecosystem Goods and Services (FEGS), and the Common International Classification of Ecosystem Services (CICES) untangle the complexities of capturing and valuing ESS. However, scholars and practitioners grapple with these frameworks and classifications when applying them on the ground (Harrison et al. 2018).

While Ecosystem Services are intuitively understood, even the youngest school child can understand that "the cow gives us milk" or the "sea gives us fish" or that "a flower is pretty". As illustrated in Table 1, the formal language used to describe ecosystem services is very precise, but not particularly accessible. While anybody can understand the concept of "a beautiful view" this, same concept is perhaps not optimally expressed as "*Intellectual and representative interactions with abiotic components of the natural environment*". These difficulties with the communication of Ecosystem services have resulted in a suite of different terms being used, for example "natures benefits", "natural capital", "natures' services".

2.4 *Scale and Polycentric Governance*

While philosophical battles have been won and lost over the ethics, as well as practical and theoretical considerations of valuing nature, these debates over valuation have tended to obscure the broader applicability of ecosystem service concepts to the field of environmental management. Flows of ecosystem services from location to location establish transactional relationships between different jurisdictions. Your country may benefit from the ecosystem services produced in my country. The most obvious current example is that of the Amazon rainforest. The Amazon provides climate regulation services (as well as a wealth of biodiversity) for the entire earth, but is largely under the jurisdiction of Brazil, therefore Brazilian management practices have the potential to increase or reduce the supply of ecosystem services to all of us. This situation sets up a power dynamic between Brazil and the rest of the global community. Systematic considerations of ecosystem service flows and the geographic characteristics of supply and demand can enable the development of institutional arrangements accounting for such natural flows.

3 The Power of the Word "Biodiversity" to Communicate with the Public

In light of these issues we propose that an important component of increasing the uptake and application of ESS or BES in everyday decision-making is through mainstreaming into policy as well as the public consciousness. Here, we will concentrate on educating and informing the public consciousness. A good place to start is by looking at the awareness of different populations to the concept of biodiversity, as a foundation for understanding of ES.

A 2015 Eurobarometer captures and presents the attitudes of Europeans towards biodiversity (European Commission 2015). More than half of the 27,718 respondents agree that the European Union (EU) should better inform its citizens about the importance of biodiversity (61%), that the EU should ensure that biodiversity concerns are taken into account when planning new infrastructure investments (55%), and that it should better implement existing nature and biodiversity conservation rules (55%) (Ibid.). Furthermore, of the 60% of respondents who have heard of the term 'biodiversity', only half of them have an understanding of what it means. Those living in western and southern areas of the EU are more likely to have heard of the term 'biodiversity' and know what it means. Two thirds (66%) of Europeans do not feel informed about the loss of biodiversity, with 22% saying that they do not feel informed at all (Ibid.).

The Union for Ethical Bio Trade carried out a six-year survey (from 2009 to 2015) to determine the levels of knowledge of the term biodiversity from 47,000 consumers in 16 countries across the globe (UEBT 2015). The key findings revealed

Table 2 Biodiversity awareness around the world (Adapted from UEBT 2015) numbers indicate percentages

Region/ Country	Have heard of biodiversity	Correct definition of biodiversity	With partial definition of biodiversity
Europe			
UK	68	26	17
Netherlands	59	27	16
France	91	34	25
Germany	38	18	9
Switzerland	83	37	18
Americas			
USA	58	22	18
Mexico	90	46	20
Colombia	93	44	18
Ecuador	82	14	30
Brazil	92	44	19
Peru	52	7	37
Asia			
China	94	64	22
India	40	1	25
South Korea	73	47	16
Japan	62	29	21
Vietnam	95	36	6

that one out of three respondents could provide a current definition of biodiversity. Table 2 provides an overview of the results.

4 Recommendations

It has been suggested that combining a number of different tools and methods can help strengthen assessment (Harrison et al. 2018; Barton et al. 2018). Recent reviews point to a persistent gap in the promise of ESS to provide easily usable information for decision-makers (Ruckelshaus et al. 2015). Based on significant field-based experience of application of ESS frameworks and classifications 6 emerging lessons have been identified (Ibid.):

1. Include Biodiversity and Ecosystem Services (BES) information as part of an iterative Science-Policy Process;
2. Keep it simple – no matter how much interdisciplinary scientists think they are over-simplifying biophysical or socio-economic processes, decision-makers typically ask for simpler, easy-to-use and understandable decision support tools that can be readily incorporated into science-policy processes (Ruckelshaus et al.

2015). Even simple tools are complicated enough for parameterizing and interpreting at early stages of applying BES information;

3. It's not always about the money – having the ability to follow biophysical ecosystem service estimates through to economic values has proven to be an important conceptual advance that has opened many decision makers to discussions they previously did not consider. However, conceptually, considering values of biodiversity for its own sake, in addition to ecosystem services, is completely consistent with an ecosystem services approach;
4. Relate BES change to livelihoods and other wellbeing metrics; and

Furthermore, combining tools and methods can yield significant benefits such as (Dunford et al. 2018):

Individual tools are unlikely to address all the needs of a given context, but a range of approaches can be used to assess different aspects of ES, such as different types of green infrastructure, different groups of services, different geographic scales or time-scales, and different types of value (e.g. biophysical, socio-cultural and monetary).

This chapter has made the case for urgent action to protect our ecological systems (biodiversity) from catastrophic decline. ESS can help shine a light on what we will lose if we fail protect valuable (and indeed invaluable) ecosystems and earth's flora and fauna in general. We have documented the development of ESS, from its foundations in environmental and ecological economics to its evolution into FEGS and CICES classifications and valuation frameworks. In doing so we have highlighted the challenges of valuation (including issues of scale, biodiversity awareness or literacy, and polycentric governance), and provided lessons and recommendations going forward for the successful application of ESS.

References

Adamowicz, W. L., Naidoo, R., Nelson, E., Polasky, S., & Zhang, J. (2011). Nature-based tourism and recreation. In P. Kareiva, G. Daily, T. Ricketts, H. Tallis, & S. Polasky (Eds.), *Natural capital: Theory and practice of mapping ecosystem services*. New York: Oxford University Press.

Barton, D. N., Kelemen, E., Dick, J., Martin-Lopez, B., Gómez-Baggethun, E., Jacobs, S., Hendriks, C. M. A., et al. (2018). (Dis) integrated valuation – Assessing the information gaps in ecosystem service appraisals for governance support. *Ecosystem Services, 29*, 529–541.

Boyd, J., & Banzhaf, S. (2007). What are ecosystem services? The need for standardized environmental accounting units. *Ecological Economics, 63*(2–3), 616–626. https://doi.org/10.1016/j.ecolecon.2007.01.002.

CBD Secretariat. (2009). Connecting biodiversity and climate change mitigation and adaptation: Report of the Second Ad Hoc technical expert group on biodiversity and climate change. Montreal, Technical Series No. 41, 126 pages. Retrieved April 3, 2019, from https://www.cbd.int/doc/publications/cbd-ts-41-en.pdf.

CBD Secretariat. (2018). Biodiversity and climate change working Group II. CBD/COP/14/L.23. Retrieved April 3, 2019, from https://www.cbd.int/conferences/2018/insession.

Costanza, R., de Groot, R., Sutton, P., van der Ploeg, S., Anderson, S. J., Kubiszewski, I., Farber, S., & Turner, R. K. (2014). Changes in the global value of ecosystem services. *Global Environmental Change, 26*, 152–158.

DeWitt, T. H., Berry, W. J., Canfield, T. J., Fulford, R. S., Harwell, M. C., Hoffman, J. C., Johnston, J. M., Newcomer-Johnson, T. A., Ringold, P. L., Russel, M. J., Sharpe, L. A., & Yee, S. J. H. (2020). The final ecosystem goods and services (FEGS) approach: A beneficiary-centric method to support. In T. O'Higgins, M. Lago, & T. H. DeWitt (Eds.), *Ecosystem-based management, ecosystem services and aquatic biodiversity: Theory, tools and applications* (pp. 127–148). Amsterdam: Springer.

Dunford, R., Harrison, P., Smith, A., Dick, J., Barton, D. N., Martín-López, B., Keleman, E., Jacobs, S., Saarikoski, H., Turkelboom, F., Verheyden, W., Hauck, J., Antunes, P., Aszalós, R., Badea, O., Baró, F., Berry, P., Carvalho, L., Conte, G., Czúcz, B., Garcia Blanco, G., Howard, D., Giuca, R., Gomez-Baggethun, E., Grizzetti, B., Izakovicova, Z., Kopperoinen, L., Langemayer, J., Luque, S., Lapola, D. M., Martinez-Pastur, G., Mukhopadhyay, R., Roy, S. B., Niemelä, J., Norton, L., Ochieng, J., Odee, D., Palomo, I., Pinho, P., Priess, J., Rusch, G., Saarela, S.-R., Santos, R., van der Wal, J. T., Vadineanu, A., Vári, Á., Woods, H., & Yli-Pelkonen, V. (2018). Integrating methods for ecosystem service assessment: Experiences from real world situations. *Ecosystem Services, 29*, 499–514.

European Commission. (2015). *Attitudes of Europeans towards biodiversity*. Special Eurobarometer 436. ISBN: 978-92-79-50788-5.

Fisher, B., Turner, R. K., & Morling, P. (2009). Defining and classifying ecosystem services for decision making. *Ecological Economics, 68*(3), 643–653. https://doi.org/10.1016/j.ecolecon.2008.09.014.

Flood, S. (2012). *Climate change and potential economic impacts in Ireland: The case for adaptation*. PhD Thesis, Maynooth University, Ireland.

Fulford, R. S., Heymans, S. J. J., & Wu, W. (2020). Mathematical modelling for ecosystem-based management (EBM) and ecosystem goods and services (EGS) assessment. In T. O'Higgins, M. Lago, & T. H. DeWitt (Eds.), *Ecosystem-based management, ecosystem services and aquatic biodiversity: Theory, tools and applications* (pp. 275–290). Amsterdam: Springer.

Hanley, N., MacMillan, D., Wright, R. E., Bullock, C., Simpson, I., Parsisson, D., & Crabtree, B. (1998). Contingent valuation versus choice experiments: Estimating the benefits of environmentally sensitive areas in Scotland. *Journal of Agricultural Economics, 49*(1), 1–15. https://doi.org/10.1111/j.1477-9552.1998.tb01248.x.

Harrison, P. A., Dunford, R., Barton, D. N., Keleman, E., Martín-López, B., Norton, L., Termansen, M., Saarikoski, H., Hendriks, K., Gómez-Baggethun, E., Czúcz, B., García-Llorente, M., Howard, D., Jacobs, S., Karlsen, M., Kopperoinen, L., Madsen, A., Rusch, G., van Eupen, M., Verweij, P., Smith, R., Toumasjukka, D., & Zuilian, G. (2018). Selecting methods for ecosystem service assessment: A decision tree approach. *Ecosystem Services, 29*, 481–498.

IPBES. (2019). Summary for policymakers of the global assessment report on biodiversity and ecosystem services of the intergovernmental science-policy platform on biodiversity and ecosystem services. In S. Díaz, J. Settele, E. S. Brondizio, H. T. Ngo, M. Guèze, J. Agard, A. Arneth, P. Balvanera, K. A. Brauman, S. H. M. Butchart, K. M. A. Chan, L. A. Garibaldi, K. Ichii, J. Liu, S. M. Subramanian, G. F. Midgley, P. Miloslavich, Z. Molnár, D. Obura, A. Pfaff, S. Polasky, A. Purvis, J. Razzaque, B. Reyers, R. Roy Chowdhury, Y. J. Shin, I. J. Visseren-Hamakers, K. J. Willis, & C. N. Zayas (Eds.). Bonn: IPBES Secretariat.

IPCC. (2014). Climate change 2014: Synthesis report. Contribution of Working Groups I, II and III to the Fifth Assessment Report of the Intergovernmental Panel on Climate Change [Core Writing Team, R. K. Pachauri & L. A. Meyer (Eds.)]. Intergovernmental Panel on Climate Change, Geneva, Switzerland, 151 pp.

IPCC. (2018). Global Warming of 1.5 °C: An IPCC special report on the impacts of global warming of 1.5 °C above pre-industrial levels and related global greenhouse gas emission pathways, in the context of strengthening the global response to the threat of climate change, sustainable development, and efforts to eradicate poverty.

IPCC. (2019). Special report on climate change, desertification, land degradation, sustainable land management, food security, and greenhouse gas fluxes in terrestrial ecosystems. Summary for policymakers approved draft. Retrieved August 29, 2019, from https://www.ipcc.ch/site/assets/uploads/2019/08/Edited-SPM_Approved_Microsite_FINAL.pdf.

Knoche, S., & Lupi, F. (2007). Valuing deer hunting ecosystem services from farm landscapes. *Ecological Economics, 64*(2), 313–320. https://doi.org/10.1016/j.ecolecon.2007.07.023.

Maes, J., Fabrega, N., Zulian, G., Barbosa, A., Vizcaino, P., Ivits, E., Polce, C., Vandecasteele, I., Marí Rivero, I., Guerra, C., Perpiña Castillo, C., Vallecillo, S., Baranzelli, C., Baranco, R., Batista e Silva, F., Jacobs-Crisioni, C., Trombetti, M., & Lavalle, C. (2015). Mapping and assessment of ecosystems and their services: Trends in ecosystems and ecosystem services in the European Union Between 2000 and 2010. Joint Research Centre-Institute for Environment and Sustainability. Luxembourg Publications Office of the European Union, p. 131.

Millennium Ecosystem Assessment (MEA). (2005). *Ecosystems and human well-being: Synthesis*. Washington, DC: Island Press.

MIT Press. (1970). *Man's impact on the global environment: Assessment and recommendations for action: Study of Critical Environmental Problems (SCEP)*. ISBN: 10: 0262690276.

Philip, L. J., & MacMillan, D. C. (2005). Exploring values, context and perceptions in contingent valuation studies: The CV market stall technique and willingness to pay for wildlife conservation. *Journal of Environmental Planning and Management, 48*(2), 257–274. https://doi.org/10.1080/0964056042000338172.

Rehdanz, K., & Maddison, D. (2008). Local environmental quality and life-satisfaction in Germany. *Ecological Economics, 64*(4), 787–797. https://doi.org/10.1016/j.ecolecon.2007.04.016.

Ruckelshaus, M., McKenzie, E., Tallis, H., Guerry, A., Daily, G., Kareiva, P., Polasky, S., Ricketts, T., Bhagabati, N., Wood, S. A., & Bernhardt, J. (2015). Notes from the field: Lessons learned from using ecosystem service approaches to inform real-world decisions. *Ecological Economics., 115*, 11–21.

Scheffers, B. R., et al. (2016). The broad footprint of climate change from genes to biomes to people. *Science, 354*, 6313. https://doi.org/10.1126/science.aaf7671.

Schumacher, E. F. (1973). *Small is beautiful: A study of economics as if people mattered* (288 p.). Blonde and Briggs. ISBN: 978-0-06-091630-5.

Silvertown, J. (2015). Have ecosystem services been oversold. *Trends in Ecology and Evolution., 30*(11). https://doi.org/10.1016/j.tree.2015.08.007.

Spangenberg, J. H., & Settele, J. (2010). Precisely incorrect? Monetising the value of ecosystem services. *Ecological Complexity, 7*(3), 327–337. https://doi.org/10.1016/j.ecocom.2010.04.007.

Spash, C. L. (1997). Ethics and environmental attitudes with implications for economic valuation. *Journal of Environmental Management, 50*, 403–416.

Spash, C. L. (2001). Broadening democracy in environmental policy processes. *Environment and Planning C: Government and Policy, 19*, 475–481.

Spash, C. L. (2008). How much is that ecosystem in the window? The one with the bio-diverse trail. *Environmental Values, 17*(2), 259–284. https://doi.org/10.3197/096327108X303882.

Stern, N. (2006). *Stern review on the economics of climate change*. London: HM Treasury.

Swinton, S. M., Lupi, F., Robertson, P. G., & Hamilton, S. K. (2007). Ecosystem services and agriculture: Cultivating agricultural ecosystems for diverse benefits. *Ecological Economics, 64* (2), 245–252. https://doi.org/10.1016/j.ecolecon.2007.09.020.

TEEB (The Economics of Ecosystems and Biodiversity). (2012). *The economics of ecosystems and biodiversity: Ecological and economic foundations* (p. 456). London: Routledge. https://doi.org/10.4324/9781849775489.

Tietenberg, T., & Lewis, L. (2007). *Environmental and natural resource economics* (5th ed.). London: Pearson Publishing.

UEBT (Union for Ethical Bio Trade). (2015). UEBT biodiversity barometer 2009–2015.

USEPA (United States Environmental Protection Agency). (2015). *National Ecosystem Services Classification System (NESCS): Framework design and policy application*. EPA-800-R-15-002. Washington, DC: United States Environmental Protection Agency.

Van Asselt, M. B. A., & Rotmans, J. (2002). Uncertainty in integrated assessment. Modelling: From positivism to pluralism. *Climatic Change, 54*, 75–105.

Wilson, M. A., & Howarth, R. B. (2002). Discourse-based valuation of ecosystem services: Establishing fair outcomes through group deliberation. *Ecological Economics, 41*(3), 431–443. https://doi.org/10.1016/S0921-8009(02)00092-7.

WWF (World Wildlife Fund). (2018). Living planet report 2018: Aiming higher. In M. Grooten & R. E. A. Almond (Eds.). Gland: WWF.

Yee, S., Cicchetti, G., DeWitt, T. H., Harwell, M. C., Jackson, S. K., Pryor, M., Rocha, K., Santavy, D. L., Sharpe, L., & Shumchenia, E. (2020). The ecosystem services gradient: A descriptive model for identifying thresholds of meaningful change. In T. O'Higgins, M. Lago, & T. H. DeWitt (Eds.), *Ecosystem-based management, ecosystem services and aquatic biodiversity: Theory, tools and applications* (pp. 291–308). Amsterdam: Springer.

10

Combining Methods to Establish Potential Management Measures for Invasive Species *Elodea nutallii* in Lough Erne Northern Ireland

Timohty G. O'Higgins, Fiona E. Culhane, Barry O'Dwyer, Leonie A. Robinson, and Maneul Lago

Abstract Lough Erne (Northern Ireland) is a heavily modified water body in a transnational catchment, straddling Northern Ireland (in the U.K.) and the Republic of Ireland. The lake has a long history of human modification from hydro-electrification, eutrophication and the introduction of non-native species. Most recently the proliferation of the non-native pond weed *Elodea nutalli* has adverse implications for recreational users of the lake. In order to establish management measures which might be acceptable to a range of lake, a number of methods, using a mixture of disciplines were combined. Fuzzy cognitive mapping exercises combined with formal goal identification surveys were conducted to establish consensus on the main environmental problems and conflicts. GIS was used to visualise potential management scenarios. Management scenarios were costed and presented to lake users to establish the preferred measures. The overall process promoted discussion and awareness of different uses and user perspectives to enable development of consensus.

Lessons Learned

- Loose coupling of models provided a useful means of analysing the system in a data poor situation
- Co-design of the models enabled the development of consensus
- Visual representation through maps and graphs enabled the communication of complexity
- Including the full suite of stakeholders is ideal but we were unable to compel unwilling stakeholders

T. G. O'Higgins (✉) · B. O'Dwyer
MaREI, Environmental Research Institute, University College Cork, Cork, Ireland
e-mail: tim.ohiggins@ucc.ie

F. E. Culhane · L. A. Robinson
School of Environmental Sciences, University of Liverpool, Liverpool, UK

M. Lago
Ecologic Institute, Berlin, Germany

- Mechanisms to promote improved farm management have the potential to enable solutions where both farmers and the environment benefit.

Needs to advance EBM

- In the context of the Erne as a transboundary system the importance of developing inclusive stakeholder fora to enable transboundary cooperation is one important requirement.
- More integrated quantitative modelling of the hydrological system and the nutrient fluxes could enhance the evidence base for action.
- Development of socially inclusive stakeholder processes.

1 Introduction

Ecosystem-Based Management (EBM) has been defined as an approach to management which "*integrates the connections between land air water and all living things, including human beings and their institutions*" (Mee et al. 2015), as such EBM in a particular location must incorporate considerations of both social and ecological aspects of that Social-Ecological System (SES). For any given system, what may be ecologically desirable to some, may be socially unacceptable to others, and different management options may result in costs or benefits to different sectors in society resulting in trade-offs or conflicts. In transboundary systems these costs and benefits may accrue in different jurisdictions, resulting in governance challenges and adding a further layer of complexity to effective management of a particular problem. In addition, achievement of socially desirable management end points may be constrained by the ecological properties of a given system. Very often there is a great degree of uncertainty about how effective a particular management measure might be, resulting in the need for adaptive management or a learning by doing approach (Holling 1978). Thus developing realistic and appropriate management strategies and targets relies on knowledge of both the ecological functioning of a system and the objectives of different users of a system, including the legislative objectives for environmental state and economic development.

The spread of Invasive Alien Species is considered a major threat to biodiversity the UN Convention on Biodiversity Strategic Plan for biodiversity (CBD 2014). In Europe, under the regulation of Invasive Alien Species (IAS) (EC 2014), a list of IAS of union concern has been drawn. The regulation forbids the transport or trade of the listed species. Member States are required to set out action plans on the pathways of invasion and to put in place surveillance systems and effective management measures for those species found to be widespread. Where an ecosystem is deemed to be "degraded, damaged or destroyed" the ecosystem must be restored unless costs of restoration are disproportionately high compared to the benefits.

Lough Erne, Co. Fermanagh Northern Ireland (NI), is comprised of Upper Lough Erne and Lower Lough Erne, both widened channels of Ireland's second largest river, the Erne. The lakes are in Northern Ireland but a substantial part of the catchment is situated within the Republic of Ireland (Fig. 1) and the region has a legacy of social and cultural division and conflict. The Erne is also connected to the

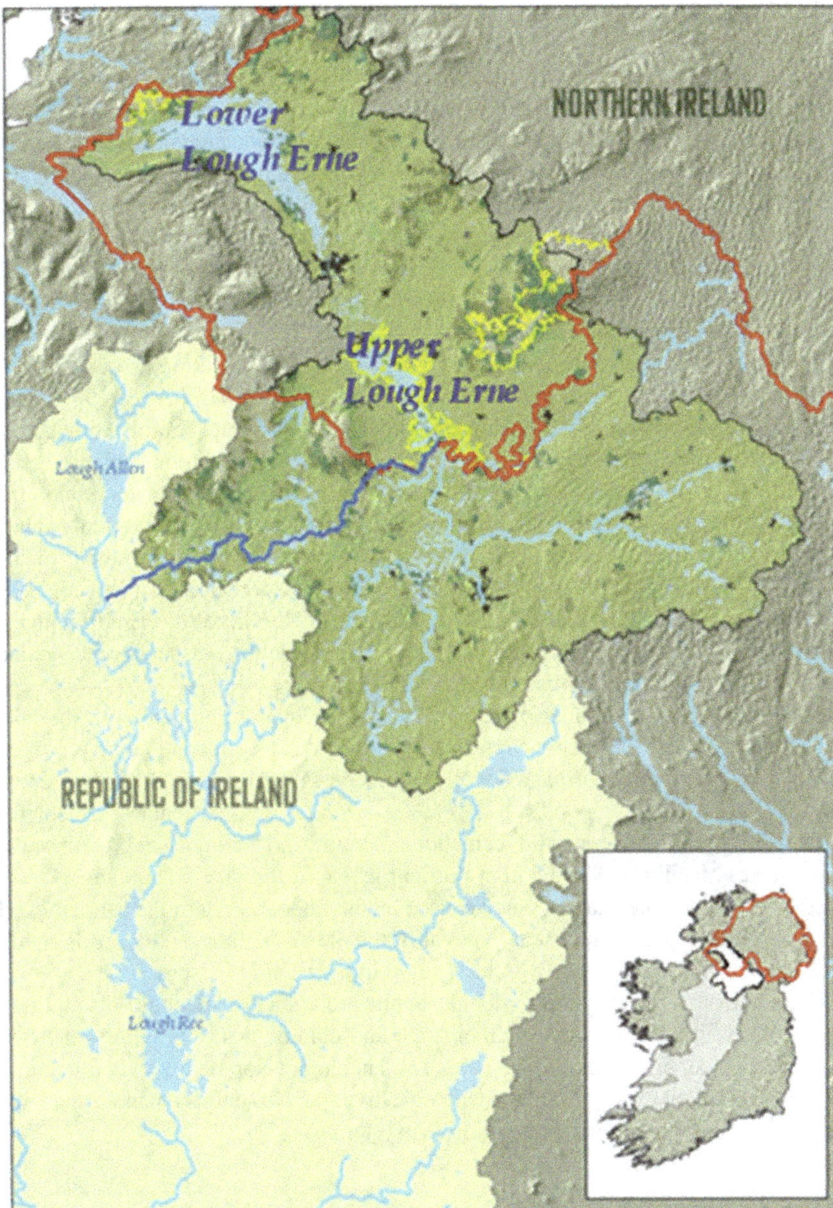

Map of Lough Erne showing the international border (red) as well as the Catchment (inset white) and the its connection with the Shannon catchment (inset light green)

Fig. 1 Map of the study area, showing Upper Lough Erne SAC and international border

Shannon river basin (the largest on the island) by the Shannon Erne Waterway. Due to this international connectivity, management of invasive species in Lough Erne is critical to the management of freshwater biodiversity across the entire island of Ireland.

The Lough Erne system has been settled since neolithic times and the ecology of the lake has been shaped by human society for millennia (Lafferty et al. 2006). References to the Erne fisheries date back to mythology and the oral tradition of early Christian times (Went 1945). In so far as a natural state can be determined for the Erne system, the fish fauna of the lake is naturally depauperate, comprised of post-glacial relic species (Salmonids, pollan and eel) supplemented by historically introduced species, including, bream, perch and pike (Rosell 2001).[1] Scientific records of non-native species date back to the late nineteenth century when Canadian pond weed (*Elodea canadenis*) was first recorded, (Moore and More 1866). Upper Lough Erne is particularly prized for its flora and fauna having several national and international environmental designations. Agriculture is also vitally important in the surrounding catchment and the lake has a history of eutrophication associated with agriculture in the catchment and associated fertilizer runoff (Battarbee 1986). It is currently considered moderately eutrophic under the EU Water Framework Directive (EC 2000). The lake supports a wide range of recreational activities, in particular boating and fishing are major contributors to the local tourist industry. Since the 1950s, the lake has been harnessed for hydro-electricity production. When constructed, the hydro power stations provided benefits to the government of the Republic of Ireland in their drive for rural electrification and to the people of Northern Ireland in the management of flooding. At the time, the Erne Drainage and Development Act (1950) which enabled the construction was an unprecedented example of cross-border cooperation.

There have been a number of recent invasive species introductions to the lake (Gallagher et al. 2015, Minchin et al. 2016) with the most significant being the zebra mussel (*Dreisenna polymorpha*) (Rosell et al. 1999; Maguire et al. 2006). The recent proliferation of Nutall's pond weed (*Eldoea nuttalli*) is a particular problem for tourism because it interferes with recreational boating and fishing. The growth of the pond weed is facilitated by the high nutrient levels of the lake waters and exacerbated by the high water transparency caused by the filter feeding of the zebra mussel (Kelly et al. 2015). This physical removal of the already established weed is costly (Kelly et al. 2013) and has proved ineffective in controlling its spread.

The aim of this paper is to describe the application of a flexible multi-disciplinary methodology in Lough Erne to demonstrate and communicate the utility of EBM practices to meet the needs of lake users and enable development and communication of Ecosystem-Based Management objectives and trade-offs which meet the needs of lake users and mangers within the system.

---

[1]Pike are known in the Irish language as "Gall Iasc" translating to "French" or "Foreign Fish" suggesting a Norman origin in Ireland's waters (ca. 1000 bp).

2 Materials and Methods

The overall approach to EBM was based on the "butterfly" assessment framework which is detailed in Elliott and O'Higgins (2020). An initial scoping meeting was held with stakeholders in the Lough Erne Invasive Species Working Group (LEISWG) (an informal collective of interested parties including several government agencies) and a second stakeholder workshop was held with participation from a variety of organisations from both Northern Ireland and the Republic of Ireland, including state agencies, national and local government as well as non-governmental organisations. The formal analysis of stakeholder goals at this meeting is described in Robinson et al. (2019) and the goals identified through that analysis formed the basis of a Fuzzy Cognitive Mapping exercise.

Fuzzy Cognitive Maps (FCM) are semi-quantitative models of system operation based on an individual's/individual stakeholder group perception of the structure and function of a given problem or system. An FCM is a diagraph or directed graph made up of variables (points, nodes or concepts), and relationships between these concepts (links or edges), or, put simply boxes and weighted arrows. Positive or negative values are assigned to these relationships and expressed as a fraction of one based on the perceived strength of the relationship. FCM has been widely applied to a range of situations; it can be used to build models of system behaviour based on expert opinion and can be used to build consensus amongst stakeholders, as well as to develop predictions for system function based on scenarios. Özesmi and Özesmi (2004) describe the mathematical aspects as well as a range of different approaches to developing FCMs.

Five separate Lough Erne stakeholder groups participated at a workshop to produce FCMs; these included, environmental NGOs and conservation groups, water managers, hydro-electricity producers and wildfowling groups as well as local government organisations. While there was high level participation by the Northern Ireland Department of Agriculture, despite invitation, farmers groups themselves chose not to attend. The DPSIR (see Elliott and O'Higgins 2020) was used as an organisational frame to elicit concepts and connections from participants. Each FCM was generated by starting with a particular Driver within the SES which interfered with the objectives of specific groups (see Robinson et al. 2019), the specific components were agreed by the groups and acted as a starting point for the FCMs. Each FCM was written on a whiteboard and relationships between all concepts identified were considered and assigned a positive or negative weight. Following the workshop the FCMs were photographed and then rendered electronically using Mental Modeller software (http://www.mentalmodeler.org/) before export. Matrices, output from Mental Modeller, from each group were combined (in Microsoft Excel) to develop a joint matrix representing the overall FCM of the whole group (called the JOINT FCM). The open source software GEPHI (https://gephi.org/) was used to visualise the data. Analysis of the FCM was carried out in R using the FCM modeller library (https://CRAN.R-project.org/package=FCMapper). In order to develop a consensus map, it was necessary to harmonise concepts within the maps. For example, concepts such as "*fish stock salmonids*" and "*game fish*" were amalgamated as were "*fish stock cyprinids*" and "*coarse fish*". To generate a final consensus map, the weight of each connection was determined by summing the weights of all connections from each contributing map. Only consensus connections from one or more groups contributed to the final map, reducing the complexity of the model from the 55 nodes of the original combined model to a final consensus model containing just eleven nodes. The baseline conditions general direction of change based on stakeholder understanding were determined by allowing the model to run to steady state. To assess the sensitivity of the systems to changes in different components 11 different model runs were performed, for each, one of the eleven model components was fixed to its initial state and the effects on all other components were examined (Table 1).

Based on the consultation process with the LEISWG, specific measures to manage the pond weed by manipulation of lake levels were identified. These measures were designed to manage the impact of pond weed proliferation on recreational activities within the lake, by controlling the amount of light available to the weed as well as increasing the draft between recreational vessels and the weed (Fig. 2). The spatial consequences of these measures were assessed using GIS. The EURODEM Digital Elevation Model (horizontal resolution 25 m) GIS was used to identify cells within a 5 km distance of the Lough with elevations marginally greater than the lake level. The Erne Drainage and Development Act (1950) stipulates strict limits for the lake levels during the summer season, between 150 ft. (45.7 m) and 154 ft. (46.9 m) above sea level), a range of approximately 1.2 m. The location and extent of potentially flooded lands was simulated at 5 increments from 0.2 m to a level of 1.2 m above the lake levels of the Digital Elevation Model, corresponding to the legally determined limits of lake water levels which are legally determined.

Table 1 Model components showing positive or negative changes >1% in the sensitivity analysis, affector components and the size of effect

Component	Affector	Effect
	Components	%
Water quality	Agriculture	−18
	Forestry	−15
	IAS	−14
	Habitats	11
	Tourism	−1
Habitat	Water quality	9
	Conservation	9
Biodiversity	Water quality	6
	Habitat	4
Tourism	Water quality	2
	Conservation	1
	IAS	−1
Agriculture	Conservation	2
	Flood management	2
Flood management	Habitat	3

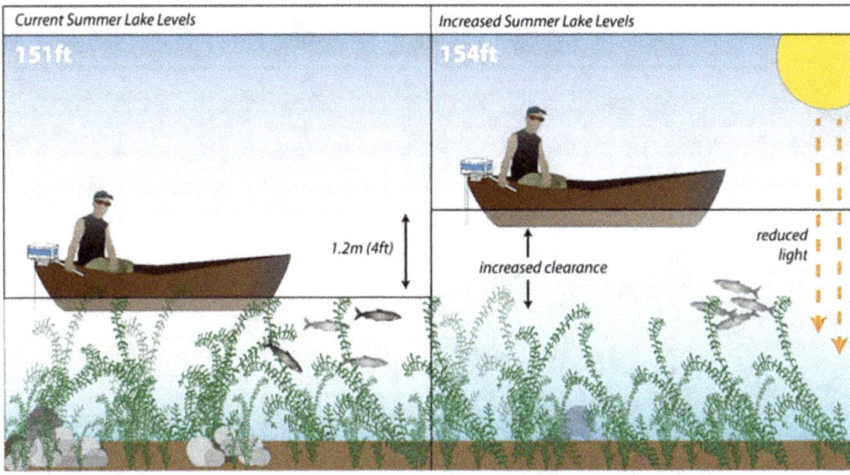

Fig. 2 Illustration of how raised lake levels could result in potential improvements for recreational boating

Economic valuation was used to appraise the costs of potential management measures. The costs to agriculture of raising lake levels were estimated in terms of annual standard output from the NI annual farm census and in terms of land value based on compulsory purchase price.

Based on the results of the fuzzy cognitive mapping exercise, valuation of a variety of agricultural measures to reduce nutrients was also performed, benefits transfer was based on Cuttle et al. (2007), these authors reviewed a range of farm Best Management Practices (BMPs) to reduce diffuse water pollution from agriculture, describing in detail the costs and technical effectiveness of each measure. Using a Cost Effectiveness Analysis (CEA) method, BMPs that could be implemented at least cost for the farmer while maximising potential Phosphorus (P) reductions were identified. The cost curve method was subsequently applied (Lago 2009) to estimate levels of abatement that could be potentially achieved as BMPs are sequentially added at farm level while considering their financial costs. The costs were calculated for two targets, 30% reduction in nutrient concentrations and 70% reduction in nutrients at the farm level under the assumption that the reductions would translate proportionally into improvements in water quality.

3 Results

Figure 3 illustrates the results of the joint FCM showing all 55 Drivers, Pressures and Ecosystem components considered by stakeholders to contribute to the functioning of the system. The concepts are scaled by the number of connections (known as density in the language of FCM). It is immediately apparent that agriculture, tourism

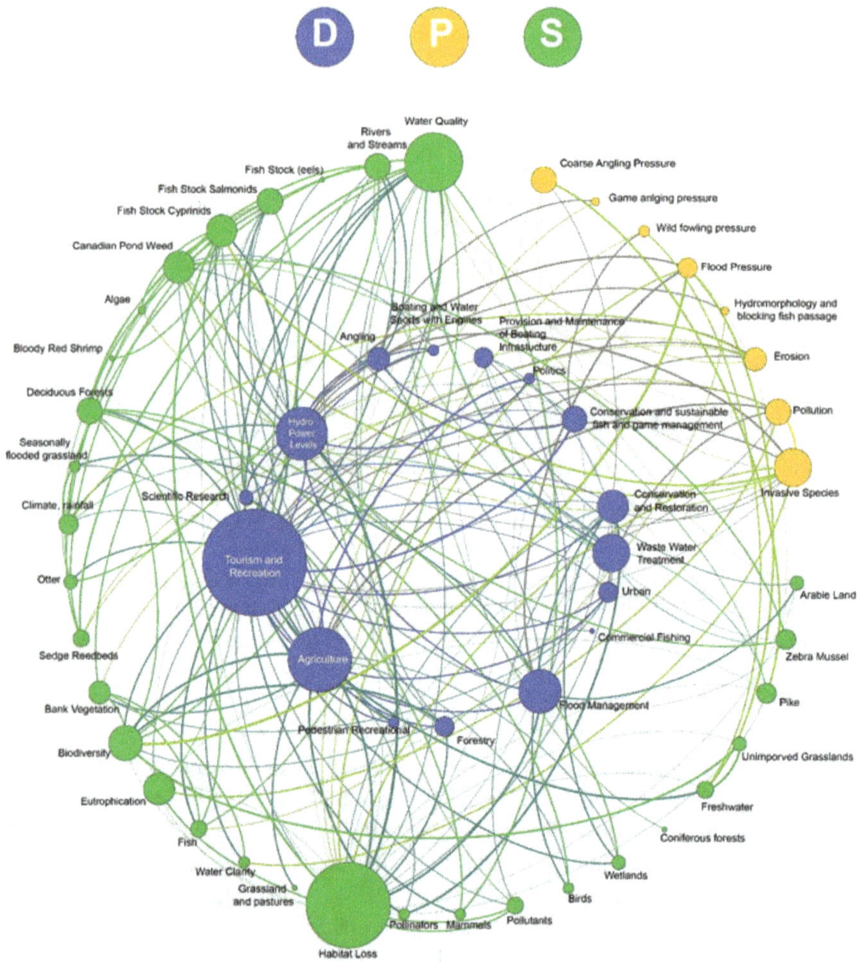

Fig. 3 Combined Fuzzy Cognitive map showing all concepts (*nodes*, *circles*) identified and relationships (*lines*, *edges*) between them based on the stakeholder workshop. Concepts are scaled according to the number of relationships with other concepts. Colours represent different components of the DPSIR framework. D = Drivers, P = Pressures, S = State (See Elliott and O'Higgins 2020)

and recreation and water quality are highly connected. Flood management, hydropower and invasive species are also seen to be highly connected to other components. Table 1 summaries the results of the sensitivity analysis conducted on the consensus FCM. Overall the model component which was most sensitive to alterations in the system was water quality showing relatively large responses to agriculture, forestry, IAS and tourism. The negative relationship between agriculture and water quality represents the strongest interaction in the whole model. Habitats had the strongest positive affect on water quality. In turn water quality had positive effects on habitat and biodiversity. The model illustrates the central importance of water quality to stakeholders, and the range of effects that water quality plays including positive effects on Habitat, Biodiversity and tourism, as well as its sensitivity to a range of drivers, but in particular agriculture. On this basis, the strong negative relationship between agriculture and water quality was used as the basis for identification of potential mitigation measures and the economic costings.

The adjustment of lake levels, which emerged as a potential management option for control of *Elodea*, would result in costs to farmers due to the inundation of productive agricultural lands and does not address the water quality of the Lough specifically. Inundation of agricultural land may also produce co-benefits in terms of biodiversity by increasing the area of semi natural riparian habitats. Maintaining the Lough at higher levels during summer may also result in benefits to the hydro-production sector enabling increased generation capacity. Figure 4 shows the area of land inundated by raising water levels by 1.2 m as well as the marginal changes in area of land inundated. The overall cost of compulsory purchase of the potentially inundated areas was just over £2 m. In the absence of compulsory purchase the total costs to farmers in terms of lost annual productivity due to inundation was under £0.5 m.

A range of nutrient abatement measures were also considered. There are many potential mechanisms to decouple agricultural activity from water quality impacts. The economic analysis summarised in Table 2 illustrates how a number of cost saving BMPs (negative costs) can save money to individual farms while also contributing to reduced nutrient loading. The target of 30% reduction in P loading can be met by implementing the first 6 measures sequentially with an overall cost of £15 m for the whole catchment.

4 Discussion

A range of techniques were employed to understand the Lough Erne SES, these included fuzzy cognitive mapping to combine individual stakeholder groups perception of the systems into an agreed and dynamic model of system behaviour, and more mechanistic GIS-based modelling approach to understand the effects of specific management measures on other activities within the catchment, as well as valuation of potential measures. In combination, these methods revealed a system which is highly complex and where incomplete knowledge is the rule. Nevertheless, the combination of simple techniques in co-design with stakeholders enabled the development of a consensus view of the system identifying eutrophication as a priority problem, with a range of potential measures for management of pond weed in the system also being identified. Ultimately the potential measures were reduced to a single metric of cost effectiveness to enable stakeholders in Lough Erne to consider the relative merits of the measures identified.

The problem of management of *Elodea* in Lough Erne is not a simple one. The problems of eutrophication are well understood to be an underlying cause for the

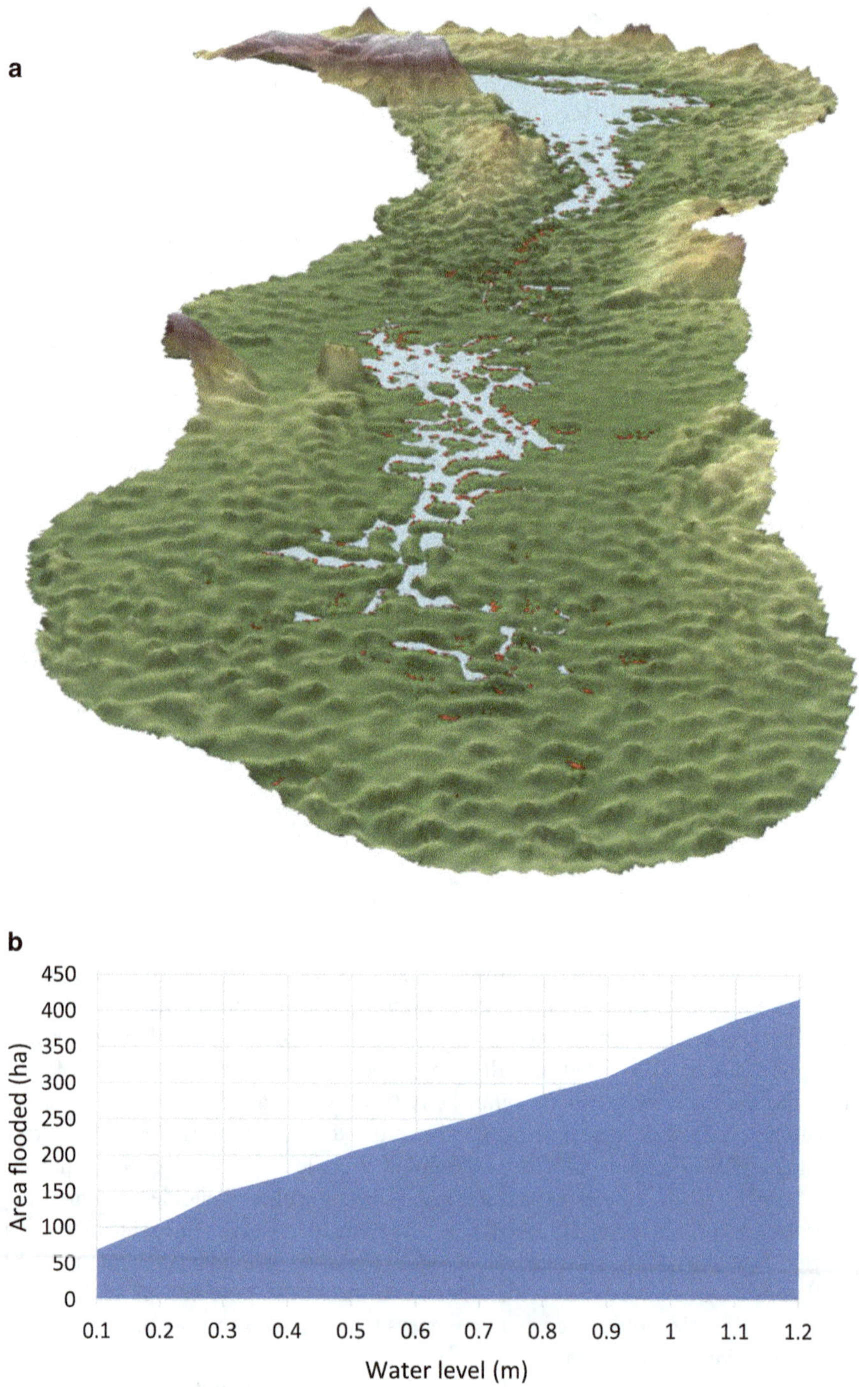

Fig. 4 (**a**) 3D visualisation of modelled areas of inundation based on raising the lake level by 1.2 m (**b**) cumulative area of inundation with increasing water level

Table 2 Cost Efficiency (CE) and percentage reduction in Phosphorus from a range of agricultural Best Practice Measures

Nutrient reduction measure		CE ratio[a]	P Loss[b]
1	Integrate fertiliser and manure nutrient supply	−472.44	0.10
2	Reduce fertiliser application rates; 20% Reduction P	−5.25	3.10
3	Do not apply P fertilisers to high P index soils	−2.62	6.00
4	Do not spread farmyard manure to fields at high-risk times	0.98	21.04
5	Do not apply manure to high-risk areas	2.25	26.57
6	Transport manure to neighbouring farms 5 km	2.69	56.68
7	Establish and maintain artificial (constructed) wetlands	2.86	74.87
8	Re-site gateways away from high-risk areas	3.17	75.63
9	Use a fertiliser recommendation system	5.25	76.36
10	Site solid manure heaps away from watercourses and field drains	5.25	77.07
11	Transport manure to neighbouring farms 20 km	5.57	86.47
12	Move feed and water troughs at regular intervals	5.85	88.36
13	Do not apply fertiliser to high risk areas	7.87	88.83
14	Site solid manure heaps on concrete and collect the effluent	22.87	89.16
15	Avoid spreading fertiliser to fields at high-risk times	27.36	89.38
16	Reduce field stocking rates when soils are wet	28.43	90.34
17	Fence off rivers and streams from livestock	38.40	90.53
18	Establish riparian buffer strips	38.40	90.72
19	Reduce overall stocking rates on livestock farms	54.05	94.15
20	Loosen compacted soil layers in grassland fields	85.04	94.21
21	Allow field drainage systems to deteriorate	177.85	94.27
22	Reduce the length of the grazing day or grazing season	255.90	94.33

[a]£/% Reduction in P loss/ha, NPV/ha over an 8-year period. Discount rate 3.5%
[b]Farm level per ha

proliferation of *Elodea* in the Erne system. Eutrophication is directly related to excess nutrients (P particularly) loading to the lake associated with fertilizer use by agriculture, one of the main socio-economic sectors in the region; fortunately, there are many well-known management measures which can be taken to reduce excess nutrient loading. The fact that some improvement in water quality can be achieved while also saving money for farmers provides a powerful justification for taking measures such as identified in items 1–6 of Table 2, thereby improving the efficiency of farms while also yielding environmental benefits for other stakeholders using or enjoying Lough Erne. However effecting change in agricultural practices is beset by governance challenges relating to the implementation of the European Common Agricultural Policy in two separate jurisdictions, where nutrient emissions occur in both the Republic and Northern Ireland but the environmental and social impacts of eutrophication effects are experienced disproportioantely in Northern Ireland. While there are clearly efficiencies to be achieved, the structuring of the Common Agricultural Policy and its single farm payments do not necessarily effectively promote this efficiency. While improved and cost-saving farming practices may reduce nutrient emissions to the Lough, the legacy effects of historic pollution combined with the recycling of nutrients by zebra mussel are likely to cause time-lags between measures and their effects.

By contrast, the proposal to manage lake levels, represents a relatively "quick-fix". However this proposal also represents an unknown quantity in terms of effectiveness and projected economic costs due to the consequent loss of farm land and productivity. The response of *Elodea* may not be as simple as reduced growth due to light limitation, and there is also potential for the existing weed to float to the surface resulting in continued nuisance for recreational activity. In reality the effectiveness of water-level management on *Elodea* are still uncertain, while the consequences to farm productivity are quite certain. Any future alteration of lake levels to control pond weed will therefore be an exercise in adaptive management or learning-by-doing.

While FCM highlighted clear and shared priorities for stakeholders, concerns for co-benefits, in terms of multiple ecosystem services that the lake ecosystem provides, did not emerge explicitly using the methods here. For example, many regulation and maintenance services are inherently valued by people but are often not prioritised compared to services linked to commercial concerns. However, participants did clearly value habitats and biodiversity. Increasing water levels, leading to an increase in the area of temporally flooded riparian habitats, or reducing nutrient inputs from agriculture, could enhance biodiversity and the supply of associated ecosystem services. The addition of holistic methods, such as the linkage framework approach proposed in (Robinson and Culhane 2020), could help to highlight other important services or elements of the SES that might influence the way management measures are evaluated.

5 Conclusions (Learning by Doing)

The current state of the Lough Erne SES results from a very long history of human use, conflict and alteration of lake and watershed ecosystems dating back for millennia. The modern Lough is a highly valued ecosystem which provides multiple benefits to humans yet also suffers from a range of chronic and acute environmental problems. European environmental legislative requirements for the Lough are not fully integrated into the management practices of the Lough, and the recent regulation on Invasive Alien Species adds an additional burden of management. Of the aquatic species listed in the regulation and found in the Lough Erne catchment, only one, *Elodea nutalli* (Nutall's pond weed) has had significant economic impacts to date.

A mixture of common and popular software as well as a range of free and open access tools was used to develop a bespoke methodology suited to a very specific local problem based on the perspectives, perceptions and using the language of interested stakeholders. Graphical output of the FCM was useful in communicating with stakeholders and in developing consensus on the main causes of the weed problem, while the resulting simplified model provided justification to focus the analysis on nutrients and eutrophication. Loose coupling of models allowed flexibility. Ultimately more complex ecological models including detailed biogeochemical components would be required to accurately predict the outcome of management measures. However, the utility of our simple approach is that it can promote understanding of trade-offs and represents a forum for interested stakeholder to contribute their knowledge to the management process. This is in itself a vital component of environmental management—since it is people who make the decision on what measures are to be taken. What our approach lacks in analytical complexity or data-driven robustness of more complex modelling techniques (many examples of which can be found in this book) it makes up for in terms of low-cost and ease of application with stakeholders when addressing a specific problem in a specific location. The co-design of the FCM with stakeholder enabled effective communication of complexity of the Lough Erne SES and trade-offs among environmental management options.

Integrated, ecosystem-based management approaches to the management of Lough Erne enable consideration of multiple primary activities and their pressures and provide a basis to meet multiple environmental as well as social and economic objectives. The transboundary nature of the Lough Erne catchment is a barrier to truly integrated management of the catchment, and the political boundaries between the two jurisdictions appear to be becoming more pronounced as the UK is set to leave the European Union. The Erne Drainage and Development Act (1950) (EDDA) was an early example of cross-border cooperation and succeeded because there were mutual benefits to be gained in the two jurisdictions. Changes to the management regime of the lake levels (within the legal limits of the EDDA) offer one opportunity for the management of the system which could continue to provide benefits to users of the Lough Erne SES on both sides of the border and could act as a focus for continued cross-border cooperation. While the inclusion of farmers within any of these management decisions is paramount, and these were notably missing from our stakeholder groups, our stakeholder approach did provide a valuable opportunity for cross border cooperation and collaboration which is one vital element in future management of the Lough Erne system.

While current EU environmental regulation provides a common cross-border framework for environmental management in the Lough Erne catchment across an international boundary, the future basis for such cooperation is unclear. The UK is currently presently in the process of leaving the European Union and the current political and economic basis for environmental regulation, as well as for enabling and subsidising agricultural production, is unlikely to remain as it is, while the potential future alternatives are largely unknown. Major changes in the social system comprising primary activities as well as the norms and values enshrined in environmental laws and regulations may be on the way. The effects of these changes on a social ecological system already characterised by overwhelming complexity cannot be foretold.

References

Battarbee, R. W. (1986). The Eutrophication of Lough Erne inferred from changes in the diatom assemblages of 210Pb and 137Cs—dated sediment cores. *Proceeding of the Royal Irish Academy Section B: Biological, Geological and Chemical Science, 86b*, 141–168.

CBD. (2014). Aichi biodiversity targets. Retrieved from https://www.cbd.int/sp/targets/.

Cuttle, S. P., Macleod, C. J. A., Chadwick, D. R., Scholefield, D., Haygarth, P. M., Newell-Price, P., Harris, D., Shepherd, M. A., Chambers, B. J., & Humphrey, R. (2007). *An inventory of methods to control diffuse water pollution from agriculture*. DEFRA-Project ES0303.

EC. (2000). Directive 2000/60/EC of the European Parliament and of the Council of 23 October 2000 establishing a framework for community actions in the field of water policy. *Official Journal of the European Communities,* L327, 1.22.12.2000.

EC. (2014). Regulation (EU) no 1143/2014 of the European Parliament and of the council of 22 October 2014 on the prevention and management of the introduction and spread of invasive alien species. *Official Journal of the European Union*, 2014, L317/35.

Elliott, M., & O'Higgins, T. G. (2020). From the DPSIR, the D(A)PSI(W)R(M) emerges... a butterfly-'protecting the natural stuff and delivering the human stuff'. In T. O'Higgins, M. Lago, & T. H. DeWitt (Eds.), *Ecosystem-based management, ecosystem services and aquatic biodiversity: Theory, tools and applications* (pp. 61–86). Amsterdam: Springer.

Gallagher, K., Rosell, R., Vaughan, L., McElarney, R., Campbell, W., O'Kane, E., & Harrod, C. (2015). Hemimysis anomala G.O. Sars, 1907 expands its invasive range to Northern Ireland. *BioInvasions Records, 4*, 43–46.

Holling, C. S. (1978). *Adaptive environmental assessment and management.* Chichester, UK: Wiley.

Kelly, J., Tosh, D., Dale, K., & Jakcson, A. (2013). The economic cost of invasive and non-native species in Ireland and Northern Ireland. Report prepared for the Northern Ireland Environment Agency and the National Parks and Wildlife Service as part of Invasive Species Ireland, 86 pp.

Kelly, R., Harrod, C., Maggs, C. A., & Reid, N. (2015). Effects of Elodea nuttallii on temperate freshwater plants, microalgae and invertebrates: Small differences between invaded and uninvaded areas. *Biological Invasions, 17*, 2123–2138.

Lafferty, B., Quinn, R., & Breen, C. (2006). A side-scan sonar and high resolution Chirp sub-bottom profile study of the natural and anthropogenic sedimentary record of Lower Lough Erne, Northwestern Ireland. *Journal of Archaeological Science, 33*, 756–766.

Lago, M. (2009). An investigation of regulatory efficiency with reference to the EU water framework directive: An application to Scottish agriculture. PhD thesis, University of Edinburgh.

Maguire, C., Rossell, R., & Roberts, D. (2006). Management of the impacts of zebra mussels in Northern Ireland and determination of the effects on fish populations in Lough Erne through alteration of food web. Environment and Heritage Service Research and Development Series 06/25, pp. 51.

Mee, L., Cooper, P., Kannen, A., Gilbert, A. J., & O'Higgins, T. (2015). Sustaining Europe's seas as coupled social-ecological systems. *Ecology and Society, 20*(1), 1. https://doi.org/10.5751/ES-07143-200101.

Minchin, D., Caffrey, J. M., Haberlin, D., Germain, D., Walsh, C., Boelens, R., & Doyle, T. K. (2016). First observations of the freshwater jellyfish Craspedacusta sowerbii Lankester, 1880 in Ireland coincides with unusually high water temperatures. *BioInvasions Records, 5*, 67–74.

Moore, D., & More, A. G. (1866). *Cybele Hibernica* (p. 538). Dublin University Press.

Özesmi, U., & Özesmi, S. L. F. (2004). Ecological models based on people's knowledge: A multi-step fuzzy cognitive mapping approach. *Ecological Modelling, 176*, 43–64.

Robinson, L., & Culhane, F. (2020). Linkage frameworks: An exploration tool for complex systems. In T. O'Higgins, M. Lago, & T. H. DeWitt (Eds.), *Ecosystem-based management, ecosystem services and aquatic biodiversity: Theory, tools and applications* (pp. 213–234). Amsterdam: Springer.

Robinson, L. A., Blincow, H. L., Culhane, F. E. L., & O'Higgins, T. (2019). Identifying barriers, conflict and opportunity in managing aquatic ecosystems. *Science of the Total Environment, 651*, 1992–2002.

Rosell, R. (2001). Monitoring fish populations in lower Lough Erne, Northern Ireland: Applicability of current methods and implications for future monitoring under the EC Water Framework Directive. *Freshwater Forum, 16*, 65–81.

Rosell, R. S., Maguire, C. M., & McCarthy, T. K. (1999). First reported settlement of zebra mussels Dreissena polymorpha in the Erne system, Co. Fermanagh, Northern Ireland. *Proceedings of the Royal Irish Academy, 98*(B), 191–193.

Went, E. J. (1945). Fishing weirs of the River Erne. *The Journal of the Royal Society of Antiquaries of Ireland., 75*(4), 213–223.

11

The Ecosystem Services Gradient: A Descriptive Model for Identifying Levels of Meaningful Change

Susan Yee, Giancarlo Cicchetti, Theodore H. DeWitt, Matthew C. Harwell, Susan K. Jackson, Margherita Pryor, Kenneth Rocha, Deborah L. Santavy, Leah Sharpe, and Emily Shumchenia

Abstract Characterization of ecosystem services can be a valuable element of Ecosystem-Based Management (EBM) in identifying meaningful measures of ecosystem change, understanding the natural resource gains or losses associated with changing ecosystem conditions, and communicating those benefits and tradeoffs to stakeholders in an intuitive way. Here, we introduce a descriptive model of the Ecosystem Services Gradient (ESG) that can be paired with the Biological Condition Gradient (BCG). The BCG is a conceptual framework that allows scientists and managers to characterize the status of an aquatic ecosystem along an anthropogenic disturbance gradient by describing and quantifying changes in biological or ecological condition with increasing levels of stressors. The ESG descriptive model builds upon the BCG approach by linking changes in ecosystem condition to effects on human health and well-being via changes in ecosystem goods and services. This involves identifying priority ecosystem services, defining them with metrics and indicators, and applying ecological production functions to translate levels of

---

S. Yee (✉) · M. C. Harwell · D. L. Santavy · L. Sharpe
Gulf Ecosystem Measurement and Modeling Division, US Environmental Protection Agency, Gulf Breeze, FL, USA
e-mail: yee.susan@epa.gov

G. Cicchetti · K. Rocha
Atlantic Coastal Environmental Sciences Division, US Environmental Protection Agency, Narragansett, RI, USA

T. H. DeWitt
Pacific Ecological Systems Division, US Environmental Protection Agency, Newport, OR, USA

S. K. Jackson
Health and Ecological Criteria Division, Office of Water, US Environmental Protection Agency, Washington, DC, USA

M. Pryor
Water Division, Region 1, US Environmental Protection Agency, Boston, MA, USA

E. Shumchenia
E&C Enviroscape, LLC, Ashaway, RI, USA

ecological condition to ecosystem services production. The ESG, through its structured approach to defining and enumerating potential changes in ecosystem services, allows decision makers to clearly assess and monitor the potential benefits, or related co-occurring benefits, of EBM, and significantly enhance how scientists and decision makers communicate these benefits to stakeholders.

Lessons Learned

- An Ecosystem Services Gradient (ESG) is introduced to describe the complete range of potential ecosystem services along a gradient of changing environmental condition
- The ESG approach leverages the concept of Final Ecosystem Goods and Services (FEGS) to identify metrics that are directly relevant to human beneficiaries
- An ESG can allow decision makers to describe meaningful and unambiguous measures that clearly communicate the potential gains or losses in ecosystem services
- The ESG facilitates a consideration of potential tradeoffs, or co-benefits, across multiple stakeholder objectives as part of EBM planning and implementation

Needs to Advance EBM

- Additional scientific research is needed to move from a narrative description of an ESG to a quantitative description that enumerates ecosystem services production with changing levels of condition
- Further development of the ESG approach is needed through case study examples across a range of ecosystem types and EBM applications

1 Ecosystem-Based Management Objectives and Tradeoffs

Ecosystem-Based Management (EBM) aims to maintain ecosystems in a healthy and resilient condition while providing the services that humans want and need (McLeod et al. 2005). However, ecosystems are complex, and layering on social and economic considerations can make operationalizing EBM seem intractable (Arkema et al. 2006; Link and Browman 2017). For successful implementation of EBM, there is a specific need to bound the scope of the problem by clarifying what really matters about a decision, including explicitly articulating how objectives will be measured and characterizing values-based tradeoffs among them (Gregory et al. 2012). To address this need, we propose a science-based descriptive model of ecosystem services production in response to changing environmental condition, the Ecosystem

Services Gradient (ESG). Scientific tools and approaches, like the ESG, can help to operationalize EBM in the decision-making process by identifying meaningful measures, defining reference points, communicating and monitoring the relevant social and economic impacts of actions, and evaluating tradeoffs across multi-sector objectives (Arkema et al. 2006; Cormier et al. 2017).

The conceptual foundation for an ESG follows that of the Biological Condition Gradient (BCG), developed over a decade ago in response to growing need to assess and effectively communicate levels of biological condition in a meaningful way (Davies and Jackson 2006; U.S. EPA 2016). The BCG leverages expert knowledge and biomonitoring data to describe ecological condition along a gradient from undisturbed to severely altered conditions. Our goal in creating an ESG framework is to build upon the original goals in developing the BCG: to create a common framework, based on measurable ecologically important attributes, that can be used to describe the complete range of condition, and provide a rational and consistent means for setting targets and communicating the consequences of different management choices.

The ESG leverages a number of practical strategies for integrating ecosystem services into decision-making, including: (1) prioritizing information and analysis to what is most important; (2) using the concept of *final* ecosystem goods and services (FEGS) to identify metrics that are unambiguous and directly relevant to human beneficiaries; (3) applying ecosystem services production functions (EPFs) to link changing condition to changes in ecosystem services; (4) understanding the range of potential outcomes; and (5) considering values-based tradeoffs across multiple, often competing, objectives (Yee et al. 2017). In this chapter, we present the conceptual foundation for the ESG as an analogy to the BCG and provide examples of how it is being developed to facilitate EBM.

2 Conceptual Foundation: The Biological Condition Gradient (BCG)

The BCG is a descriptive model that describes how attributes of biological condition change in response to increasing levels of anthropogenic stress (Fig. 1a; Davies and Jackson 2006). The BCG approach was developed to address a need for science-based approaches to more precisely and effectively communicate the existing and potential condition of aquatic resources for water quality management purposes under the U.S. Clean Water Act (United States Code title 33, sections 1251–1387). The biological characteristics, defined in the BCG as "attributes," include aspects of community structure, non-native taxa, organism condition, ecosystem function, and inter-habitat connectivity. The highest level of biological condition is Level 1, which represents natural or undisturbed biological communities and anchors the best condition for defining five levels of change or departure from this condition. Level 6 represents conditions most severely altered by anthropogenic

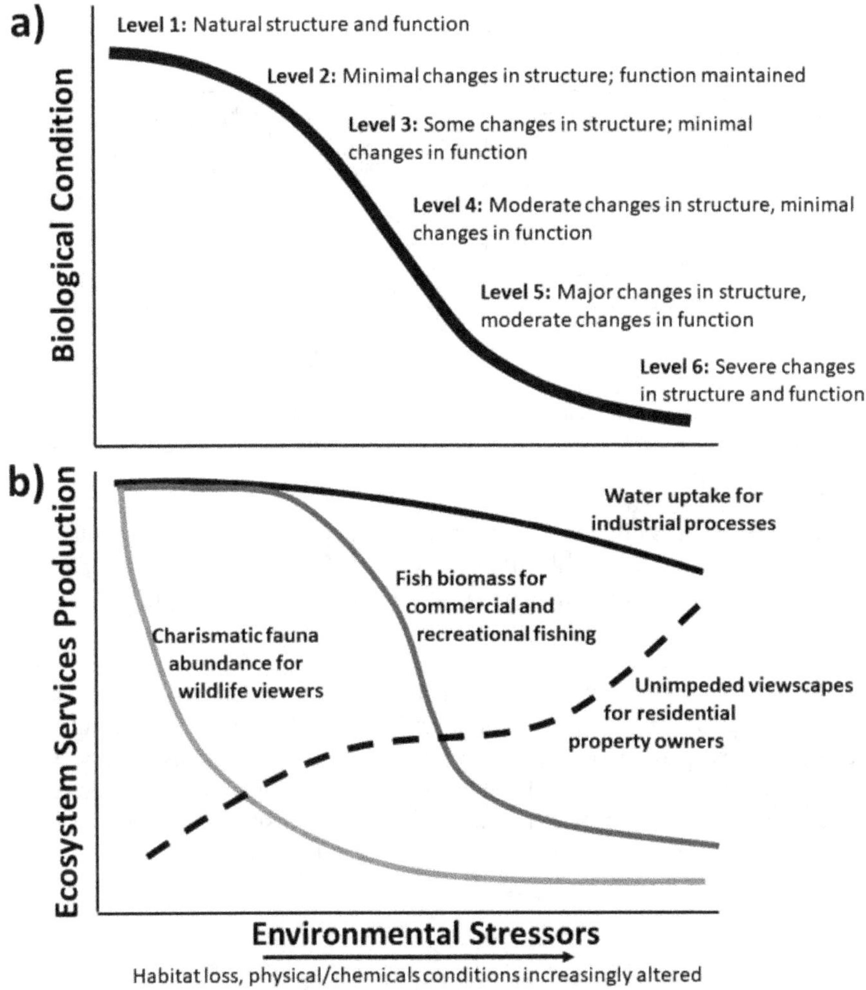

Fig. 1 The BCG model (**a**; top panel) of incremental changes in biological condition along a stressor gradient, and hypothetical changes in select ecosystem services (**b**; bottom panel) along the same gradient

stress. Each level is defined by an empirically-derived narrative description that can be consistently interpreted regardless of biology, location, or sampling method. A quantitative model is derived from narrative descriptions for each level and translated using metrics and measurable indicators to develop quantitative decision rules to identify thresholds to discriminate between BCG levels (for details see U.S. EPA 2016; Cicchetti et al. 2017).

In BCG development, a specific sequence of steps is undertaken for a given ecosystem to develop the BCG components (Table 1; U.S. EPA 2016). Because this structured approach is generalizable, BCG models can and have been developed for

Table 1 Steps in the process of developing and using a BCG

Biological Condition Gradient framework	Process
What biological attributes are relevant?	Identify and prioritize attributes
How will we measure them?	Identify metrics and indicators
What biological condition did we have?	Establish reference (natural) condition
What biological condition do we have now?	Collect and review bioassessment data
What biological condition do we want?	Set targets
How do we get there?	Identify management actions
What are the social and economic consequences?	*Conduct and communicate ecosystem services assessment (ESG)*

Adapted from Cicchetti et al. 2017
An additional step (in *italics*) indicates where an assessment of ecosystem services could supplement the process, and in conjunction with stressor and other data, inform management decisions

different regions and different ecosystems, including streams (Davies and Jackson 2006), estuaries (Cicchetti et al. 2017), and coral reefs (Bradley et al. 2014; Santavy et al. 2016). Though originally developed for aquatic ecosystems, the approach is applicable in terrestrial ecosystems as well.

The BCG can help precisely define biological condition, identify and protect high quality waters, evaluate the potential for improvement of degraded waters, select restoration targets, and clearly communicate the likely impacts of management decisions to the public. The additional step of assessing how ecosystem services change with corresponding levels of BCG (Fig. 1b) can help to communicate the social and economic benefits of protecting or restoring a site, or potential tradeoffs between different management scenarios (Cicchetti et al. 2017).

3 The Ecosystem Services Gradient (ESG)

3.1 Interpreting the ESG

Building on the conceptual foundation of the BCG, the ESG describes the complete range of ecosystem services along a gradient of biological condition from natural to severely altered. In environmental management situations where protecting biological integrity is the primary goal, directly pairing an ESG with a BCG can help decision-makers understand the potential co-occurring benefits and tradeoffs of management activities and communicate them to the public (Fig. 1). Furthermore, if a waterbody is designated for a particular use, such as recreational fisheries or contact recreation, an assessment of ecosystem services in conjunction with BCG can help identify the levels of biological condition that can be protected while still supporting desired levels of services (Davies and Jackson 2006).

Along a gradient of declining biological condition, ecosystem services may decline at different rates depending on the biological attributes providing those services (Fig. 1b). The quality of a recreational fishery, for example, may depend on the presence of uncommon taxa that are particularly vulnerable to stressors, whereas commercially-important fish species may be able to persist with some moderate degree of habitat degradation. Other ecosystem services may remain relatively unaffected along the gradient if the attributes that define biological condition are disconnected ecologically from the attributes providing the service. For example, the ecosystem service of water availability for use in industrial applications may be only partially influenced by ecosystem condition. Other ecosystem services might increase with declining biological condition. For example, depending on what local residents or recreational users consider to be aesthetically pleasing, charismatic species or unimpeded viewscapes may increase in value as presence of habitat or condition declines.

Operationally, an ESG may be directly paired with a BCG (Fig. 1). Table 1 describes an approach where an added step to the BCG process, ideally occurring during BCG development, could involve an assessment of ecosystem services. However, because the attributes that define biological condition in the BCG may not be the same attributes providing ecosystem services, EBM practitioners may prefer to develop an ESG independently of a BCG. Analogous to a BCG, the ESG would describe the full range of potential ecosystem services provisioning along a stressor gradient. Moreover, different biological attributes contribute to different ecosystem services, such that a suite of ESG curves may be needed for describing a range of different ecosystem services in a given system. However, the underlying approach in building the ESG is the same regardless, differing only in how the descriptive model is presented as either a gradient of decreasing biological condition or a gradient of changing service production (Fig. 1).

3.2 *Steps for Developing an ESG*

The steps to building and using an ESG closely parallel the steps to develop and use a BCG (Table 2). An important first step is working with decision makers and stakeholders to identify the relevant ecosystem services for the specific environmental management problem. A FEGS approach can help reduce ambiguity by explicitly and *directly* connecting biophysical indicators to the people that benefit from them (Fig. 2; Boyd et al. 2015; DeWitt et al. 2020). Along a continuum of ecological production, FEGS are distinguished from *intermediate* regulating and supporting ecological functions (e.g., habitat quality, water quality) that require additional steps to reach the ecological features (e.g., harvestable fish) directly experienced by human beneficiaries (Landers and Nahlik 2013).

Monitoring data on FEGS metrics form the basis for quantitatively describing levels of production from highest potential production to severely altered production along a gradient of changing environmental condition. While environmental

Table 2 Steps in the process of developing and using an ESG

Ecosystem Services Gradient Framework	Process
What *final* ecosystem goods and services (FEGS) are relevant?	Identify and prioritize FEGS
How will we measure them?	Identify FEGS metrics and indicators, and the bio-physical attributes that provide them
What FEGS could we have?	Establish potential availability under a range of bio-physical conditions
What FEGS do we have now?	Measuring, mapping, and ecological production functions (EPFs)
What FEGS do we want?	Evaluate co-occurring benefits and tradeoffs
How do we get there?	Identify management actions
What are the social and economic consequences?	Conduct and communicate benefits assessment using ecological benefit functions (EBFs)

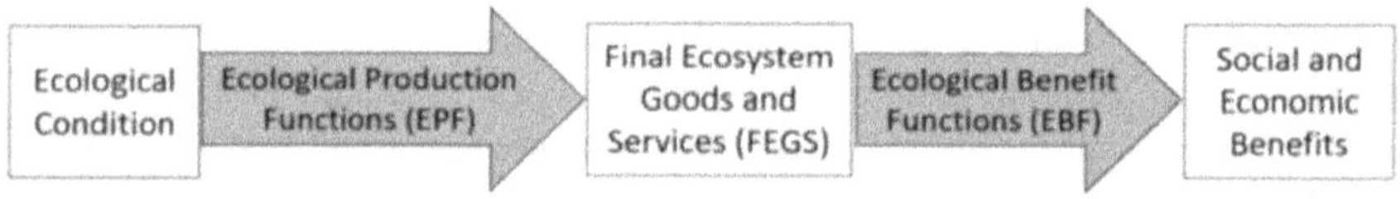

Fig. 2 Conceptual model illustrating the relationships between ecological condition, FEGS, and socio-economic benefits

assessments and monitoring often focus on collecting data on ecological condition, development of an ESG may rely on reasonable proxies where direct data or models are not available. Models, known as ecological production functions (EPFs), may be needed to translate environmental condition data to FEGS metrics (Fig. 2; Wainger and Mazzotta 2011; Bruins et al. 2017). This combination of expert judgment on meaningful metrics, collection of field data, and application of EPFs is used to first narratively and then, ideally, numerically describe incremental changes in ecosystem services provisioning along a stressor gradient to form the ESG. If ESGs are developed for more than one ecosystem service (e.g., Fig. 1b), then potential co-occurring benefits or tradeoffs can be examined alongside changes in ecosystem condition.

While FEGS represent the end product of what the environment provides to human beneficiaries, they require human input and interaction (e.g., a boat to collect the fish) for those services to be realized as actual benefits (Mazzotta et al. 2016). As such, a benefits assessment, using ecological benefits functions (EBFs) to translate ecosystem service supply into monetary, health, or other measures of benefit, could be an additional step for characterizing and communicating the benefits of EBM decisions (Fig. 2).

4 Example ESG Applications

4.1 Communicating Benefits of Coral Reef Protection

Healthy coral reef ecosystems supply multitudes of benefits on which many economies and societies rely (Wilkinson 2008; van Beukering et al. 2011), including recreation such as fishing, tourism, boating, SCUBA diving; education; coastal protection; and bioprospecting for novel pharmaceuticals and biochemicals (Moberg and Folke 1999; Principe et al. 2012). Marine coastal areas, including coral reefs, are exposed to increasing loads of nutrients, sediments, pollutants, and other materials originating from terrestrial sources that can deleteriously impact the ecosystem goods and services they provide and place them at risk of being lost (Harborne et al. 2017). Consequently, there is continuing urgency to develop tools to effectively communicate this information to improve public awareness of reef condition; understand what actions are most likely to protect these irreplaceable ecosystems; and provide a more robust process to inform management of the biological condition of coral reefs to ensure protection of high quality marine waters and their biological communities, and to develop restoration targets.

The framework used to develop the BCG model for freshwater streams, rivers, and lakes was adapted to incorporate coral reef attributes judged important to protect the biological integrity of tropical Caribbean and Western Atlantic waters, including marine coastal habitats such as mangroves, seagrasses, and coral reefs (Bradley et al. 2014; Santavy et al. 2016). From a preliminary narrative model of all coral reef assemblages (Table 3), two narrative BCG models were developed for Caribbean coral reefs, one for coral reef fish and a second for sessile marine assemblages, built primarily using attributes from scleractinian coral communities, but including algae, sponges, and octocorals (Santavy et al. 2016). For each, a numeric BCG model is being developed by eliciting expert knowledge in combination with bioassessment data and underwater videos, using mathematical fuzzy set theory to define decision rules for BCG levels (U.S. EPA 2016).

In order to build upon these coral reef BCGs and develop an ESG for coral reefs, the first step would be to identify the most relevant ecosystem goods and services for the particular decision context (Table 2), recognizing that ESGs would need to be developed on a site-by-site basis so that each ESG properly reflects the makeup of human beneficiaries at that site. For example, a coral reef that is situated within a Marine Protected Area that limits or bans many consumptive activities will provide a very different set of FEGS than a coral reef that has fewer use restrictions. Once the stakeholder groups associated with the reef have been clearly identified, the ways in which each of those groups benefit from the reef can be identified and clearly articulated. At this point, the FEGS necessary to achieve those benefits can be identified. This can be done in an ad hoc fashion or using a more structured approach such as the National Ecosystem Services Classification System (U.S. EPA 2015) or the Common International Classification of Ecosystem Services (CICES; Haines-Young and Potschin 2018) as a starting point to identify potential FEGS. Because

Table 3 Examples of narrative condition levels and associated attributes in coral reef BCG

Condition level	Physical structure	Corals	Fish, other vertebrates and invertebrates	Gorgonians, sponges, algae	Condition
Very Good Excellent BCG Level 1–2	High rugosity; very clear water; no sediment	High species diversity; includes rare, large, and old colonies	Balanced species abundances and sizes; large, long-lived species present	Low abundance of fleshy algae; sensitive species present	Low prevalence of disease
Good BCG Level 3	Moderate to high rugosity; water slightly turbid; low sediment	Moderate coral diversity; rare species absent	Noticeable decline in apex predators; large, long lived species absent locally	More fleshy algae than Level 1–2; highly sensitive species missing	Disease slightly above background level; some irregular tissue loss
Fair BCG Level 4	Low rugosity; water turbid; sediment accumulation	Reduced coral diversity; few or no living, large, old colonies; emergence of tolerant species	Absence of small reef fishes; large, long lived species absent locally; sensitive species conspicuously absent	Gorgonians replace sensitive coral and sponge species; abundant and diverse fleshy algae	Higher prevalence of disease and tissue loss
Poor BCG Level 5	Very low rugosity; very turbid water; thick sediment	Absence of colonies except highly tolerant species	No large fish; only tolerant species remain; high abundance of sediment dwelling invertebrates	Small and few colonies; highly tolerant species; high cover of fleshy algae	High prevalence of disease and high tissue loss

Simplified from Bradley et al. 2014

the potential list of beneficiaries and associated FEGS can get quite long, some prioritization may be needed to generate a manageable set for further consideration, based on greatest relevance to stakeholders or likelihood of impact by management (see the FEGS Scoping Tool; Sharpe et al. 2020).

Once the FEGS for a given coral reef have been identified, the next step is to develop metrics and indicators for each of the FEGS or the biophysical attributes that provide them. Table 4 below steps through this process for three common groups of human coral reef beneficiaries. The process begins by asking those benefiting from the coral reef what matters directly to them. That information can help identify the type of FEGS necessary to receive that benefit as well as the type of metrics that would be useful. The examples in Table 4 are generic, but more site-specific FEGS and metrics can, and should, be developed for site-specific ESGs. For example, fish diversity and abundance are suggested metrics for the beneficiary category of SCUBA divers. However, alternative or more specific metrics might be more

Table 4 FEGS metrics for a set of generic coral reef beneficiaries

Beneficiary	What matters directly?	FEGS type	Example metric
SCUBA Divers and Snorkelers	Is there sufficient visibility to be pleasurable to divers?	Water	Secchi disk depth (m)
	Is water quality safe for diving?	Water	Pathogen, contaminant, toxin concentrations
	Do these species attract the beneficiary?	Fauna	Fish diversity and abundance
	Do these species attract the beneficiary?	Fauna	Coral diversity and abundance
Recreational Anglers	Will I catch what I am expecting?	Fauna	Edible species abundance
	Will I catch something interesting?	Fauna	Charismatic species abundance
	Is it safe for boating?	Water	Wave intensity, surge height
	Is this reef aesthetically enjoyable?	Site appeal	Viewscape, sounds, smells
Coastal Property Owners	Will my property be damaged?	Water	Flood risk and coastal protection by the reef

appropriate, such as the presence or abundance of sea turtles, depending on site-specific factors such local ecology or cultural significance. Expert judgment can be used to determine the most appropriate substitute, depending on local factors as well as data availability. For example, it may not be feasible to collect daily Secchi disk depth readings, but local dive reports on water visibility may be easily collected.

After a complete set of beneficiaries and their associated metrics have been developed, these metrics can be compiled into models like those in Fig. 1 that demonstrate how changes in the level of environmental stress impact these prioritized FEGS. As shown in our coral reef example, there are multiple types of FEGS that may combine with one another to provide the overall benefit to a given beneficiary. For SCUBA diver beneficiaries, metrics related to coral diversity, fish abundance, and water visibility are all important for measuring the level of ecosystem service provided (Fig. 3a). Ecological production functions (EPFs) can be applied (e.g., Yee et al. 2014) to quantitatively link shifts in the level of environmental stress to shifts in the levels of FEGS provided to beneficiaries. One or more FEGS could be assessed for a single beneficiary (Fig. 3a), or those individual FEGS can be combined into an overall measure of realized benefit using ecological benefit functions (EBFs) (Fig. 3b, solid line). Either approach allows for a clear description as to how changes in environmental stressors directly impact different coral reef stakeholder groups (Fig. 3b, dashed lines).

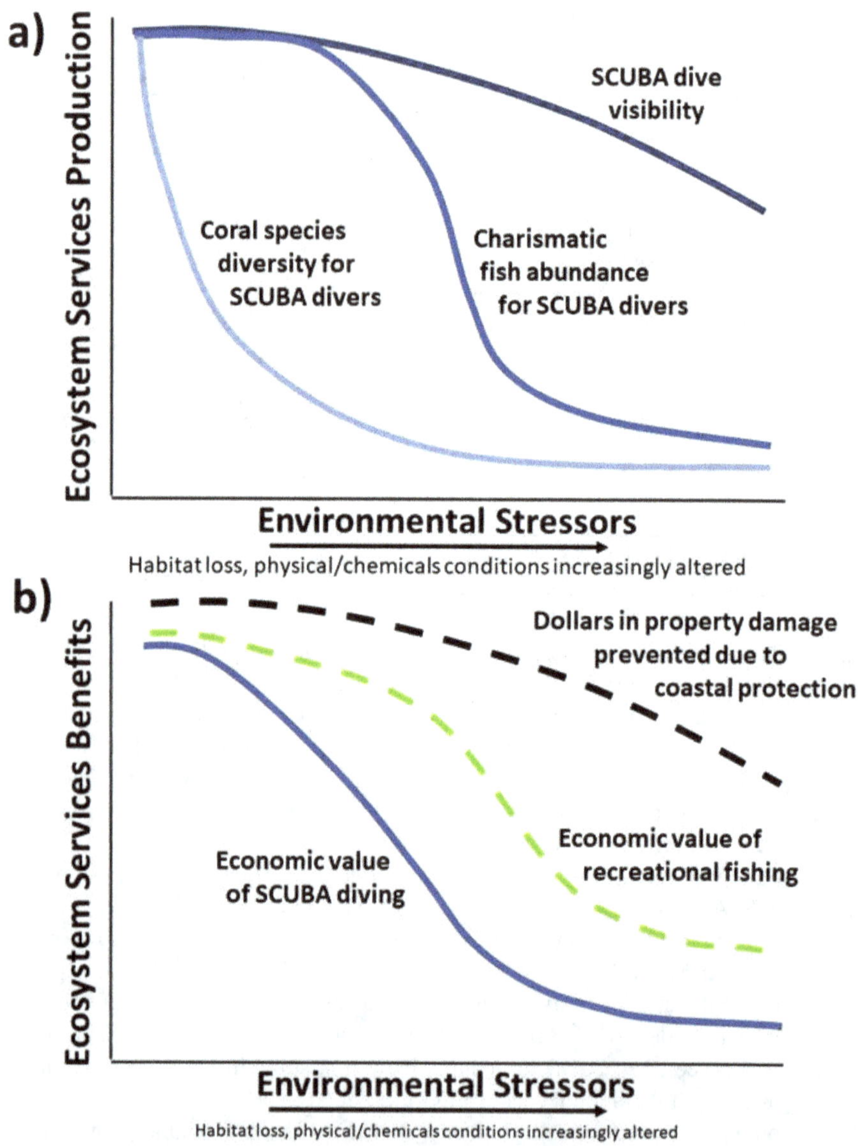

Fig. 3 A hypothetical ESG model showing changes in ecosystem service production and realized benefits in response to increases in environmental stressors. Top panel (**a**) shows changes in three individual FEGS for a single beneficiary (SCUBA divers) as developed through application of EPFs to changes in environmental attributes. Bottom panel (**b**) shows realized benefits to SCUBA divers (solid line) in addition to benefits to two other types of coral reef beneficiaries (dashed lines) as developed through application of EBFs to changes in EPFs

4.2 *Measuring Benefits of Habitat Restoration in Massachusetts Bay*

The Massachusetts Bays National Estuary Program (MassBays) is one of 28 National Estuary Programs (NEPs) across the United States, charged with developing and implementing a long-term plan to improve the waters, habitats, and living resources of their estuary(s). MassBays turned to the BCG framework as part of their plan to identify target conditions to manage 47 embayments along 1770 km of shoreline. MassBays is planning to use the BCG to communicate condition of key biological components that resonate with the public, including relevant invertebrate, fish, and habitat indicators with which local decision-makers can set public-supported targets. MassBays further realized that presenting community members with socio-economic information together with BCG biological information would greatly strengthen outreach and lead to better-informed decisions. In the course of updating their long-term plan (MassBays 2019), MassBays sought input on how people use the estuaries. Responses suggest the public cares about estuarine health, clean water and water quality, with favored activities that include shellfishing, swimming, and fishing. Developing a BCG classification scheme and a BCG/ESG approach (Fig. 1) will help to communicate the potential benefits of environmental improvement for both nature and people and to set appropriate, measurable, community-supported targets for restoration and protection.

As a proposed example of how MassBays could combine BCG and ESG, we focus on seagrass (*Zostera marina*), a habitat of restoration importance to MassBays. Seagrass beds throughout the MassBays system (and on the entire U.S. coastline) have lost significant area and function due to stressors including nutrient pollution, increasingly extreme weather events, and disease. Seagrasses support a diverse fauna and provide ecosystem services for many beneficiaries. To illustrate the approach, we selected recreational anglers, shellfishers, and birdwatchers as example beneficiaries. Fishing and shellfishing in seagrass beds are popular activities for those seeking striped bass, bluefish, scallops, crabs, and other species. Seagrasses are nursery habitats for many valued species and seagrass beds reliably sustain diving waterfowl for birdwatchers.

The BCG/ESG seen in Fig. 1 as two stacked graphs can also be presented as a Table to better show qualitative and quantitative thresholds of measures that define both the BCG levels of biological condition and the ESG measures of social and economic benefits. Table 5 provides an example of this for seagrass habitat, where the first column identifies each row with BCG level and the second column characterizes the seagrass biology defined by that level as narrative (which could most easily be quantified using seagrass acres as a proxy). The third column lists possible FEGS measures of valued fauna, and the fourth column shows measures of benefits to people. The last three columns align with the three boxes of the conceptual model of Fig. 2. To illustrate management application of Table 5, consider a hypothetical seagrass survey that shows only a few acres of sparse seagrass in a managed area. Sparse seagrass is a Level 5 (fourth row) BCG narrative measure (second column)

Table 5 A hypothetical example of a seagrass BCG/ESG set up as a table, with possible BCG measures (second column) linked to FEGS measures (third column) linked to benefit measures (fourth column)

BCG level	BCG narrative measures	Possible FEGS measures	Possible benefit measures
Level 1 to 2	Managed area has a large extent of abundant, dense, and healthy seagrass that supports diverse and abundant fauna	Fish/shellfish surveys and eBird (Sullivan et al. 2009) show valued fish species and scallops are very abundant and bird populations are very diverse	Observations and recreational fishing surveys show many people are fishing, scalloping, or birdwatching
Level 3	Some loss of acres from Level 1 to 2 (above); Abundant, dense, and healthy seagrass in some places; Thin and/or poor quality seagrass elsewhere; Diverse and abundant fauna in dense beds	Surveys and eBird show valued fish species and scallops are abundant and bird populations are very diverse	Observations and recreational fishing surveys show many people are fishing, scalloping, or birdwatching, comparable to BCG Level 1 to 2 above
Level 4	Moderate loss of acreage from Level 1 to 2 in managed area; Thin and/or poor quality seagrass in most places supports fewer and less diverse seagrass fauna	Surveys and eBird show valued fish species and scallops are moderately abundant and bird populations are diverse	Observations and recreational fishing surveys show a moderate number of people are fishing, scalloping, or birdwatching
Level 5	Major loss of acres from Level 1 to 2 in managed area; Sparse seagrass supports sparse fauna	Surveys and eBird show valued fish species and scallops are scarce and bird population diversity is only slightly elevated from adjacent non-vegetated areas	Observations and recreational fishing surveys show few people fishing, scalloping, or birdwatching
Level 6	No seagrass, shift to less diverse and productive non-vegetated faunal communities in managed area	Fish, scallop, and bird populations are comparable to those in non-vegetated areas	No more people are fishing, scalloping, or birdwatching than in local non-vegetated areas

with scarce valued fauna, and few people enjoying benefits (third and fourth columns). Presenting this information to the public together with descriptions of better environmental and socio-economic conditions at higher BCG levels could inspire a long-term vision of achieving, say, Level 3 conditions (second row) with abundant seagrass and fauna and many fishers and birdwatchers in some but not all places within the managed area. Once quantitative targets are set (e.g., for Level 3 acres) public-supported management actions (perhaps significant nutrient reductions) can be developed and implemented, then changes in BCG, FEGS, and benefit measures can be quantitatively monitored and reported back to the public.

The success of this potential approach for MassBays depends on the data and effort utilized to create a working BCG/ESG gradient that could apply at several scales: an overarching application to all 47 embayments; application to groups of embayments classified based on specific characteristic conditions; and to embayments at scales most relevant to management. Key elements of the effort are acquiring data and determining appropriate reference narratives and values for biological, FEGS, and benefit indicators. Comparability of the approach among embayments relies on a consistent identification of Level 1 or combined Level 1 / Level 2 condition.

To address data needs, MassBays worked with partners to synthesize a large amount of environmental, social, and economic data for these embayments. Reference condition for seagrass acres may be available through historic maps, charts, and surveys, or (as with all measures) by using best available current data, which may not represent Level 1 or 2, but can be interpreted in the BCG construct as Level 3 (or lower). It is likely that other existing MassBays data might serve as proxies for FEGS measures (here fishing, shellfishing, and birdwatching) based on methods in the literature, such as Rapid Benefits Indicators (e.g., Mazzotta et al. 2016).

Combining the BCG with an assessment of ecosystem services allows communication of environmental condition directly linked to the socio-economic benefits of environmental improvement (Cicchetti et al. 2017). This approach can resonate with people whose belief systems run the entire spectrum from those who most appreciate nature for its own sake to those who most appreciate the socio-economic benefits that nature provides to humans. Engaging a range of stakeholders invests more people in the value of environmental protection and is an important tenet of ecosystem-based management (Arkema et al. 2006). The BCG/ESG framework captures stakeholder input to develop goals using the approach of "what did we have, what do we have, what do we want, and how do we get there" to communicate both nature and benefits (Tables 1 and 2). This allows managers to set appropriate, measurable environmental targets that are supported by a diverse public.

5 Role of an ESG in Ecosystem-Based Management

The introduction of ecosystem goods and services advances the utility and applicability of the BCG framework for ecosystem-based management activities. Important action items within an EBM approach to decision making can include: the identification of objectives and performance measures to describe what really matters to stakeholders about a decision; the identification of management alternatives; the articulation of potential user conflicts or tradeoffs between management alternatives; and the articulation of potential direct ecosystem services benefits, or related co-occurring benefits, for a given decision context (Cormier et al. 2017). The ESG, through its structured approach to defining and enumerating potential changes in ecosystem services, allows decision makers to clearly articulate the elements feeding into each of these steps.

An important foundational principle of the BCG is science communication through BCG visualization and the accompanying tables that describe the technical aspects of each BCG level. Likewise, an ESG framework also lends itself to strong science communication served up in a strategic manner (*sensu* Harwell et al. 2020), allowing for communication of key messages to targeted audiences. Paired with a BCG, the ESG can allow decision makers to describe meaningful and unambiguous environmental objectives and their measures and clearly communicate the potential gains or losses in ecosystem services. The quantitative measures defined by the levels of an ESG might also indicate where biomonitoring can be used to assess whether actions are having the desired outcomes and what adjustments can be made to future actions as part of targeted adaptive management efforts (e.g., LoSchiavo et al. 2013).

Finally, the ESG approach can be helpful for identification of critical gaps in knowledge, helping EBM practitioners identify where resources may be needed to fill those gaps—in particular, what future scientific research is needed to move a narrative description for a given ESG level to a quantitative description. Future development of ESG principles include in-depth application to a suite of case study examples across a range of ecosystem types, both related to information and regulatory needs, such as condition assessments conducted for water quality management purposes under the U.S. Clean Water Act, and to broader EBM questions such as ecological protection, restoration or fisheries management.

Disclaimer This chapter has been subjected to Agency review and has been approved for publication. The views expressed in this paper are those of the author(s) and do not necessarily reflect the views or policies of the U.S. Environmental Protection Agency.

References

Arkema, K. K., Abramson, S. C., & Dewsbury, B. M. (2006). Marine ecosystem-based management: from characterization to implementation. *Frontiers in Ecology and the Environment, 4*, 525–532.

Boyd, J. W., Ringold, P. L., Krupnick, A. J., Johnston, R. J., Weber, M., & Hall, K. (2015). *Ecosystem services indicators: Improving the linkage between biophysical and economic analyses. RFF DP 15-40*. Washington, DC: Resources for the Future.

Bradley, P., Santavy, D. L., & Gerritsen, J. (2014). *Workshop on biological integrity of coral reefs August 21–22, 2012, Caribbean Coral Reef Institute, Isla Magueyes, La Parguera, Puerto Rico. EPA/600/R-13/350*. Narragansett, RI: U.S. Environmental Protection Agency, Office of Research and Development, Atlantic Ecology Division.

Bruins, R. J. F., Canfield, T. J., Duke, C., Kapustka, L., Nahlik, A. M., & Schäfer, R. B. (2017). Using ecological production functions to link ecological processes to ecosystem services. *Integrated Environmental Assessment, 13*(1), 52–61.

Cicchetti, G., Pelletier, M. C., Rocha, K. J., Bradley, P., Santavy, D. L., Pryor, M. E., Jackson, S. K., Davies, S. P., Deacutis, C. F., & Shumchenia, E. J. (2017). *Implementing the biological condition gradient framework for management of estuaries and coasts. EPA/600/R-15/287*. Narragansett, RI: U.S. Environmental Protection Agency, Office of Research and Development, Atlantic Ecology Division.

Cormier, R., Kelble, C. R., Anderson, M. R., Allen, J. I., Grehan, A., & Gregersen, O. (2017). Moving from ecosystem-based policy objectives to operational implementation of ecosystem-based management measures. *ICES Journal of Marine Sciences, 74*, 406–413.

Davies, S. P., & Jackson, S. K. (2006). The biological condition gradient: A descriptive model for interpreting change in aquatic ecosystems. *Ecological Applications, 16*, 1251–1266.

DeWitt, T. H., Berry, W. J., Canfield, T. J., Fulford, R. S., Harwell, M. C., Hoffman, J. C., Johnston, J. M., Newcomer-Johnson, T. A., Ringold, P. L., Russel, M. J., Sharpe, L. A., & Yee,

S. J. H. (2020). The final ecosystem goods and services (FEGS) approach: A beneficiary-centric method to support ecosystem-based management. In T. O'Higgins, M. Lago, & T. H. DeWitt (Eds.), *Ecosystem-based management, ecosystem services and aquatic biodiversity: Theory, tools and applications* (pp. 127–148). Amsterdam: Springer.

Gregory, R., Failing, L., Harstone, M., Long, G., McDaniels, T., & Ohlson, D. (2012). *Structured decision making: A practical guide to environmental management choices*. Chichester, UK: Wiley-Blackwell.

Haines-Young, R., & Potschin, M. B. (2018). *Common International Classification of Ecosystem Services (CICES) V5.1 and guidance on the application of the revised structure*. Nottingham, UK: Fabis Consulting.

Harborne, A. R., Rogers, A., Bozec, Y.-M., & Mumby, P. J. (2017). Multiple stressors and the functioning of coral reefs. *Annual Review of Marine Sciences, 9*, 445–468.

Harwell, M. C., Molleda, J. L., Jackson, C. A., & Sharpe, L. (2020). Establishing a common framework for strategic communication in ecosystem-based management and the natural sciences. In T. O'Higgins, M. Lago, & T. H. DeWitt (Eds.), *Ecosystem-based management, ecosystem services and aquatic biodiversity: Theory, tools and applications* (pp. 165–188). Amsterdam: Springer.

Landers, D., & Nahlik, A. (2013). *Final ecosystem goods and services classification system (FEGS-CS). EPA/600/R-13/ORD-004914*. Washington, DC: U.S. Environmental Protection Agency.

Link, J. S., & Browman, H. I. (2017). Operationalizing and implementing ecosystem-based management. *ICES Journal of Marine Sciences, 74*, 379–381.

LoSchiavo, A., Best, R., Burns, R., Gray, S., Harwell, M., Hines, E., McLean, A., St. Clair, T., Traxler, S., & Vearil, J. (2013). Lessons learned from the first decade of adaptive management in comprehensive Everglades restoration. *Ecology and Society, 18*, 70–85.

MassBays, Massachusetts Bay National Estuary Program. (2019). *Comprehensive conservation and management plan: A blueprint for the bays*. Boston, MA: Massachusetts Bay National Estuary Program.

Mazzotta, M., Bousquin, J., Ojo, C., Hychka, K., Druschke, C. G., Berry, W., & McKinney, R. (2016). *Assessing the benefits of wetland restoration: A rapid benefit indicators approach for decision makers. EPA/600/R-16/084*. Narragansett, RI: U.S. Environmental Protection Agency.

McLeod, K. L., Lubchenco, J., Palumbi, S. R., & Rosenberg, A. A. (2005). Scientific consensus statement on marine ecosystem-based management. Communication Partnership for Science and the Sea. Retrieved October 22, 2019, from http://www.coml.us/wp-content/uploads/2010/06/Scientific-Consensus-Statement-on-Marine-Ecosystem-Based-Management.pdf.

Moberg, F., & Folke, C. (1999). Ecological goods and services of coral reef ecosystems. *Ecological Economics, 29*, 215–233.

Principe, P., Bradley, P., Yee, S., Fisher, W., Johnson, E., Allen, P., & Campbell, D. (2012). *State of the science on linkages among coral reef condition, functions, and ecosystem services. EPA/600/R-11/206*. U.S. Research Triangle Park, NC: Environmental Protection Agency, Office of Research and Development.

Santavy, D. L., Bradley, P., Gerritsen, J., & Oliver, L. (2016). The biological condition gradient, a tool used for describing the condition of US coral reef ecosystems. *Proceedings of the 13th International Coral Reef Symposium*, Honolulu, pp. 548–559.

Sharpe, L., Hernandez, C., & Jackson, C. (2020). Prioritizing stakeholders, beneficiaries and environmental attributes: A tool for ecosystem-based management. In T. O'Higgins, M. Lago, & T. H. DeWitt (Eds.), *Ecosystem-based management, ecosystem services and aquatic biodiversity: Theory, tools and applications* (pp. 189–212). Amsterdam: Springer.

Sullivan, B. L., Wood, C. L., Iliff, M. J., Bonney, R. E., Fink, D., & Kelling, S. (2009). eBird: A citizen-based bird observation network in the biological sciences. *Biol Conserv, 142*(10), 2282–2292.

U.S. EPA. (2015). *National ecosystem services classification system (NESCS): Framework design and policy application. EPA-800-R-15-002*. Washington, DC: United States Environmental Protection Agency.

U.S. EPA. (2016). *A practitioner's guide to the biological condition gradient: A framework to describe incremental change in aquatic ecosystems. EPA-842-R-16-001*. Washington, DC: U.S. Environmental Protection Agency.

van Beukering, P., Brander, L., van Zanten, B., Verbrugge, E., & Lems, K. (2011). *The economic value of the coral reef ecosystems of the United States Virgin Islands. Report R11/06*. Amsterdam: IVM Institute for Environmental Studies.

Wainger, L., & Mazzotta, M. (2011). Realizing the potential of ecosystem services: A framework for relating ecological changes to economic benefits. *Environmental Management, 48*, 710–733.

Wilkinson, C. (2008). *Status of coral reefs of the world: 2008*. Townsville, Australia: Global Coral Reef Monitoring Network and Reef and Rainforest Research Centre.

Yee, S. H., Dittmar, J. A., & Oliver, L. M. (2014). Comparison of methods for quantifying reef ecosystem services: A case study mapping services for St. Croix, USVI. *Ecosystem Services, 8*, 1–15.

Yee, S., Bousquin, J., Bruins, R., Canfield, T. J., DeWitt, T. H., de Jesús-Crespo, R., Dyson, B., Fulford, R., Harwell, M., Hoffman, J., Littles, C. J., Johnston, J. M., McKane, R. B., Green, L., Russell, M., Sharpe, L., Seeteram, N., Tashie, A., & Williams, K. (2017). *Practical strategies for integrating final ecosystem goods and services into community decision-making. EPA/600/R-17/266*. Gulf Breeze, FL: U.S. Environmental Protection Agency.

12

Establishing a Common Framework for Strategic Communications in Ecosystem-Based Management and the Natural Sciences

Matthew C. Harwell, Jeannine L. Molleda, Chloe A. Jackson, and Leah Sharpe

Abstract There is a need for a generalized framework and guidance for developing strategic communication efforts for interdisciplinary practitioners of ecosystem-based management to ensure engagement and communication efforts focus on effective science-society dialogue. Too often, however, developing and implementing communication strategies is viewed as separate from the research and not undertaken until the research is complete. Developing a strategic communication plan involves outlining and articulating a project's goals and objectives, identifying communication goals, defining messages, audiences, and vehicles, characterizing the different types of communication flow paths (both internal and external), and developing clear metrics that will allow for evaluating the success of the communication plan. A strategic communication matrix provides an organizational and operational structure for implementing a strategic communication plan. Here, we offer specific guidance tailored to scientists, stakeholders, and decision makers for developing strategic communication efforts. This tailored framework is then examined through a case study application in the field of ecosystem-based management.

Lessons Learned

- There is a lack of peer-reviewed literature on the development and implementation of strategic communications (focusing on *message*, *audience*, *vehicle*) with the ecosystem-based management and natural sciences literature.
- Efforts to develop and implement strategic communication plans are more effective when there is holistic buy-in from organizations.

M. C. Harwell (✉) · J. L. Molleda · L. Sharpe
Gulf Ecosystem Measurement and Modeling Division, U.S. EPA, Gulf Breeze, FL, USA
e-mail: Harwell.matthew@epa.gov; jmolleda@pgrenewables.com; Sharpe.leah@epa.gov

C. A. Jackson
ORISE Research Participant, Pacific Ecological Systems Division, Newport, OR, USA
e-mail: Jackson.chloe@epa.gov

- It is key to have a proactive communication effort from the beginning of a scientific effort, and it is just as important to develop evaluation and feedback methods to understand the effectiveness of the specific *messages* presented to particular *audiences* using the chosen communication *vehicles*.
- An organized framework for strategic communication moves general science communication from a tactical "list of tasks" to a more comprehensive strategy to communicate the relevance of the science.
- A strategic communication matrix can be used to effectively organize *messages* for specific *audiences* using specific *vehicles* to address overall communication goals for a given effort.

Needs to Advance EBM

- There is a need to implement strategic communication frameworks, ideally from the beginning of a project life-cycle, to advance both principles of EBM and case-study applications of EBM into future studies.
- There is a need to study previous EBM communication efforts in both the peer-reviewed and grey literature to determine their effectiveness and help improve communication efforts moving forward.
- Documentation of strategic communication efforts can inform assessing the effectiveness and success of communication and be used to help inform future communication efforts.

1 Introduction

Over the last few decades, the use of strategic communication has become increasingly prominent and valued across many disciplines, including Ecosystem-Based Management (EBM). Strategic communication is a three-element process that involves specific efforts to get the proper *message* delivered using the correct form (*vehicle*) of communication, to the intended *audience*, at the appropriate time (Braus 2009). A strategic communication approach is a useful tool to tackle sensitive topics, define and prioritize target groups, and standardize communication processes (Ekebom et al. 2008). In addition to communicating information, principles of strategic communication have been used to achieve a variety of goals ranging from persuasion (Halloran 2007) to coordination (Murphy 2008) to behavioral changes (Cabanero-Verzosa and Elaheebocus 2008; Mortenius 2014).

A strategic communication approach asks decision makers and stakeholders to think holistically about their communication efforts, looking beyond information sharing to think purposefully about what they want to achieve by sharing information. For practitioners of EBM, the importance of a meaningful science-policy dialogue is paramount to the effectiveness of using an EBM approach to bring science into the discussion and decision-making process for socio-ecological decisions (e.g., Long et al. 2015; Mattheiß et al. 2018; O'Higgins et al. 2020).

A recent case study analysis of multiple EBM efforts by Mattheiß et al. (2018) concluded that better communication strategies are needed to improve effectiveness of EBM efforts. In this chapter, we demonstrate why this approach is worth taking and articulate a generalizable framework for scientists, decision makers, stakeholders, and the larger EBM community of practitioners. We argue that it is not just the field of science that should embrace a strategic communication philosophy, but the full suite of EBM practitioners, including decision makers and stakeholders, and that this approach should not begin once results are ready for dissemination or publication but be incorporated throughout the decision process.

Ecosystem-based managers who develop conservation plans, projects, and policies work to understand people's perceptions while promoting habitat conservation (Goldberg et al. 2016). A number of chapters in this text acknowledge the important role of strong communication to maximize effectiveness of EBM efforts (e.g., Myer and Johnston 2020; Williams and Hoffman 2020). Long et al.'s (2015) analysis of core principles of EBM recognise the importance of communications as part of EBM implementation (Long et al. 2015).

Stakeholder engagement and involvement, a focused effort on bringing the appropriate groups of people together to discuss, and engage, on aspects of a given decision, is recognized as another core principle in EBM (Long et al. 2015). The goals of a given communication, or dialog, effort often go beyond simple transfer of information. For example, the Uganda Nutrition and Early Childhood Development Project identified behavior change as their ultimate goal and incorporated two-way dialog with stakeholders to ensure project objectives aligned with needs and demands of their project's beneficiaries (Cabanero-Verzosa and Elaheebocus 2008). By understanding beneficiaries' attitudes, beliefs, and practices, they improved the health and nutritional status of their clients, improved stakeholders' knowledge and practices, increased a demand for community health services and schooling, and enhanced local and social capacity within the community (Cabanero-Verzosa and Elaheebocus 2008). Including audiences' attitudes and perceptions in management considerations increases the likelihood of success (Goldberg et al. 2016).

A strategic approach to communication recognizes that true communication is a two-way process and that it is important to understand what the identified audiences are looking to get out of these interactions (USFWS 2016). A strategic communication program moves beyond limitations of most common communication models (e.g., "one size fits all," "presenting everything and letting the audience decide what is important," or "thinking that communication ends once information has been presented") and specifically focuses on building a communication framework composed of three interlinked pillars—*message*, *audience*, and *vehicle*—resting on the common foundation of clearly articulated communication goals. In addition to serving as an organizational framework, the physical structure of a strategic communication plan shows an audience where they fit into the larger picture. From an EBM context, this aligns with the core principle of "recognise coupled social-ecological system" (Long et al. 2015). Additionally, a robust strategic communication plan incorporates context-specific metrics for determining communication effectiveness and success. As described later, metrics can focus on one or more communication elements to inform whether the communication goals were met. As a whole, these important elements are relevant to applied science, stakeholder engagement, and decision making.

The National Academies of Sciences, Engineering, and Medicine (NASEM) find that it is important to communicate science effectively because people need to be able to integrate accurate scientific information into their personal values and life decisions (NASEM 2017). Traditional science communication approaches often assume that the main goal is to address the deficit of scientific understanding among non-scientists (Groffman et al. 2010; NASEM 2017). This approach assumes that once an audience is educated on the topic, the work of communicators is done (Kellstedt et al. 2008; Bubela et al. 2009; Groffman et al. 2010; NASEM 2017). However, studies show that while audiences do not lack the knowledge to understand, they interpret and use information they receive in different ways (e.g., Hansen et al. 2003). The difference between a science communication deficit model and a strategic communication model is that the latter is "a purposeful use of communication by an organization to fulfil its mission" (Hallahan et al. 2007) that does not make inherent assumptions about the audience's level of knowledge about a topic and encompasses goals beyond information transfer.

Liang et al. (2018) propose the term "Strategic Environmental Communication" (SEC), which combines the concepts of environmental communication, strategic communication, and persuasion research. Practitioners and scientists can use SEC to design strategic environmental campaigns using well-designed messages and science-based strategies (Liang et al. 2018). Strategic environmental communication takes the concept of strategic communication and applies it to increase the effectiveness of environmental campaigns used to promote pro-environmental attitudes, behaviors, and investments (Liang et al. 2018). Whether focusing solely on strategic communication, or adopting the concept of SEC, practitioners and scientists can benefit from a strategic approach to communication research.

Scientific results are not always available or accessible to the public and decision makers and questions addressed by scientists do not always speak directly to stakeholder concerns. We argue that strategic communication is a way to help bridge the gap between the two (de Bruin and Bostrom 2012; Winterfeldt 2012; Jones et al. 2013). By taking a strategic communication approach throughout the lifecycle of a research project, scientists can tailor their approach to respond to stakeholder priorities or develop buy-in and participation from stakeholders and collaborators, making communication a well thought-out, long-term process rather than a reactive exercise or one constructed after the project's completion (Odugbemi and Mozammel 2005). This helps ensure that the science meets the need of management within communities, and vice versa (DeLauer et al. 2012). We argue that the use of a strategic communication approach thus can be useful in efforts to achieve another core EBM principle, "decisions reflect societal choice" (Long et al. 2015), or at the minimum, a societal-choice-informed decision adequately considers the science elements involved. By studying and researching strategic science communication, one can better understand how effective communication might influence audience interpretations and reactions, as well as how audience responses might influence next steps in science programs and management decisions (Barker 2006; Jones et al.

2013). Through an analysis of strategic science communication research to date, we have developed a generalizable framework that can be applied across a range of scientific disciplines.

There is wide consensus that natural sciences would benefit from putting more effort into strategic communication efforts (Barker 2006; Hobbs 2006; Groffman et al. 2010), but most natural scientists, decision makers, and stakeholders have not been trained in strategic communication, and traditional science communication training has typically focused on information transfer clarity after the research has been completed (e.g., Turbek et al. 2016). In order to present a framework based on analysis of the available literature and tailored guidance for facilitating a strategic communications approach for natural scientists, decision makers, and stakeholders, this chapter is organized in the following manner. First, we conduct a literature review on strategic communication in the natural sciences. Second, we discuss the eight steps of developing a generalized framework for strategic communication in the natural sciences. Third, we discuss how to develop, implement, and track a strategic communication plan through a hypothetical ecosystem services case study.

2 Literature Review

A natural sciences literature review was conducted to characterize published efforts to organize, develop, and utilize strategic communication. The search focused on examples of strategic communication within the natural sciences to identify examples that included a clearly articulated strategic communication plan or framework. The primary search was on peer-reviewed literature related to the natural sciences. Additionally, we examined other science disciplines (e.g., health, political, and social sciences), non-science examples (business, military, and education), and communications field examples (public relations, technical/social media, and customer service). Utilizing multiple literature search engines, including Web of Science, Google Scholar, and Publish or Perish, the term "strategic communication" was searched by itself, as well with the following terms: "natural resources management," "ecosystem restoration," "adaptive management," "structured decision-making," "habitat conservation," and "ecosystem based management." While "strategic communication" approaches were not explicitly identified in many EBM examples, a large EBM case study analysis by Mattheiß et al. (2018) concluded that "the better the communication strategy the likelier the demand for scientific knowledge from the social system."

Potential articles were identified and subsequently filtered by title and abstract to determine whether papers were relevant with applicable information looking for title key words such as "communication" and "strategic" or articulation of strategic communication elements in the abstract. All remaining articles were then examined for information applicable to identifying elements of strategic communication yielding articles for further, in-depth analysis. The remaining articles were examined for three criteria: key words throughout the text such as message, audience, or vehicle; a clearly articulated strategic communication plan, goal, or a framework; and, whether examples of implementing a strategic communication plan were included (e.g., case studies). A separate, additional literature analysis examined specific journals such as *Science Communication*, the *International Journal of Strategic Communication*, and the *Proceedings of the National Academy of Science* for any missing sources or materials.

Additionally, we identified examples from U.S. federal agencies and environmental non-governmental organizations (ENGOs) in the grey literature on strategic communication efforts in the natural sciences. The Google search engine was used to identify relevant natural science federal agencies and ENGOs using the keywords "strategic communication plan," and websites for each relevant source were then searched using the same keywords. The natural sciences peer-reviewed publications and ENGO sources are cited in the Appendix, along with strategic communication elements identified for each source.

Common elements and important process steps in developing a strategic communication plan were identified through this literature review. As no generalizable framework for developing a strategic communication plan was identified from this literature review, we compiled elements of strategic communication and developed a strategic communication framework presented here (Fig. 1). While none of the individual framework elements are novel within the field of science communication, developing a generalizable overall framework tailored to the needs of scientists represents an important advancement in strategic communication for EBM practitioners, and the larger suite of natural sciences in general. In addition to outlining this framework, we present one approach for operationalizing strategic communication through development of a Strategic Communication Matrix.

3 Generalized Framework Development for Strategic Communication in Ecosystem-Based Management and the Natural Sciences

Three interlinked pillars of message, audience, and vehicle, built on clearly articulated goals, form the core of a generalizable framework (Fig. 1). The first step in the generalized framework is establishing project goals and objectives (Step 1). Once project goals are identified, communication goals (Step 2) can be derived and help drive the communication process and aid in accomplishing project goals and objectives. Many natural science papers identified establishing communication goals as one of the first steps in a strategic communication plan (Bronson 2004; Dayer and Meyers 2012; Timm et al. 2016). When identifying communications goals, decision makers and stakeholders must ask themselves what they are trying to achieve with their communication efforts. All too often, decision makers and stakeholders fall into the trap of communicating information for the sake of the information itself. In the framework presented here, when setting a communication goal, the question

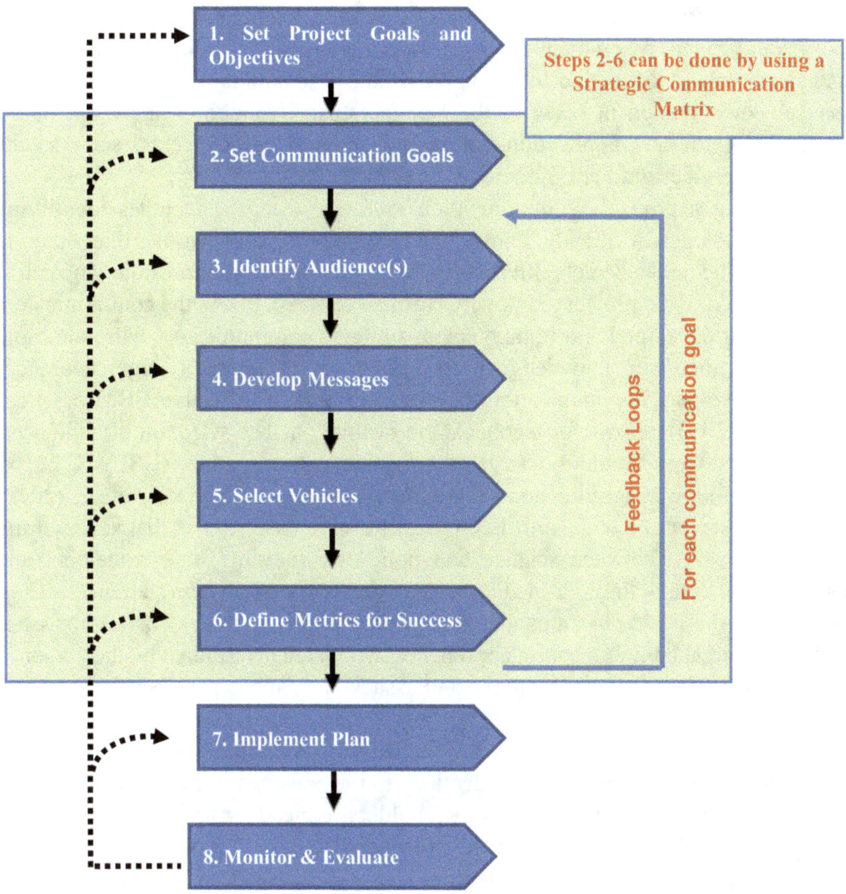

Fig. 1 Generalizable Strategic Communication Conceptual Framework using a Strategic Communication Matrix

decision makers and stakeholders must ask themselves is, "*What* are they trying to achieve?" The decision makers and stakeholders in this example may have several goals obscured under the more nebulous idea of "communicating the importance" of a science topic or decision. These goals could range from wanting to share their results or decisions with fellow stakeholders, to wanting to garner support for future work from stakeholders, to wanting to encourage external decision makers to incorporate scientific findings into their decision-making processes.

Using an example of ecosystems services to inform EBM, a more outcome-focused version of a communication goal could be "demonstrate how using the concept of ecosystem services could improve decisions related to human health and well-being." This type of goal could be identified as part of a larger effort to introduce the concepts of applying ecosystem services to inform decision making in general (e.g., DeWitt et al. 2020) as a way to present the desire to apply a similar approach in another decision context. In this text, examples include the use of ecosystem services assessment mapping for communicating future decision scenarios in Long Island (Myer and Johnston 2020), the articulation of ongoing ecosystem services consideration in EBM of the Vouga estuary (Lillebø et al. 2020), or the development of new applications of ecosystem services concepts, such as the ecosystem services gradient framework (Yee et al. 2020).

Identifying a project's communication *audience* (Step 3) includes identifying what the audience(s) already knows, what communicators would like them to know, how the audience gets information, and what the communicators would like those involved in the project to know (Groffman et al. 2010). As the communication plan is being developed and implemented, strategic communicators will gain more information about the knowledge base and values of different target audiences because a two-way communication approach, critical for effective EBM practices (Long et al. 2015), allows for feedback and evaluation. The selection of audiences, those groups who should be targeted to achieve those goals (USFWS 2016), organically follows identification of communication goals. Driscoll et al. (2012) examined a series of Long Term Ecological Reserve case studies that successfully built relationships between science and policy by focusing on engagement and distillation of results through media that met the needs of diverse audiences. Their case studies showcase programs that engage with decision makers on various issues through different communication approaches, which are determined by the audience of decision makers and the context and issues (i.e., messages) being addressed (Driscoll et al. 2012).

In an EBM context, the importance of two-way communication has been long established, including historical connections to the conservation literature, focusing on communication among relevant stakeholders as needing to be interactive and continuous. To learn more about the role of stakeholder engagement in socio-ecological decision-making contexts such as EBM, the reader is directed to examples such as Newton and Elliott (2016) and Lillebø et al. (2020). Recent advances in approaching prioritization of stakeholders in natural sciences are identified in Sharpe et al. (2020).

Developing appropriate *messages* for each goal and audience (Step 4) is more detailed and less organic than audience identification. Audiences approach a message with their own backgrounds, ideas, attitudes, and beliefs that must be acknowledged in developing successful messages (Barker 2006; Halpern et al. 2012). Developing clear messages for different stakeholders or audiences is important for promoting the responses intended by communicators and ensuring that identified goals and objectives are being met (USFWS 2016). Several natural science papers highlighted establishing effective, multi-party communication among decision makers, scientists, and community members, which led to more successful science and management (Leong et al. 2008; DeLauer et al. 2012). Message development must be done with the audiences' perspectives in mind. For example, if a communications program goal is to get recreational fishermen to comply with a licensing requirement, messaging that urges compliance should include language on how compliance helps achieve fishing community goals (e.g., licensing leads to more accurate estimates of fish populations which leads to more sustainable fishing, allowing fishing to be available to future generations).

Identifying specific communication *vehicles* (Step 5) will depend on what is known about the audience and is selected based on communication goals. The USFWS (2016) discusses selecting tactics, tools, and channels based upon the target audience, and Hobbs (2006) reminds the reader to "choose your medium carefully." The potential effectiveness of a communication vehicle can also be used to steer a science communication effort. Within the Tonle Sap Region of Cambodia, the Participatory Natural Resources Management Team recognized their target communication audiences were made up of different ethnic groups, languages, and religions, and they found meetings, workshops, posters, and environmental educators were the most effective communication vehicles for success of their communication goals (Thompson 2006). Additionally, Thompson (2006) recognized that vehicles which did not involve community members in development or interpretation process had very limited impact.

When outlining the messages/audiences/vehicles for each communication goal, *metrics of success* (Step 6) can be identified (e.g., NOAA 2009; Sea Grant 2003; NASEM 2006; USFWS 2014; NPS 2016) and used during implementation of a strategic communication plan and in efforts to monitor and evaluate the strategic communication plan. Metrics can focus on a specific aspect of communication (e.g., effectiveness of a given presentation to a particular audience), or a more comprehensive aspect (e.g., metrics that should be tracked to learn whether overall communication goals are met).

The approach for implementing a strategic communication plan should be tied back to overarching communication and project goals and may include establishing a shared understanding between science communicators and the audience. Operationally, the strategic communication plan is implemented with both content development and delivery in different forums (Step 7). Recognizing that what is persuasive for one audience may not be for another, a single project can have multiple messages, but targeted differently. Targeted messages could be as simple as having different sets of fact sheets for different regions, each highlighting regionally specific work. Additionally, different aspects of a larger project may have targeted messages for a particular audience. A Strategic Communication Matrix is one way to organize elements of a communication plan to help track and implement more sophisticated approaches to communications (e.g., more components or more complex sets of messages/audiences/vehicles) around a larger project. We present a hypothetical example in Table 1 demonstrating using a matrix to understand how different elements within the plan relate to, and rely upon, one another. More complex Strategic Communication Matrices can be operational in nature, including capturing additional metadata and tracking information.

Evaluations and feedback loops are critical to a strategic communication plan (Step 8). Interacting with intended audiences and/or stakeholders on a regular basis is important to ensure that messages are being received and having the intended impact. Further, it is important that communicators understand how audiences are responding to messages and whether adjustments are needed. These feedback loops

Table 1 Example template of a Strategic Communication Matrix

<table>
<tr><td>Project goal</td><td colspan="4">Insert Project Goal 1 here. This template can be adjusted to fit your project needs based on the identified project goal.</td></tr>
<tr><td>Project sub-goals</td><td colspan="4">Insert Project Sub-goal 1 here. This is the first sub-goal necessary in aiding and accomplishing project goal 1.</td></tr>
<tr><td>Communication goals</td><td colspan="2">Insert Communication Goal 1 here. This is the first communication goal necessary in aiding and accomplishing Sub-Goal 1. Ask “What are you trying to achieve?”</td><td colspan="2">Insert Communication Goal 2 here. This is the second communication goal necessary in aiding and accomplishing sub-goal 1. Ask “What are you trying to achieve?”</td></tr>
<tr><td>Audiences</td><td>Insert Audience 1 here. This is the first group targeted to achieve Communication Goal 1.</td><td>Insert Audience 2 here. This is the second group targeted to achieve Communication Goal 1.</td><td>Insert Audience 1 here. This is the first group targeted to achieve Communication Goal 2.</td><td>Insert Audience 2 here. This is the second group targeted to achieve Communication Goal 2.</td></tr>
<tr><td>Messages</td><td>Insert list of messages here. These messages are appropriate in aiding and accomplishing Communication Goal 1 and are specific to the targeted group identified as Audience 1.</td><td>Insert list of messages here. These messages are appropriate in aiding and accomplishing Communication Goal 1 and are specific to the targeted group identified as Audience 2.</td><td colspan="2">Insert list of messages here. These messages are appropriate in aiding and accomplishing Communication Goal 2 for both Audience 1 and Audience 2 and are specific to the targeted groups identified as Audience 1 & 2.</td></tr>
<tr><td>Vehicles</td><td>Insert a list of vehicles here that is specific to Audience 1 and their messages.</td><td>Insert a list of vehicles here that is specific to Audience 2 and their messages.</td><td colspan="2">Insert a list of vehicles here that is specific to Audience 1 & 2 and their messages.</td></tr>
<tr><td>Metrics</td><td>Insert a list of metrics for success. These metrics aid in monitoring and evaluating the success of communicating Communication Goal 1 with Audience 1.</td><td>Insert a list of metrics for success. These metrics aid in monitoring and evaluating the success of communicating Communication Goal 1 with Audience 2.</td><td colspan="2">Insert a list of metrics for success. These metrics aid in monitoring and evaluating the success of communicating Communication Goal 2 with Audience 1 & 2.</td></tr>
</table>

This matrix can be expanded or collapsed based on project needs to include as many communication goals, audiences, messages, vehicles, and metrics are necessary to aid in accomplishing a project goal are iterative in nature and ideally occurring throughout the project (e.g., getting feedback on each communication effort as it occurs). It is only by providing mechanisms to learn how a given audience responds that practitioners can refine and improve efforts (Hartman and Lenk 2001; LTER Network 2010; Okaka 2010; USFWS 2013; Ferguson 2015). The framework in Fig. 1 presents the need for both iterative development (e.g., NOAA 2016) and feedback loops throughout the strategic communication process (Table 1).

4 Ecosystem Services in an Ecosystem-Based Management Case Study

Ecosystem services, also referred to as the "benefits of nature," involve the identification and valuation of ecosystem attributes that benefit humans. Changes in *final* ecosystem goods and services (FEGS, or Final EGS), those services that directly benefit people (Landers and Nahlik 2013; DeWitt et al. 2020), can translate into changes in human health and well-being. An example Strategic Communication Matrix illustrates how to implement and track communication goals and messages (Tables 2 and 3) for an ecosystem services application in an EBM context.

In our example, the project goal is to examine and quantify how the supply and benefits of FEGS are delivered to different populations within a community as it relates to informing a specific EBM context (de Jesus Crespo and Fulford 2018). The studies within this goal involve identifying community-based preferences and values for natural resources, conducting quantitative modeling of FEGS and their benefits for human health, and exploring relationships between ecosystem services and human health, all in the context of the EBM decision at hand. This overarching project goal has multiple sub-goals, each having its own set of communication goals (Table 2). For example, the first project sub-goal focused on demonstrating the value of the concept of beneficiaries to stakeholders. To achieve this goal, several separate communications goals have been articulated, each aimed at achieving a different purpose with their respective target audiences (Table 2):

1. Clearly explain the concept of beneficiaries (those that receive the benefits provided by the ecosystem good or service) in EBM decision contexts. This goal is one of information transfer, in particular, supporting the establishment of a shared understanding of terminology.
2. Demonstrate how using the concept could influence existing EBM decision-making processes. This goal is aimed at building support for the work, getting buy-in among target audiences (in this case, the EBM stakeholders), and recognition of the coupled socio-ecological system (Long et al. 2015).
3. Understand what, if any, concepts are currently being used in place of the one proposed. This goal is aimed at expanding scientists' understanding of how their work is being received; the use of scientific knowledge in an integrated management context are both core EBM principles (Long et al. 2015).

Table 2 A hypothetical example of a Strategic Communication Matrix—project goals, sub goals, communication goals, and audience identification

Project goal	Goal X: Examine how the supply and benefits of Final EGS (DeWitt et al. 2020) are delivered to different populations (beneficiaries) through studies involving community-based preferences and values for natural resources, conducting quantitative modeling of Final EGS and their benefits to human health outcomes, and providing a beneficiary perspective on advancing ecosystem to human health relationships.					
Project sub goal	Demonstrate the value of the concept of beneficiaries to stakeholders.					
Communication goals (EBM principle)	Increase awareness, understanding, and appreciation for the concept of beneficiaries in EBM decision contexts. (stakeholder involvement; Interdisciplinarity)			Demonstrate how using the concept of beneficiaries can change decisions/reframe EBM decision contexts. (coupled socio-ecological system)		Elicit and understand what concept(s) and methods are being used currently in place of a beneficiary concept. (use of scientific knowledge; Integrated management)
Audiences	*Agency leadership*	*Scientific collaborators*	*Community-level decision makers*	Scientific collaborators	Regulatory agencies/community-level decision makers	Regulatory agencies/community-level decision makers

The three audiences in *italics* are expanded upon in Table 2. To aid the EBM practitioner, we identified those core EBM principles from Long et al. (2015) most related to each of these example communication goals (in underlined text)

Table 3 A hypothetical example of a Strategic Communication Matrix—Audience identification, messages, vehicles, and metrics

Audiences	Agency leadership	Scientific collaborators	Community-level decision makers
Messages	• Beneficiaries are "the interests of an individual (i.e., person, organization, household, or firm) that drive active or passive consumption and/or appreciation of ecosystem services resulting in an impact (positive or negative) on their welfare." • Identifying beneficiaries allows researchers and decision makers to solicit input from groups that may be affected by changes in ecosystem goods and services, and to target beneficiary groups of interest. • Future work on beneficiaries will be used to identify how the benefits of Final EGS are delivered to different populations through EBM-related studies involving community-based preferences and modeling of Final EGS and their benefits to human health outcomes.		• Beneficiaries are "the interests of an individual (i.e., person, organization, household, or firm) that drive active or passive consumption and/or appreciation of ecosystem services resulting in an impact (positive or negative) on their welfare." • The concept of beneficiaries is useful in EBM decision contexts because it directly connects those who benefit from the environment to the ways in which they benefit
Vehicles	• Internal reports • Newsletters • Presentations to leadership	• Peer-reviewed journal articles • Conference presentations	• Presentations or webinars to community planning groups • Plain language web page
Metrics	• Sharing project results up the chain • Publicizing project results • Continued support for the research to continue	• Citations of work on beneficiaries • Publications of work using the beneficiaries concept	• Recognition and comprehension of the concept of beneficiaries • Agreement that the concept of beneficiaries would be useful in EBM-related decision making

This example is for the first communication goal identified in Table 2

In this example, the first goal focuses on providing foundational information. While the first goal does not directly focus on the science, communications may have to go beyond the bounds of the specific research in order to set the stage properly. The importance of setting the stage is underscored in the ecosystem services example, where the general public may not be aware—or supportive—of ecosystem services per se but does see the value in the concepts (and language) of the "benefits of nature" (Metz and Weigel 2013). The second goal focuses on messages specifically related to the science. The third goal gives an example of two-way communication that should be fostered in a strategic communication plan so that scientists are able to refine their scientific research and messaging to reflect the audience's needs and values. As these are done for a specific EBM context, communication goals may change as the EBM context changes.

For the first of these communication goals, three specific audience types were identified—agency leadership, scientific collaborators, and community-level decision makers (Table 2). In developing the actual plan, these audiences may be broken down further for more specific identification and targeting of messages. In our example Strategic Communication Matrix, we built upon the first communication goal in Table 2 and different messages were identified for the various audiences (Table 3). In this example, for this communication goal, one set of messages was aimed at agency leadership and scientific collaborators, and another set aimed at community-level decision makers. Although there is overlap between the sets of messages, they are differentiated by what they are attempting to achieve. When focusing on agency leadership and scientific collaborators, project communicators hope to build an understanding of the concept of beneficiaries, so it is considered in future planning efforts. With community-level decision makers, the messaging is aimed at showing how the concept of beneficiaries would specifically be useful to them in their work.

All messages are aimed at clearly explaining the concept of beneficiaries, but they are crafted with targeted audiences in mind. In this scenario, incorporating ecosystem services into EBM decision processes is done through the use of a "FEGS approach"—one that focuses on beneficiaries and their role in defining and articulating relevant ecosystem services for a given decision context (DeWitt et al. 2020). This represents only one of several ways to present ecosystem services information into the decision process. Other approaches, such as those focused on capturing the supply of ecosystem services (e.g., Lillebø et al. 2020; Myer and Johnston 2020) might have different communication goals related to identifying stakeholders or recipients of nature's benefits. A strategic communication approach can be useful for capturing information about those different approaches, focused on different target audiences, allowing for different target audiences to learn more to answer their different questions. For a discussion on the overall FEGS approach, the reader is directed to DeWitt et al. (2020); for more discussion on the value and approach of engaging with beneficiaries with this approach, the reader is directed to Sharpe et al. (2020).

Different sets of vehicles and metrics are identified for each targeted audience (Table 3). Even though the same set of messages is being targeted at agency leadership and scientific collaborators, the strategic communication plan recognizes that peer-reviewed journal articles and conference presentations that are successful at spreading their message to scientific collaborators would be far less effective in communicating to agency leadership. Similarly, success in communicating the concept of beneficiaries for consideration in future work looks very different to these two audiences. For agency leadership, successful communication is measured, in part, by continued support for the science, whereas successful communication with scientific collaborators is measured by seeing them use the beneficiaries concept in their own work on EBM decisions. When designing metrics for specific communication goals, the strategic communication plan approach asks the planner to consider what outcomes they are hoping to see overall and with each targeted audience.

It is easy to see how the complexity of strategic communication planning can increase exponentially when considering a large project in its entirety, underscoring the importance of a strategic communication plan and the value of a well-designed Strategic Communication Matrix. In our example, communicating the science on ecosystem services and its role with targeted beneficiaries within a larger EBM decision context through the use of a Strategic Communication Matrix allows advances in strategic communications for: (1) considering the full set of communication goals in a coordinated fashion; (2) finding opportunities for coordination across project goals; (3) finding ways to combine messaging associated with separate communication goals aimed at the same audience; and (4) leveraging limited resources into a more focused and outcome-driven communication effort.

5 Conclusion

Traditionally, the work of strategic communications has been done by individuals and organizations other than those conducting the research and making the decisions. We suggest that research projects and management plans benefit when scientists and decision makers proactively engage in the communication process from the beginning. This engagement allows them to focus the science and decisions in ways that resonate with targeted audiences and to share the work in ways more likely to have an impact; this aligns strongly with the core EMB principle of recognizing the coupled socio-ecological nature of the system. Although scientists and managers may have not traditionally been expected or trained to participate in communication work, the strategic communication framework laid out here (the three interlinked pillars of *message*, *audience*, and *vehicle*, resting on a common foundation of *communication goals*) can be used as a template. To be most effective, there needs to be buy-in, from an organizational perspective, on the effort required to build and implement a plan. The core EBM principle of stakeholder involvement calls for development of a strategic communication element in an EBM program.

An example in the interdisciplinary field of ecosystem services science was presented to demonstrate how to develop a clear and simple matrix to provide a visual roadmap for communication and to help coordinate efforts across a project. The use of a generalizable framework and a strategic communication matrix allows science and decision communications to be pursued using a systems-thinking approach. The EBM core principle of the use of scientific knowledge calls for practitioners to share their results or decisions with fellow scientists and stakeholders, garner support for future work from funding institutions, or encourage external decision makers to incorporate findings into decision-making processes. All scientists, decision makers, and stakeholders have project goals related to multiple audiences. The framework and matrix laid out here provide a pathway to help those scientists, decision makers, and stakeholders take a strategic approach in using communication efforts to help achieve their communication goals for their audiences. This approach expands scientific communication from the old deficit model, in which the primary goal was to make information available, towards a model that recognizes communication can be an invaluable tool in achieving science and management goals.

While the idea of strategic communication exists in the literature, the practice seems to be underutilized by scientists, decision makers, and stakeholders in general, including EBM practitioners. Those scientists, decision makers, and stakeholders who might be using the concept of strategic communication are not discussing their efforts in the peer-reviewed literature. Those that are publishing it in the literature are typically proposing strategic communication plans for future projects, rather than following up with the results of implementation and analysing whether or not communication improved. Therefore, our assumption of a lack of strategic communication plans/frameworks used among scientists, decision makers, and stakeholders may be overly simplistic. We hope that this chapter can encourage EBM practitioners to be more open/transparent about their science, decision making, management, and communication efforts in order to help the scientific community better communicate results. For strategic communication efforts that may have already taken place, but are unacknowledged in the peer-reviewed literature, a useful next step could be an effort to survey scientists, decision makers, and stakeholders to examine the question of their communication practices. This approach could give a more accurate picture of the state of strategic communication work being done by EBM practitioners.

Disclaimer This chapter has been subjected to Agency review and has been approved for publication. The views expressed in this paper are those of the author(s) and do not necessarily reflect the views or policies of the U.S. Environmental Protection Agency.

Appendix

Search results from both the literature (top table) and government and ENGO reports (bottom table)

Citation	Elements (G = Goals; M = Message; A = Audience; V = Vehicle; T = Two-way Communication; S = Metrics for Success)
Bronson, D. (2004). *Engaging Canadians: Building Professional Communications in Parks Canada. Communicating Protected Areas* pp. 61–68.	G, M, A, V
Day, B.A. and M.C. Monroe. (2000). *Environmental Education and Communication for a Sustainable World: Handbook for International*	G, M, A, V

(continued)

Citation	Elements (G = Goals; M = Message; A = Audience; V = Vehicle; T = Two-way Communication; S = Metrics for Success)
Practitioners. Academy for Educational Development, Washington, DC. 141 pp.	
Dayer, A. and R.M. Meyers. (2012). *Appalachian Mountains Joint Venture Strategic Communications Plan 2013–2017*. Communications Report 2012–01, Skaneateles, NY. 62 pp.	G, M, A, V, T
DeLauer, V., S. Ryan, I. Babb, P. Taylor, and P. Di-Bona. (2012). *Linking Science to Management and Policy through Strategic Communication. Advancing an Ecosystem Approach in the Gulf of Maine*. Stephenson, R.L., J.H. Annala, J.A. Runge and M. Hall-Arber (eds.). American Fisheries Society, Symposium 79:89–101.	M, A, V, T
Driscoll, C.T., K.F. Lambert, F.S. Chap–in III, D.J. Nowak, T.A. Spies, F.J. Swanson, D.B. Kittredge, and C.M. Hart. (2012). Science and society: The role of long-term studies in environmental stewardship. *BioScience, 62*(4):354–366.	G, M, A, V, T, S
Ekebom, J., J. Jäänheimo, J. Reker, M. Kindström, C. Lindblad, A. Mattisson, A. Sandstrom, and V. Jermakovs. (2008). *Towards Marine Spatial Planning in the Baltic Sea. BALANCE Technical Summary Report 4*(4).	G, M, A, V, S
Ferguson, D.B. (2015). *Linking Environmental Research and Practice: Lessons from the Integration of Climate Science and Water Management in the Western United States*. 2015 AGU Fall Meeting.	M, T
Goldberg, J., N. Marshall, A. Birtles, P. Case, E. Bohensky, M. Curnock, M. Gooch, H. Parry-Husbands, P. Pert, R. Tobin, C. Villani, and B. Visperas. (2016). Climate change, the Great Barrier Reef and the response of Australians. *Palgrave Communications, 2*:15046.	M, A, V, T
Groffman, P.M., C. Stylinski, M.C. Nisbet, C.M. Duarte, R. Jordan, A. Burgin, M.A. Previtali, and J. Coloso. (2010). Restarting the conversation: Challenges at the interface between ecology and society. *Frontiers in Ecology and the Environment, 8*(6):284–291.	A, V, T
Halpern, B.S., J. Diamond, S. Gaines, S. Gelcich, M. Gleason, S. Jennings, S. Lester, A. Mace, L. McCook, K. McLeod, N. Napoli, K. Rawson, J. Rice, A. Rosenberg, M. Ruckelshaus, B. Saier, P. Sandifer, A. Sholtz, and A Zivian. (2012). Near-term priorities for the science, policy and practice of Coastal and Marine Spatial Planning (CMSP). *Marine Policy, 36*(1):198–205.	M, A, V, T

(continued)

Citation	Elements (G = Goals; M = Message; A = Audience; V = Vehicle; T = Two-way Communication; S = Metrics for Success)
Hesselink, F., W. Goldstein, P.P. van Kempen, T. Garnett, and J. Dela. (2007). *Communication, Education, and Public Awareness (CEPA): A toolkit for National Focal Points and NBSAP Coordinators.* Secretariat of the Convention on Biological Diversity and IUCN, Montreal, Canada. 310 pp.	M, A, V, T
Kellam, D. (2004). *New Hampshire Estuaries Project: Strategic Communication Plan.* PREP publications. 17 pp.	G, M, A
Lawas, T.P., M.S.C. Tirol, V.R. Cardenas, and S.B. Jamias. (2010). Communication resource mapping for coastal resources management of Barangay Malabrigo, Lobo, Batangas, Philippines. *Journal of Environmental Science and Management, 12*(2):38–56.	G, M, A, V
Leong, K.M., K.A. McComas, and D.J. Decker (2008). Formative coorientation research: A tool to assist with environmental decision making. *Environmental Communication, 2*(3):257–273.	A, S
Long-Term Ecological Research (LTER) Network. (2010). *LTER Strategic Communication Plan: Bridging to Broader Audiences.* 55 pp.	G, M, A, V, S
Okaka, W. (2010). Developing regional communications campaigns strategy for environment and natural resources management policy awareness for the East African community. *Research Journal of Environmental and Earth Sciences* 2(2):106–111.	G, M, A, V, T
Smith, D. C., Smith, A. D. M., Dichmont, C., Steele, W., & Webb, H. (2014). *Towards a strategic relationship between CSIRO and FRDC.* Fisheries Research and Development Corporation. 36 pp.	A, V, S
Thompson, J.S. (2006). Strategic Communication in Community-Based Fisheries and Forestry: A Case from Cambodia. In G. Bessette (Ed.), *People, Land, and Water: Participatory Development Communication for Natural Resource Management.* International Development Research Centre (IDRC), Ottawa, Canada.	G, M, A, V, T, S
Timm, K., R. Hum, and M. Duckenmiller. (2016). *Using Communication Theory and Strategy to Communicate Science and Build Stakeholder Relationships in the Arctic.* 10 pp.	G, M, A, V, T
Velasco, M.T.H. (2006). *Management and Implementation of Communication Programs for Natural Resources Management in Agriculture.* In: Information and Communication for Natural Resource Management in Agriculture: A Training Sourcebook. College of Development Communication, University of the Philippines Los Banos pp. 85–96.	G

(continued)

Citation	Elements (G = Goals; M = Message; A = Audience; V = Vehicle; T = Two-way Communication; S = Metrics for Success)
Vidal, R.M. and G. Lucia (2004). *Strategic Communication Planning for a National System of Protected Areas, Mexico. Communicating Protected Areas*. Commission on Education and Communication, Switzerland and Cambridge, UK, pp. 69–86.	A, V, T
Wiggill, M.N. (2014). Communicating for organizational legitimacy: The case of the Potchefstroom Fire Protection Association. *Public Relations Review 40*(2):315–327.	G, A
Winterfeldt, D.V. (2012). Bridging the gap between science and decision making. *Proceedings of the National Academy of Sciences 110*(3):14055–14061.	M, A, T

Citations are included and strategic communication elements are identified for each of the citations. This literature review identified the extent to which strategic communication is studied and incorporated into projects within the field of natural sciences, and allowed published efforts to organize, develop, and utilize a strategic communication plan in ecosystem-based management to be characterized

Citation	Elements (G = Goals; M = Message; A = Audience; V = Vehicle; T = Two-way communication; S = Metrics for Success)
Africa Biodiversity Collaborative Group (ABCG). (2004). *Working together to help conserve Africa's biodiversity—revised communications strategy*. 13 pp.	G, M, A, V
International Union for Conservation of Nature. (2016). *Communications strategy Sri Lanka—global forest governance project: Strengthening voices for better choices*. 38 pp.	G, M, A, V, T
National Academies of Sciences, Engineering, and Medicine. (2006). *Review of the Marine Recreational Information Program (MRIP)*. The National Academies Press, Washington, DC. https://doi.org/10.17226/24640.	M, A, V, T, S
National Marine Sanctuaries. (2003). *Cordell Bank, Gulf of the Farralones and Monterey Bay National Marine Sanctuaries—strategic communication plan*. 10 pp.	G, M, A, V
National Oceanic and Atmospheric Administration (NOAA) Office of Chief Information Officer. (2009) *Communications plan: Reliable and consistent information exchange across NOAA's information technology community*. 21 pp.	G, M, A, V, S

(continued)

Citation	Elements (G = Goals; M = Message; A = Audience; V = Vehicle; T = Two-way communication; S = Metrics for Success)
National Oceanic and Atmospheric Administration (NOAA) Fisheries. (2016) *NOAA Fisheries communications implementation plan for the Pacific Islands.* U.S. Department of Commerce, National Oceanic and Atmospheric Administration, National Marine Fisheries Services. 8 pp.	G, M, A, V, T
National Park Service (NPS). (2016). *Wildland fire communication plan centennial edition: 2016–2020.* 18 pp.	G, M, A, T, S
Sea Grant (2003) *Positioning Sea Grant—An Integrated National Communications Plan.* 2003–2006. 24 pp.	G, M, A, V, S
US Fish and Wildlife Service. (2011). *North Atlantic LCC DRAFT communications strategy.* 14 pp.	G, M, A, V, T
US Fish and Wildlife Service. (2013). *Desert Landscape Conservation Cooperative communications plan.* 13 pp.	G, M, A, V
US Fish and Wildlife Service. (2014). *Climate change communications and engagement strategy for the National Wildlife Refuge System.* 20 pp.	G, M, A, V, S
US Fish and Wildlife Service. (2016). *National Wildlife Refuge System Communications Strategy—final,* February 1, 2016. 25 pp.	G, M, A, V, T

References

Africa Biodiversity Collaborative Group (ABCG). (2004). Working together to help conserve Africa's biodiversity—Revised communications strategy. In Department of Environmental Affairs, *South Africa's National Biodiversity Strategy and Action Plan—2015–2025.* Retrieved October 21, 2019, from https://www.environment.gov.za/sites/default/files/docs/publications/SAsnationalbiodiversity_strategyandactionplan2015_2025.pdf.

Barker, S. (2006). Environmental communication in context. *Frontiers in Ecology and the Environment, 4*(6), 328–329.

Braus, J. (2009). *Tools of engagement: How education and other social strategies can engage people in conservation action.* Free Choice Learning and the Environment. Lanham, MD: AltaMira Press, pp. 87–104.

Bronson, D. (2004). *Engaging Canadians: Building professional communications in parks Canada* (pp. 61–68). Gland, Switzerland and Cambridge UK: IUCN Commission on Education and Communication.

Bubela, T., Nisbet, M. C., Borchelt, R., Brunger, F., Critchley, C., Einsiedel, E., & Jandciu, E. W. (2009). Science communication reconsidered. *Nature Biotechnology, 27*(6), 514–518.

Cabanero-Verzosa, C., & Elaheebocus, N. (2008). *Strategic communication in early childhood development programs: The case of Uganda* (pp. 331–351). Washington, DC: The International Bank for Reconstruction and Development/The World Bank.

Day, B. A., & Monroe, M. C. (2000). *Environmental education and communication for a sustainable world: Handbook for international practitioners*. Washington, DC: Academy for Educational Development.

Dayer, A., & Meyers, R. M. (2012). *Appalachian Mountains joint venture strategic communications plan 2013–2017*. Communications Report 2012-01, Skaneateles, NY.

de Bruin, W. B., & Bostrom, A. (2012). Assessing what to address in science communication. *Proceedings of the National Academy of Sciences, 110*(3), 14062–14068.

de Jesus Crespo, R., & Fulford, R. (2018). Eco-health linkages: Assessing the role of ecosystem goods and services on human health using causal-criteria analysis. *International Journal of Public Health, 63*(1), 1–12.

DeLauer, V., Ryan, S., Babb, I., Taylor, P., & Di-Bona, P. (2012). Linking science to management and policy through strategic communication. In R. L. Stephenson, J. H. Annala, J. A. Runge, & M. Hall-Arber (Eds.), *Advancing an ecosystem approach in the Gulf of Maine (Symposium), 79*:89–101. Bethesda, MD: American Fisheries Society.

DeWitt, T. H., Berry, W. J., Canfield, T. J., Fulford, R. S., Harwell, M. C., Hoffman, J. C., Johnston, J. M., Newcomer-Johnson, T. A., Ringold, P. L., Russel, M. J., Sharpe, L. A., & Yee, S. J. H. (2020). The final ecosystem goods and services (FEGS) approach: A beneficiary-centric method to support ecosystem-based management. In T. O'Higgins, M. Lago, & T. H. DeWitt (Eds.), *Ecosystem-based management, ecosystem services and aquatic biodiversity: Theory, tools and applications* (pp. 127–148). Amsterdam: Springer.

Driscoll, C. T., Lambert, K. F., Chapin, F. S., III, Nowak, D. J., Spies, T. A., Swanson, F. J., et al. (2012). Science and society: The role of long-term studies in environmental stewardship. *Bioscience, 62*(4), 354–366.

Ekebom, J., Jäänheimo, J., Reker, J., Kindström, M., Lindblad, C., Mattisson, A., et al. (2008). Towards marine spatial planning in the Baltic Sea. *Balance Technical Summary Report, 4*(4). Retrieved October 21, 2019, from http://balance-eu.org/.

Ferguson, D. B. (2015). *Linking environmental research and practice: Lessons from the integration of climate science and water management in the western United States*. 2015 AGU Fall Meeting, San Francisco.

Goldberg, J., Marshall, N., Birtles, A., Case, P., Bohensky, E., Curnock, M., et al. (2016). Climate change, the great barrier reef and the response of Australians. *Palgrave Communications, 2*, 15046.

Groffman, P. M., Stylinski, C., Nisbet, M. C., Duarte, C. M., Jordan, R., Burgin, A., et al. (2010). Restarting the conversation: Challenges at the interface between ecology and society. *Frontiers in Ecology and the Environment, 8*(6), 284–291.

Hallahan, K., Holtzhausen, D., van Ruler, B., Vercic, D., & Sriramesh, K. (2007). Defining strategic communication. *International Journal of Strategic Communication, 1*, 3–35.

Halloran, R. (2007). Strategic communication. *Parameters, 37*(3), 4.

Halpern, B. S., Diamond, J., Gaines, S., Gelcich, S., Gleason, M., Jennings, S., et al. (2012). Near-term priorities for the science, policy and practice of coastal and marine spatial planning (CMSP). *Marine Policy, 36*(1), 198–205.

Hansen, J., Holm, L., Frewer, L., Robinson, P., & Sandøe, P. (2003). Beyond the knowledge deficit: Recent research into lay and expert attitudes to food risks. *Appetite, 41*(2), 111–121.

Hartman, J., & Lenk, M. M. (2001). Strategic communication capital as an intangible asset. *International Journal on Media Management, 3*(3), 147–153.

Hesselink, F., Goldstein, W., van Kempen, P. P., Garnett, T., & Dela, J. (2007). *Communication, education, and public awareness (CEPA): A toolkit for national focal points and NBSAP coordinators*. Montreal: Secretariat of the Convention on Biological Diversity and IUCN.

Hobbs, R. (2006). Overcoming barriers to effective public communication of ecology. *Frontiers in Ecology and the Environment, 4*(9), 496–497.

International Union for Conservation of Nature (IUCN). (2016). *Communications strategy Sri Lanka—Global forest governance project: Strengthening voices for better choices*. Forest Conservation Programme, Columbo, pp 38.

Jones, K., Baker, P., Doyle, J., Armstrong, R., Pettman, T., & Waters, E. (2013). Increasing the utility of systematic reviews findings through strategic communication. *Journal of Public Health, 35*(2), 345–349.

Kellam, D. (2004). *New Hampshire estuaries project: Strategic communication plan* (227p.). Durham, New Hampshire: University of New Hampshire, PREP Publications.

Kellstedt, P. M., Zahran, S., & Vedlitz, A. (2008). Personal efficacy, the information environment, and attitudes toward global warming and climate change in the United States. *Risk Analysis, 28*, 113–126.

Landers, D. H., & Nahlik, A. M. (2013). *Final ecosystem goods and services classification system (FEGS-CS).* EPA/600/R-13/ORD-004914. Washington, DC: U.S. Environmental Protection Agency, Office of Research and Development.

Lawas, T. P., Tirol, M. S. C., Cardenas, V. R., & Jamias, S. B. (2010). Communication resource mapping for coastal resources management of Barangay Malabrigo, Lobo, Batangas, Philippines. *Journal of Environmental Science and Management, 12*(2), 38–56.

Leong, K. M., McComas, K. A., & Decker, D. J. (2008). Formative coorientation research: A tool to assist with environmental decision-making. *Environmental Communication, 2*(3), 257–273.

Liang, Y., Kee, K. F., & Henderson, L. K. (2018). Toward an integrated model of strategic environmental communication: Advancing theories of reactance and planned behavior in a water conservation text. *Journal of Applied Communication Research, 46*(2), 135–154.

Lillebø, A. I., Teixeira, H., Martínez-López, J., Genua-Olmedo, A., Marhubi, A., Delacámara, G., Mattheiß, V., Strosser, P., O'Higgins, T., & Nogueira, A. A. J. (2020). Mitigating negative unintended impacts on biodiversity in the Natura 2000 Vouga estuary (Ria de Aveiro, Portugal). In T. O'Higgins, M. Lago, & T. H. DeWitt (Eds.), *Ecosystem-based management, ecosystem services and aquatic biodiversity: Theory, tools and applications* (pp. 461–498). Amsterdam: Springer.

Long, R. D., Charles, A., & Stephenson, R. L. (2015). Key principles of marine ecosystem-based management. *Marine Policy, 57*, 53–60.

Long-Term Ecological Research (LTER) Network. (2010). *LTER strategic communication plan: Bridging to broader audiences*. Albuquerque, New Mexico.

Mattheiß V, Strosser P, Krautkraemer A, Charbonnier C, McDonald H, Röschel L, et al. (2018). *Evaluation of ecosystem-based management responses in case studies: AQUACROSS Deliverable 8.2.* European Union's Horizon 2020 Framework Programme for Research and Innovation Grant Agreement No. 642317. Retrieved October 20, 2019, from www.aquacross.eu.

Metz, D., & Weigel, L. (2013). *Language of conservation 2013: Updated recommendations on how to communicate effectively to build support for conservation*. The Nature Conservancy. Retrieved October 20, 2019, from https://www.conservationgateway.org/Files/Pages/language-conservation-mem.aspx.

Mortenius, H. (2014). Creating an interest in research and development as a means of reducing the gap between theory and practice in primary care: An interventional study based on strategic communication. *International Journal of Environmental Research and Public Health, 11*(9), 8689–8708.

Murphy, D. M. (2008). *The trouble with strategic communication(s)*. Army War College, Center for Strategic Leadership, Carlisle Barracks, Pennsylvania.

Myer, M., & Johnston, J. M. (2020). Models and mapping tools to inform resilience planning after disasters: A case study of hurricane Sandy and Long Island ecosystem services. In T. O'Higgins, M. Lago, & T. H. DeWitt (Eds.), *Ecosystem-based management, ecosystem services and aquatic biodiversity: Theory, tools and applications* (pp. 417–430). Amsterdam: Springer.

National Academies of Sciences, Engineering, and Medicine (NASEM). (2006). *Review of the marine recreational information program (MRIP)*. The National Academies Press, Washington, DC. https://doi.org/10.17226/24640.

National Academies of Sciences, Engineering, and Medicine (NASEM). (2017). *The science of science communication III: Inspiring novel collaborations and building capacity. Proceedings of a colloquium.* The National Academies Press, Washington, DC. https://doi.org/10.17226/24958.

National Marine Sanctuaries. (2003). *Strategic communication plan. Cordell Bank, Gulf of the Farallones, and Monterey Bay.* pp 10. Retrieved October 20, 2019, from https://nmssanctuaries.blob.core.windows.net/sanctuaries-prod/media/archive/jointplan/cc_outreach/cc_nmsp_com_temp.pdf.

National Oceanic and Atmospheric Administration (NOAA) Fisheries. (2016). *NOAA Fisheries communications implementation plan for the Pacific Islands* pp. 8. National Oceanic and Atmospheric Administration, National Marine Fisheries Services.

National Oceanic and Atmospheric Administration (NOAA) Office of Chief Information Officer. (2009). *Communications plan: Reliable and consistent information exchange across NOAA's information technology community*. pp. 21.

National Park Service (NPS). (2016). *Wildland fire communication plan.* Centennial edition: 2016–2020. pp. 18.

Newton, A., & Elliott, M. (2016). A typology of stakeholders and guidelines for engagement in transdisciplinary, participatory processes. *Frontiers in Marine Science, 3*, 230.

O'Higgins, T., DeWitt, T. H., & Lago, M. (2020). Using the concepts and tools of social ecological systems and ecosystem services to advance the practice of ecosystem-based management. In T. O'Higgins, M. Lago, & T. H. DeWitt (Eds.), *Ecosystem-based management, ecosystem services and aquatic biodiversity: Theory, tools, and practice* (pp. 3–14). Amsterdam: Springer.

Odugbemi, S., & Mozammel, M. (2005). *With the support of multitudes: Using strategic communication to fight poverty through PRSPs.* London: World Bank Publications.

Okaka, W. (2010). Developing regional communications campaigns strategy for environment and natural resources management policy awareness for the east African community. *Research Journal of Environmental and Earth Sciences, 2*(2), 106–111.

Sea Grant. (2003). *Positioning Sea Grant: An integrated national communications plan 2003-2006.* pp. 24.

Sharpe, L., Hernandez, C., & Jackson, C. (2020). Prioritizing stakeholders, beneficiaries and environmental attributes: A tool for ecosystem-based management. In T. O'Higgins, M. Lago, & T. H. DeWitt (Eds.), *Ecosystem-based management, ecosystem services and aquatic biodiversity: Theory, tools and applications* (pp. 189–212). Amsterdam: Springer

Smith, D. C., Smith, A. D. M., Dichmont, C., Steele, W., & Webb, H. (2014). *Towards a strategic relationship between CSIRO and FRDC.* Fisheries Research and Development Corporation, Australian Capital Territory.

Thompson, J. S. (2006). *Strategic communication in community-based fisheries and forestry: A case from Cambodia.* Ottawa: International Development Research Centre.

Timm, K., Hum, R., & Duckenmiller, M. (2016). *Using communication theory and strategy to communicate science and build stakeholder relationships in the Arctic.* 2016 Arctic Observing Summit, Fairbanks, Alaska.

Turbek, S. P., Chock, T. M., Donahue, K., Havrilla, C. A., Oliverio, A. M., Polutchko, S. K., et al. (2016). Scientific writing made easy: A step-by-step guide to undergraduate writing in the biological sciences. *Bulletin of the Ecological Society of America, 97*(4), 417–426.

U.S. Fish and Wildlife Service (USFWS). (2011). *North Atlantic LCC draft communications strategy,* pp 14. Retrieved October 9, 2018, from https://www.fws.gov/northeast/science/pdf/Handout_15_DRAFT_North_Atlantic_LCC_Comms_Plan_2011_0420.pdf.

U.S. Fish and Wildlife Service (USFWS). (2013). *Desert Landscape Conservation Cooperative communications plan,* pp 13. Retrieved October 20, 2019, from https://lccnetwork.org/resource/desert-lcc-communications-plan.

U.S. Fish and Wildlife Service (USFWS). (2014) *Climate change communications and engagement strategy for the National Wildlife Refuge System,* pp 20. Retrieved October 20, 2019, from https://www.fws.gov/refuges/vision/pdfs/ClimateChangeEngagementStrategyFinal.pdf.

U.S. Fish and Wildlife Service (USFWS). (2016). *National Wildlife Refuge System Communications Strategy—Final, February 1, 2016,* pp 25. Retrieved October 20, 2019, from https://www.fws.gov/refuges/vision/pdfs/NWRSCommunicationsStrategy.pdf.

Velasco, M. T. H. (2006). Management and implementation of communication programs for natural resources management in agriculture. In Food and Agricultural Organization (Ed.), *Information and communication for natural resource management in agriculture: A training sourcebook* (pp. 85–96). Los Banos: University of the Philippines.

Williams, K. C., & Hoffman, J. C. (2020). Remediation to restoration to revitalisation: Ecosystem-based management to support community engagement at clean-up sites in the Laurentian Great Lakes. In T. O'Higgins, M. Lago, & T. H. DeWitt (Eds.), *Ecosystem-based management, ecosystem services and aquatic biodiversity: Theory, tools and applications* (pp. 543–560). Amsterdam: Springer

Winterfeldt, D. V. (2012). Bridging the gap between science and decision-making. *The Proceedings of the National Academy of Sciences, 110*(3), 14055–14061.

Yee, S., Cicchetti, G., DeWitt, T. H., Harwell, M. C., Jackson, S. K., Pryor, M., Rocha, K., Santavy, D. L., Sharpe, L., & Shumchenia, E. (2020). The ecosystem services gradient: A descriptive model for identifying thresholds of meaningful change. In T. O'Higgins, M. Lago, & T. H. DeWitt (Eds.), *Ecosystem-based management, ecosystem services and aquatic biodiversity: Theory, tools and applications* (pp. 291–308). Amsterdam: Springer.

13

Ecosystem-Based Management for more Effective and Equitable Marine Protected Areas

Hugh McDonald, Helene Hoffman, Adriana Ressurreição, Lina Röschel, Holger Gerdes, Manuel Lago, Ben Boteler, Keighley McFarland, and Heliana Teixeira

Abstract Marine Protected Areas (MPAs) are increasingly employed as a tool to protect Europe's swiftly declining marine biodiversity. However, despite increasing coverage, MPA effectiveness and equity is considered highly variable. Concurrently, Ecosystem-Based Management (EBM)—that is, management that aims to protect, restore, or enhance the resilience and sustainability of an ecosystem to ensure sustainable flows of ecosystem services and conserve its biodiversity—is growing in prominence. We applied EBM in the Faial-Pico Channel, a 240 km^2 MPA in the Azores, Portugal, to assess whether EBM can protect biodiversity whilst meeting diverse stakeholder and policy goals. Collaborating with local stakeholders and policy-makers, this chapter documents the steps of EBM: identifying integrative policy and stakeholder objectives, understanding the social-ecological system, scenario development, and identification and evaluation of EBM measures and policies. We find that stakeholder co-creation and collaboration is a key strength of EBM and should be strengthened in the Faial-Pico Channel. We find that local stakeholders support effective and equitable EBM of MPAs by clearly identifying challenges and priorities, co-creating solutions, providing low-cost knowledge and expertise, and

H. McDonald (✉) · L. Röschel · H. Gerdes · M. Lago · K. McFarland
Ecologic Institute, Berlin, Germany
e-mail: hugh.mcdonald@ecologic.eu

H. Hoffman
Zukunft – Umwelt – Gesellschaft (ZUG) gGmbH, Berlin, Germany

A. Ressurreição
CCMAR - Centre of Marine Sciences, Faro, Portugal

B. Boteler
Institut für transformative Nachhaltigkeitsforschung /Institute for Advanced Sustainability Studies e.V. (IASS), Potsdam, Germany

H. Teixeira
DBIO & CESAM, Universidade de Aveiro, Aveiro, Portugal

through ongoing monitoring, enforcement, and evaluation of the impact of management.

Lessons Learned

- Stakeholder engagement and participation supports long-term sustainable protection of biodiversity and equitable and effective management of MPAs
- Stakeholders can contribute at each stage of EBM: identifying social objectives, understanding the social-ecological system, identifying an EBM plan, and evaluating impact
- Stakeholders contribute by clearly identifying challenges and priorities, co-creating solutions, and generally by providing low-cost knowledge and expertise, as well as increasing societal acceptance.
- EBM is an appropriate framework for increasing effectiveness and efficiency of MPAs

Needs to Advance EBM

- Clear guidance on how to effectively engage stakeholders at each stage of the EBM process
- EBM has high environmental and socio-economic data demands. Guidance on how to apply EBM in low-data environments would support uptake.

1 Introduction

Globally, marine biodiversity declined by 49% between 1970 and 2012 (Tanzer et al. 2015). This rapid decline threatens the resilience of marine ecosystems and their ability to sustainably produce ecosystem services that humans depend on to survive and thrive (Cardinale et al. 2012). Policy makers have turned to Marine Protected Areas (MPAs) as a key tool to reverse marine biodiversity loss (Gill et al. 2017). Indeed, globally, the Convention of Biological Diversity's Aichi Target 11 and the UN Sustainable Development Goal 14 aim to "efficiently and equitably" protect 10% of coastal and marine areas within MPAs (UN 2016; Secretariat of the CBD 2011). However, the efficacy and equity of MPAs is questioned and considered highly variable (Gill et al. 2017).

Researchers, policy-makers, and environment managers are increasingly interested in the ecosystem-based management concept as a promising approach to more effectively, efficiently, and equitably manage aquatic ecosystems (see, e.g., Delacámara et al. 2020). Ecosystem-based management (EBM) is a principle-based management approach that aims to protect, restore, or enhance the resilience and sustainability of an ecosystem to ensure sustainable flows of ecosystem services and conserve its biodiversity (see Gómez et al. 2017; Rouillard et al. 2017). While there is increasing interest in ecosystem-based management, there are still relatively few practical examples worldwide, especially as applied to Marine Protected Areas.

This chapter presents a summarised excerpt from the more detailed AQUACROSS project case study report (McDonald et al. 2018), documenting the application ecosystem-based management (EBM) in the richly biodiverse Faial-Pico Channel, a 240 km^2 Marine Protected Area in the Azores, Portugal. We include it in this book as it illustrates in an integrated manner how each of the concepts developed in the AQUACROSS project can be combined to practically apply EBM to manage biodiversity. To apply ecosystem-based management, we collaborated with local stakeholders and policy-makers and follow the AQUACROSS Assessment Framework (Gómez et al. 2017). The chapter aims to: (1) demonstrate how the AQUACROSS Assessment Framework can be followed to practically apply ecosystem-based management; (2) identify how ecosystem-based management can protect biodiversity and improve social welfare in the specific context of the Faial-Pico Channel social-ecological system, and (3) understand how ecosystem-based management generally can support existing MPAs to become more effective and equitable.

2 The Faial-Pico Channel Marine Protected Area: Case Study Context

The Faial-Pico Channel is rich in biodiversity, and its complex of habitats, species, and ecological processes is recognised as one the most diverse and representative complex of habitats in the Azores archipelago (MarBEF Data System 2006; OSPAR Commission 2016). However, despite a 30 year history of increasing international, Azorean, and local protection for the area (Abecasis et al. 2015), biodiversity in the MPA continues to be lost, as indicated by falling population indices of target coastal species in the channel (Afonso et al. 2014).

Numerous human activities in the Channel place pressure on the ecosystem, especially fishing and tourism. Fishers and tourism operators (including diving operators) value the biodiversity hotspots within the Channel, but have different objectives for how they should be managed. It is important to balance these objectives, as both tourism and fisheries are important local industries for the 30,000 people who live on Channel's neighbouring islands. Commercial fisheries are a historically important driver of the local economy, and still employ 1.5–3.2% of the total working population (Ojamaa 2015; Statistics Portugal 2017).[1] Tourism has swiftly become central to the local economy, with the number of tourist nights in the Azores tripling from 1995–2015; in 2016, tourists spent 228,000 nights on the islands (SREA 2017). As one indicator of the sector's importance, in 2015, the

[1]Statistics Portugal: own calculations, Fishermen registered at 31 December 2015 in Azores. This compares to a rate of 0.6% for Portugal.

accommodation sector directly employed 2% of the total Azorean workforce.[2] This has supported economic growth, with GDP per capita growing at 2.7% per year since 2000 (currently at €16,000).[3]

The increased demand by tourists (and tourism providers) for eco-tourism in the Channel and declining biodiversity is leading to conflict between commercial fishers and other stakeholders as to how the Channel should be managed (AQUACROSS 2017). Managing the Channel is complicated by multi-level and overlapping responsibilities, with policy development and enforcement split across the local-level Nature Park of Faial and Nature Park of Pico, both under the mandate of the Regional Directorate for the Environment (Direcção Regional do Ambiente, DRA). Other relevant managing authorities include the Azores-level Regional Directorate for Sea Affairs (DRAM) and the Regional Directorate for Fisheries (Direcção Regional das Pescas, DRP), all who must consider local (i.e., Faial and Pico Island), Azorean, Portuguese, and EU policy targets.

In response to falling local biodiversity and to balance stakeholder competition for space, local authorities have extended Marine Protected Area to cover the Faial-Pico Channel. Parts of the Channel have been protected under local policy as a MPA since 1980, with this extended under NATURA 2000 protection in 1995, and OSPAR coverage in 2006, and consolidated under new Azorean Island National Park regulation in 2007 (Abecasis et al. 2015).

Dovetailing this government push for increased biodiversity protection, bottom-up stakeholder demands have driven Faial-Pico Channel management, resulting in an increase in stakeholder participation in MPA management. An early, nearby example was the Condor Seamount, which in 2010 following a stakeholder participatory process was designated a temporary MPA to facilitate marine research (Ressurreição and Giacomello 2013; Ressurreição et al. 2017). Following this and other Azorean examples, local government and scientists supported Faial-Pico tourism operators when they published an open letter calling for an extension of MPA coverage in the Channel to promote non-extractive recreational activities, instigating two stakeholder meetings to gather input on MPA management revisions. While these workshops lacked sufficient representatives from the tourism sector and no recreational fishing representatives, they represent more inclusive management of the MPA by local authorities and the resulting change in law (Ordinance 53 2016) increased protection for some of the high biodiversity zones in the Channel. Within this context—of falling biodiversity, increased competition for the Channel, and at the same time more inclusive MPA management—our application of ecosystem-based management aims to build on previous policies and approaches and identify how local authorities and stakeholders can increase the effectiveness and equity of Faial-Pico Channel MPA management.

---

[2]Eurostat: own calculations, SBS data by NUTS 2 regions and NACE Rev. 2 (2014–2016). This compares to a rate of 2.3% for Portugal.

[3]EUROSTAT: GDP at current market prices by NUTS2 region.

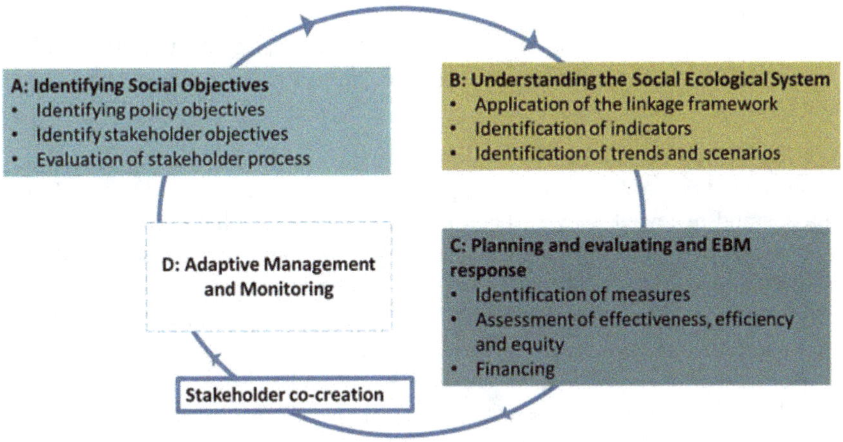

Fig. 1 AQUACROSS Assessment Framework, as applied in this case study

3 Methodology

To apply ecosystem-based management, we followed the AQUACROSS Assessment Framework (Gómez et al. 2017). As shown in Fig. 1, we applied this in three overlapping steps.[4] Below, we describe the different methodologies applied at each step, as well as how stakeholder co-creation supported the whole process.

3.1 Stakeholder Co-creation

Common to our methodology at all steps was co-creation with local stakeholders. Given EBM's ambition to reflect the complexity and multifunctionality of the Faial-Pico Channel, diverse representative stakeholder participation was required. We mapped stakeholder interest and influence, using snowball sampling to identify and recruit diverse stakeholders (following Reed 2008). Through phone and in-person semi-structured interviews and small meetings we gathered input and feedback from all key stakeholders including recreational and commercial fishers, diving operators, environmental NGOs, scientists, and representatives of all relevant policy ministries and departments (Regional Directorates). Stakeholders also identified issues, shared their views, and provided input and feedback at two workshops: (1) Stakeholder workshop 1—Horta—3rd of October, 2017*:* 31 local stakeholders discussed the current and future management of the Faial-Pico Channel MPA, and how science and local knowledge can support policy (AQUACROSS 2017);

[4]Due to the timing of the case study, we did not progress to applying the fourth step of adaptive management and monitoring.

(2) Stakeholder workshop 2—Horta—23rd of May, 2018: 18 local stakeholders collaborated on a concrete plan for stakeholder-based management of the Faial-Pico Channel MPA, and prioritised and developed measures to managed the Channel (AQUACROSS 2018).

(A) Identifying Policy and Stakeholder Objectives

To understand *policy objectives* we applied at a local level Rouillard et al.'s (2017) approach and reviewed relevant Faial-Pico and Azores regulations, laws and strategies related to the environment, fishing, and tourism, i.e., the sectors driving pressures on local biodiversity. We assessed key features, implementing measures, and governance of the most important local policies, and applied the Driver-Pressure-State-Impact-Response model to identify the expected pathway through which the management measures impact biodiversity in the Faial-Pico Channel, i.e., how the policy affects ecosystem state, pressures, or drivers. Finally, we identified synergies, conflicts, and gaps in relation to how local management and policy affects biodiversity in the Channel, and how biodiversity protection could be improved. To understand *stakeholder objectives* we relied on stakeholder interviews and the two workshops. To understand *stakeholder processes* and to identify how current stakeholders could better support MPA management, we used the development of a recent relevant policy Fishing Ordinance no. 53/2016 as a case study, evaluating how existing stakeholder processes could be adapted to the requirements for EBM.

(B) Understanding the Social-Ecological System

We applied the AQUACROSS Linkages Framework to understand the current Faial-Pico Channel socio-ecological system (Robinson and Culhane 2020). We mapped marine habitats present in the Channel and then used expert judgement, local scientific reports and economic and environmental data, and interviews with local scientists and regulators to identify drivers and activities, the pressures these place on habitats, and link these habitats to ecosystem-services production. Having identified key elements in the Channel's simplified social-ecological system (see Fig. 2), we then identified indicators and collected data on state and trends. We presented this to stakeholders at workshop 2 and co-developed future scenarios to identify future trends that would require integrative management and to identify potential trade-offs associated with different approaches for managing fishing, tourism, and biodiversity within the MPA.

(C) Identifying an EBM Plan

To identify the combination of management measures and implementing policies that make up the EBM plan, we collaborated with local stakeholders and policy-makers. They suggested a long list of potential measures/policy instruments in interviews and at stakeholder workshop 1 (AQUACROSS 2017). At stakeholder workshop 2, stakeholders selected priority management measures and implementing policies and developed how these should be implemented in the Channel (AQUACROSS 2018). We then ensured the workability of these individual measures and policies and combined them into an EBM plan. Finally, we evaluated this EBM plan relative to a baseline of current management using three criteria:

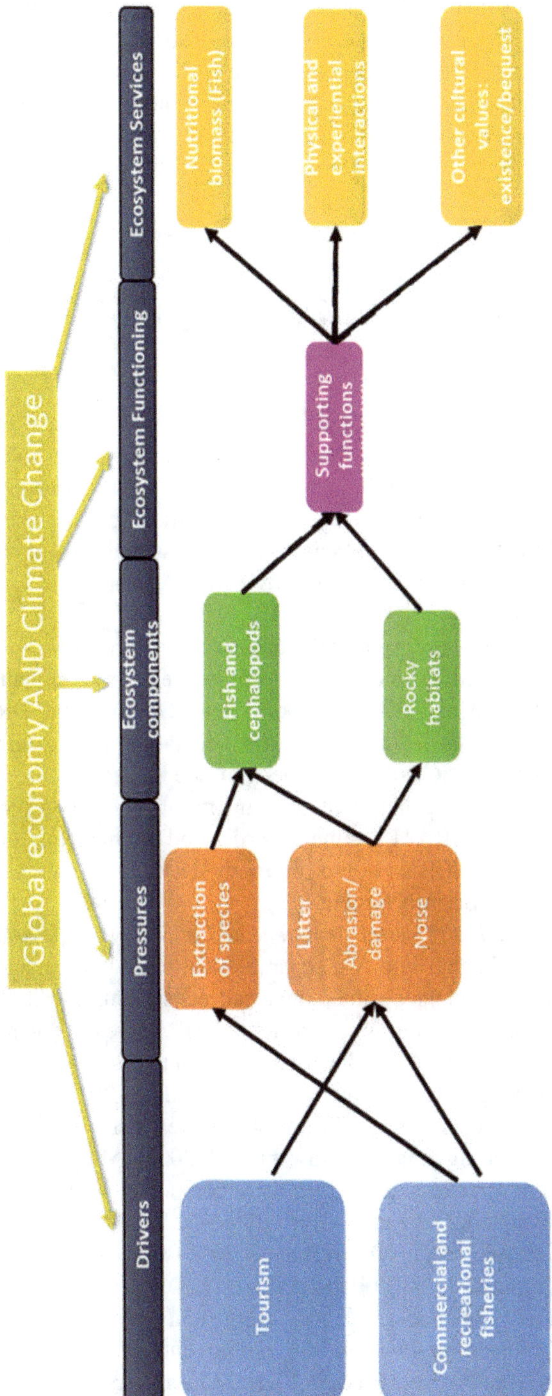

Fig. 2 Simplified linkage framework for Faial-Pico Channel, showing key drivers, pressures, ecosystem-components, ecosystem functioning, and ecosystem services

effectiveness, efficiency, and equity. Here, we drew on stakeholder and expert input and the AQUACROSS Linkage Framework to qualitatively assess direct and indirect impacts. To assess how the direct costs of the EBM plan could be financed, we interviewed participants and quantitatively assessed tax and levy impacts (following European Commission et al. 2017).

4 Results

4.1 Identifying Policy and Stakeholder Objectives

Policy Objectives

Biodiversity in the Channel is protected by environmental policies. However, as described in Rouillard et al. (2017), the positive impact of these policies can be undermined by sectoral policies, which support drivers (fishing, tourism) that place pressures on biodiversity. Together with local policy-makers and stakeholders, we concluded that, while local policy already targets sustainability, there are three policy gaps that should be priorities for improving MPA management:

- *Lack of coordinated management of the Channel limits synergies*—The current dispersion of responsibilities and management between environmental directorates (Faial and Pico Island Nature Parks, DRA, DRAM, and DRP) hinders integrated and coordinated management, implementation, monitoring, and evaluation of the Faial-Pico Channel. Leaders of the Island Nature Parks have reported lacking expertise and interest in non-terrestrial protected areas (AQUACROSS 2017). DRAM has the expertise and the mandate for coordinating and regulating the MPAs but is currently lacking operational means to implement monitoring or enforcement.
- *Issues of scale of marine resources not reflected in policy or governance*—The current split of the Channel into two separate Faial and Pico management units fails to recognise the Channel's interconnected ecosystem, and its links to the wider Azores marine ecosystem. A key benefit of MPAs are the potential positive spillover effects: MPAs elsewhere have been shown to increase species richness and catch rates in neighbouring waters (Russ and Alcala 2011). Negative spillover effects can also occur, where closure of one area increases fishing effort in boundary or neighbouring zones (Murawski et al. 2005). Managing the Channel as one integrated unit could help balance these competing spillover and network effects to meet local and Azorean biodiversity goals. In this way, the most recent MPA management regulation (Ordinance 53 2016) suggests a way forward: it was developed by DRAM in collaboration with DRP, who also manage the Azores Marine Park, thus better reflecting ecosystem scale.
- *A lack of monitoring data limits target setting and adaptive management*—Ecosystem-based management requires decision-makers to monitor policy impact and regularly revisit management tools if objectives are not being met

effectively, efficiently, and equitably (Rouillard et al. 2017). This requires monitoring and data at the appropriate spatial scale (i.e., Faial-Pico Channel), as well as clearly defined and spatially consistent policy objectives and targets. Ideally, this should include both ecological data (i.e., measures of biodiversity state, such as fish stocks) and socio-economic data (benefits and costs for society, e.g., fishing income, MPA visits). This data challenge is compounded by the issue of scale: policy objectives are set—and existing biodiversity and economic data collected—at the national (or, in some cases, island) scale, rather than at the Faial-Pico Channel-level. This makes it difficult to set and evaluate quantitative local targets. Additionally, Channel monitoring data is currently insufficient to manage biodiversity.

Stakeholder Objectives

EBM aims to maximise overall social welfare. Accordingly, it is important that as well as existing policy objectives, MPA management must consider other stakeholder goals. In the Faial-Pico Channel, there was considerable overlap between policy objectives and stakeholder priorities, but we did identify additional stakeholder objectives, some of which all stakeholder groups shared, and others where different groups were in conflict.

- *Shared stakeholder objectives*: Stakeholders all recognised that they share the Faial-Pico Channel MPA and come from the same community. Accordingly, all stakeholder groups share four central objectives: long-term sustainability, simplified and holistic management of the Channel, regular monitoring, and ongoing participatory management. (AQUACROSS 2017, 2018).
- *Conflicting stakeholder objectives*: The major stakeholder groups within the Channel also have conflicting objectives (AQUACROSS 2017, 2018). Additionally, as the Channel consists of many distinct habitats, stakeholders also place different value on different parts of the Channel (Schmiing et al. 2015; Afonso et al. 2014). For example, commercial fishers' prioritise access to fishery grounds, which can be in conflict with recreational fishers wish for extended catch limits and tourism operators' desire for expansion of the MPA to protect biodiversity and restrict extractive uses.

Enhancing cooperation and managing these conflicts relies on transparent and inclusive governance, which stakeholders believe could additionally decrease conflict, increase knowledge, and motivate greater environmental protection (AQUACROSS 2018).

Stakeholder Processes

Stakeholder processes are central to EBM, and given the gap we identified between policy objectives and stakeholder objectives and the presence of stakeholder conflicts, we evaluated existing stakeholder processes for integrating stakeholders into policy development. We found that while policy-makers' development of a non-technical scientific report (Afonso et al. 2014) and stakeholder workshops were positive steps in enabling stakeholders to contribute to policy design/

development, low participation from two key sectors—recreational fishing and tourism operators—meant the process was not representative. A second conclusion was that stakeholders should be involved throughout the policy cycle, not just in the policy development stage. Such adaptive management requires ongoing monitoring, evaluation, and, if necessary, adaptation of any management measures. This ongoing stakeholder engagement, for example through clear communication or regular workshops, would help ensure that decision-makers have full information on stakeholder objectives and priorities and feedback on whether current management is optimal or needs adjustment.

4.2 Understanding the Social-Ecological System

The second step of the AQUACROSS Assessment Framework is to understand the Faial-Pico Channel Social-Ecological System (SES). Effective management requires an understanding of how society affects the ecosystem, and how the ecosystem provides benefits to society, as well as the complex processes within the SES. We used the AQUACROSS Linkage Framework and developed indicators to understand the current state of the SES, and also used co-developed scenarios to identify potential future challenges and trends that would need managing.

Linkage Framework Analysis

Figure 2 presents a simplified social-ecological system for the Channel. We find that biodiversity in the Faial-Pico Channel is affected by the society that surrounds it: human activities like fishing and tourism place pressures on the Channel. These pressures affect the ecosystem's health and its ability to deliver valuable ecosystem services, such as fish and recreational experiences, which drive human activities and responses.

Our analysis shows that both the key sectors of fishing and tourism place many of the same pressures on the ecosystem, such as litter and noise. Unsurprisingly, fishing is most associated with the key pressure of extraction of fauna and flora. The linkage framework also assesses impacts over time: we find that fishing exerts more acute pressures, while tourism is associated with pressures that are more chronic. Accordingly, policies targeting fisheries will more swiftly decrease pressures than tourism-targeted policies.

We also used the Linkage Framework to assess which ecosystem components were most central to the Faial-Pico Channel SES. Fish are highly valued by all stakeholders. We find that rocky habitats support the most ecosystem functions and were associated with the most ecosystem services. This aligns with recent research on values of biodiversity indices around the Faial and Pico islands, which shows that the highest values were linked to rocky habitat, which provide refuge and substrate for various marine species, making rocky habitats important sites for fishing and diving (Schmiing et al. 2014). These insights suggest that management should prioritise protection of these central and valued ecosystem components.

Indicators

Our development and evaluation of indicators suggests that policy-makers can use indicators to understand the system, set quantitative targets, and monitor and evaluate trends and the impact of management measures. However, a key conclusion of this exercise was that a lack of quantitative Faial-Pico Channel data limits ability to apply EBM. The small scale and trans-boundary nature of the case study makes it difficult to use Azores-level data. Ecosystem-based management of the Channel calls for collecting and developing more specific Faial-Pico Channel data, especially to measure the current state of the ecosystem and its biodiversity, and on flows of key ecosystem services (fish for consumption, recreational experiences, and existence/bequest values).

Future Scenario Development

Scenarios are valuable as they provide a vehicle for incorporating diverse information into a comprehensive, actionable vision of the expected future (Gómez et al. 2017). Together with Azorean stakeholders and policy-makers (AQUACROSS 2018), we reflected on the understanding of the current SES, as well as our understanding of policies and stakeholder objectives, to develop identify what 2018–2050 is likely to bring to the Channel:

- **Climate change** will impact all sectors, increasing variability and uncertainty.
- The **global economy** will continue to drive ongoing—but fluctuating—growth.
- **Tourism** will continue to grow economically—with more visitors, income, and infrastructure.
- These changes mean **marine biodiversity** will be under increasing pressure in Faial-Pico Channel.
- **Commercial fisheries** and **recreational fishing** will remain central to local life, but sensitive to uncertain trends in fish stocks and biodiversity.

Developing this scenario clarified the gaps between current management (and the resulting expected future) and the future stakeholders and policymakers and stakeholders desired. Overall, we concluded that all stakeholders depend on a sustainable and resilient ecosystem. Given the large uncertainties and unknowns, stakeholders and policy-makers need to be adaptive—employing regular monitoring, evaluation, and if necessary, management changes.

4.3 Identifying an EBM Plan

Our final steps in applying ecosystem-based management in the Faial-Pico Channel was to reflect on identified objectives and policy gaps, and draw on our understanding of the current and future state of the SES to identify a set of priority management measures and implementing policies (the EBM Plan). We then evaluated the extent to which this EBM Plan would increase effectiveness, equity, and efficiency relative to a baseline of current management. We also investigated how regulators could finance the EBM Plan, which has important equity affects as well as being crucial for MPA effectiveness (Gill et al. 2017).

EBM Plan

We identified the following measures and policies as priorities for EBM management of the Faial-Pico Channel:

1. **Increased monitoring of biodiversity**
2. **Increased stakeholder participation through a Stakeholder Advisory Group** consisting of representatives of all sectors.
3. **Integrate and coordinate Channel management** through a Marine Protected Area management plan and policy coordination group.
4. **Clear communication and enforcement of existing regulations**—e.g., through simple information panels and surveillance cameras
5. **Implement a sustainability tax**—a tourism tax/diving fee.

Evaluation of the EBM Plan

Effectiveness: Due to data and methodological limitations, we are unable to decisively quantitatively assess how the EBM Plan will affect biodiversity (i.e., its environmental effectiveness). The EBM plan has direct impacts on biodiversity by increasing enforcement and awareness of existing fisheries/biodiversity regulation, which will increase compliance and decrease a key pressure on local biodiversity, extraction of species. The implementation of a sustainability tax will marginally decrease tourism and related pressures. The EBM Plan would also have indirect positive impacts on biodiversity by increasing scientific knowledge and financing to support management, policy integration, and stakeholder cooperation. Stakeholders believe that a stakeholder advisory group would result in greater environmental protection and increases in biodiversity (AQUACROSS 2018).

Efficiency: Assessing economic efficiency of the EBM Plan requires an understanding of its direct and indirect costs and benefits. However, given the indirect, supporting nature of the majority of elements of the EBM Plan, we cannot quantitatively assess this. Using the AQUACROSS Linkage Framework, we find that there is uncertain impacts on the value of fish caught to be eaten; increases in the existence/bequest value of the system; and likely increases in the value of experiential/physical interactions with the ecosystem. Alongside this qualitative assessment, evidence of efficiency is provided by the fact that each of the policy instruments that form the EBM plan were co-created with local stakeholders, whose selection of the plan, who believe that the benefits of the plan will outweigh the costs (AQUACROSS 2017, 2018).

Equity: A key focus of the EBM Plan is to increase stakeholder involvement and ownership of MPA management in such a way that the EBM Plan recognises and balances the costs and benefits to different stakeholder groups, and focusses on synergies and a shared commitment to environmental sustainability. Indeed, all stakeholders prioritised this cooperative, participatory element of the EBM Plan, arguing that it would decrease conflicts between different users and policy entities through better communication, and the promotion of multiple uses of the Marine

Protected Area (AQUACROSS 2018); all evidence of greater equity under the EBM Plan than under current management.

Financing: The first four elements of the proposed EBM plan place costs on fishers (who will face increased enforcement and compliance costs), while tourists, tourism operators, and other local stakeholders benefit (both from exclusive access to diving locations and positive environmental impacts). Financing can be used as a way to share the costs between those who benefit and those who bear cost. Our assessment of two financing options (a per dive fee levied by tourism operators and a per night occupancy tax) suggests that even at low rates of €2 per dive or €0.25 per night, either of these options could cover the likely direct costs of the EBM Plan and share the costs between different stakeholder groups to improve equity.

4.4 Local Policy Recommendations

Overall, our co-development of an EBM plan for the Faial-Pico Channel with stakeholders resulted in the following set of complementary management measures and policy instruments: (1) increase scientific monitoring, (2) implement stakeholder co-management with a Stakeholder Advisory Group, (3) increase integration and coordination of Channel management (e.g., by means of a coordination group of fishing, tourism, and environment Regional Directorates and island national parks); (4) communicate and enforce existing fishing and biodiversity regulations, and (5) finance biodiversity protection and share costs. This plan would better protect Channel biodiversity, whilst also ensuring economic and social sustainability. A key element of this plan is extending the stakeholder participation and policy cooperation that was evident in the EBM process and in existing local government stakeholder engagement efforts. In light of the Azores government's strategic goal of increasing MPA coverage, to ensure their success, we encourage continued engagement of stakeholders in planning, implementation, and evaluation. This, along with increased scientific knowledge and cross-sectoral policy coordination, will enable adaptive management in the Channel, reduce stakeholder conflict, and can improve effectiveness and efficiency of management, delivering benefits to the whole community into the future.

5 Conclusion and Discussion: How Can Ecosystem-Based Management Support Effective and Efficient Management of Marine Protected Areas?

We conclude that the Faial-Pico Channel case study provides evidence that ecosystem-based management and the AQUACROSS Assessment Framework can support decision-makers to manage Marine Protected Areas more effectively, so that they equitably meet biodiversity goals, both in the specific case of the Faial-Pico Channel and more generally in existing MPAs.

Our key conclusion is that stakeholder engagement and participation is beneficial for long-term sustainable protection of biodiversity and equitable and effective management of MPAs, and that ecosystem-based management's placing of representative stakeholder participation at the centre of ecosystem management is its key strength. Stakeholder engagement and participation has value in its own right. Reed (2008) reviewed stakeholder engagement literature and found that it promotes active citizenship, increases public trust, empowers stakeholders through co-generation of knowledge, improves public perception of policy, promotes social learning, and can reduce conflict between stakeholders and lead to creative solutions to environmental problems. In addition, stakeholder engagement is one of the defining principles of EBM (Long et al. 2015; Gómez et al. 2017). Stakeholder co-creation within this case study increased the relevance, acceptance, and quality of the management plan, and, as recognised by stakeholders, promotes synergistic solutions that provide multiple benefits, reducing stakeholder conflict, as well as improving knowledge and justifying more biodiversity protection (AQUACROSS 2018). It can be challenging involving stakeholders: for example, we found some stakeholders are harder to involve than others, and the process can be time-consuming, focused on discussion rather than action. However, on balance, we believe that the benefits of stakeholder co-creation outweigh these costs. This conclusion aligns with recent participatory management initiatives within the Azores, such as Condor seamount (Austen et al. 2019) and the Azorean fisheries regulation (Ordinance 53 2016) that increased protection for some high biodiversity areas in the Faial-Pico Channel. Our case study built on these initiatives and underlines the importance of integrated and representative management as a way to cope with the complexity and interlinkages of marine social-ecological systems.

Our experience also identified other strengths and challenges of ecosystem-based management for managing Marine Protected Areas. We found that ecosystem-based management provides a framework for integration of diverse stakeholders and objectives (biodiversity/environmental and sectoral). This integration clarifies the interconnectedness of the social-ecological system, and strengthens understanding of and arguments for collaborative, sustainability-focussed long-term ecosystem management. Key challenges that we faced were that while the interdisciplinary work of ecosystem-based management results in more useful and impactful policy, it requires diverse expertise and sometimes challenging cross-sectoral and cross-disciplinary collaboration and communication. Additionally, the newness and apparent complexity of the interdisciplinary work can make it challenging to get buy-in from sectors and policy makers. Finally, while EBM's emphasis on science-informed management are likely to support effective biodiversity protection, data and methodological limitations were a challenge in our case study.

It is too soon to evaluate the impact of the Faial-Pico MPA EBM process, though we conclude that the process had stakeholders' support and that it contributed to sustainable marine policy development in the Azores. Stakeholders demonstrated their support for the EBM process through their participation and positive comments in the workshops (AQUACROSS 2017, 2018). In particular, stakeholders supported EBM's commitment to representative stakeholder participation in policy development (AQUACROSS 2018). Alongside concurrent Azores projects and policy development, the case study and resulting EBM plan support ongoing MPA policy development and increasing stakeholder involvement in Azores marine policy, as evidenced by current processes to update Azorean MPA policy.

Overall, The Faial-Pico Channel EBM Plan, and its development and evaluation, provide evidence of how ecosystem-based management can support existing and future marine protected area management. The results are relevant in the Azores, where the government is committed to expanding MPA coverage, and globally to meet international MPA coverage targets. This study provides valuable information on how participatory management can support effective and equitable MPAs through clear identification of challenges and priorities, creative co-creation of solutions, low-cost knowledge and expertise, and ongoing monitoring, enforcement, and evaluation of the impact of management.

Acknowledgments This paper was funded as part of the AQUACROSS project, which received funding from the European Union's Horizon 2020 Programme for Research, Technological Development and Demonstration under Grant Agreement no. 642317. This study would not have been possible without the generous collaboration of all stakeholder interviews/workshop participants who took time to answer our questions and to share their views with us. We are also thankful to DRAM for their collaboration. The authors bear responsibility for any errors and omissions.

Adriana Ressurreição acknowledges Fundação para a Ciência e Tecnologia (FCT), through postdoctoral grant (SFRH/BPD/102494/2014) and the strategic project (UID/Multi/04292/2019) granted to CCMAR.

References

Abecasis, R. C., Afonso, P., Colaço, A., Longnecker, N., Clifton, J., Schmidt, L., & Santos, R. S. (2015). Marine conservation in the Azores: Evaluating marine protected area development in a Remote Island context. *Deep-Sea Environments and Ecology, 2*, 104. https://doi.org/10.3389/fmars.2015.00104.

Afonso, P., Schmiing, M., Santos, M., Diogo, H., & Fontes, J. (2014). *Áreas Marinhas Protegidas nos Parques Naturais de Ilha do Faial e do Pico, sector Canal: cenários iniciais*. Horta: IMAR - Universidade dos Açores.

AQUACROSS. (2017). *The Faial-Pico Channel stakeholder workshop - scientists, stakeholders, and policy-makers - working together to improve marine protected area management*. Proceedings of AQUACROSS. Retrieved from https://www.ecologic.eu/sites/files/event/2017/2803-faial-pico-channel-workshop-3-october-2017-proceedings-english_0.pdf.

AQUACROSS. (2018). *The Faial-Pico Channel: Future stakeholder management of the marine protected area*. Proceedings of AQUACROSS Faial-Pico Channel Stakeholder Workshop #2. http://dataportal.aquacross.eu/dataset/faial-pico-channel-workshop-2-proceedings.

Austen, M., Anderson, P., Armstrong, C., Döring, R., Hynes, S., Levrel, H., Oinonen, S., & Ressurreição, A. (2019). *Valuing Marine Ecosystems - Taking into account the value of ecosystem benefits in the Blue Economy* (Future science brief 5) (J. Coopman, J. J. Heymans, P. Kellett, A. Muñiz Piniella, V. French, B. Alexander, Eds.). Ostend, Belgium: European Marine Board. 32pp. ISBN: 9789492043696; ISSN: 4920-43696. Retrieved from https://doi.org/10.5281/zenodo.2602732.

Cardinale, B. J., Emmett Duffy, J., Gonzalez, A., Hooper, D. U., Perrings, C., Venail, P., Narwani, A., et al. (2012). Biodiversity loss and its impact on humanity. *Nature, 486*(7401), 59–67. https://doi.org/10.1038/nature11148.

Delacámara, G., O'Higgins, T., Lago, M., & Langhans, S. (2020). Ecosystem-based management: moving from concept to practice. In T. O'Higgins, M. Lago, & T. H. DeWitt (Eds.), *Ecosystem-based management, ecosystem services and aquatic biodiversity: Theory, tools and applications* (pp. 39–60). Amsterdam: Springer.

European Commission, Industry Directorate-General for Internal Market Entrepreneurship and SMEs, & PwC. (2017). *The impact of taxes on the competitiveness of European tourism.* Final Report.

Gill, D. A., Mascia, M. B., Ahmadia, G. N., Glew, L., Lester, S. E., Barnes, M., Craigie, I., et al. (2017). Capacity shortfalls hinder the performance of marine protected areas globally. *Nature, 543*(7647), 665–669. https://doi.org/10.1038/nature21708.

Gómez, C. M., Delacámara, G., Jähnig, S., Mattheiss, V., Langhans, S., Domisch, S., Hermoso, V., Piet, G., Martínez-López, J., Lago, M., Boteler, B., Rouillard, J., Abhold, K., Reichert, P., Schuwirth, N., Hein, T., Pletterbauer, F., Funk, A., Nogueira, A., Lillebø, A., Daam, M., Teixeira, H., Robinson, L., Culhane, F., Schlüter, M., Martin, R., Iglesias-Campos, A., Barbosa, A. L., Arévalo-Torres, J., & O'Higgins, T. (2017) *Developing the AQUACROSS Assessment Framework. Deliverable 3.2, AQUACROSS, European Union's Horizon 2020 Framework Programme for Research and Innovation Grant Agreement No. 642317.* Technical Report. European Union (H2020 FP Grant Agreement)-AQUACROSS.

Long, R. D., Charles, A., & Stephenson, R. L. (2015). Key principles of marine ecosystem-based management. *Marine Policy, 57*(July), 53–60. https://doi.org/10.1016/j.marpol.2015.01.013.

MarBEF Data System. (2006). Faial-Pico Channel, Azores. Retrieved from http://www.marbef.org/data/sitedetails.php?id=12909.

McDonald, H., Boteler, B., Gerdes, H., Hoffman, H., McFarland, K., & Röschel, L. (2018). *Case Study 8 report - Ecosystem-based solutions to solve sectoral conflicts on the path to sustainable development in the Azores.* D9.2. Retrieved from https://aquacross.eu/sites/default/files/D9.2_CS8_28092018_FINAL.pdf.

Murawski, S. A., Wigley, S. E., Fogarty, M. J., Rago, P. J., & Mountain, D. G. (2005). Effort distribution and catch patterns adjacent to temperate MPAs. *ICES Journal of Marine Science, 62* (6), 1150–1167. https://doi.org/10.1016/j.icesjms.2005.04.005.

Ojamaa, P. (2015). *Fisheries in Azores.* European Parliament. Retrieved from http://www.europarl.europa.eu/RegData/etudes/STUD/2015/540355/IPOL_STU(2015)540355_EN.pdf.

Ordinance 53. (2016). "Regulamento para o exercício da pesca na zona marítima das ilhas do Faial e Pico" - Portaria n.° 53/2016 de 21 de Junho de 2016. Secretaria Regional do Mar, Ciência e Tecnologia on the 1 June 2016.

OSPAR Commission. (2016). Faial-pico channel - Marine protected area (OSPAR). *OSPAR.* Retrieved from http://mpa.ospar.org/accueil_ospar/fiches_didentite_des_amp/fiche_didentite_dune_amp?wdpaid=555556986&gid=1512&lg=0.

Reed, M. S. (2008). Stakeholder participation for environmental management: A literature review. *Biological Conservation, 141*(10), 2417–2431. https://doi.org/10.1016/j.biocon.2008.07.014.

Ressurreição, A., & Giacomello, E. (2013). Quantifying the direct use value of Condor Seamount. *Deep Sea Research Part II: Topical Studies in Oceanography.* (An Integrated Approach for Studying Seamounts: CONDOR Observatory, *98*(December), 209–217. https://doi.org/10.1016/j.dsr2.2013.08.005.

Ressurreição, A., Menezes, G., & Giacomello, E. (2017). Assessing the annual revenue of marine industries operating at Condor seamount, Azores. In *Handbook on the economics and management of sustainable oceans* (UNEP). Cheltenham: Edward Elgar.

Robinson, L., & Culhane, F. (2020). Linkage frameworks: An exploration tool for complex systems. In T. O'Higgins, M. Lago, & T. H. DeWitt (Eds.), *Ecosystem-based management, ecosystem services and aquatic biodiversity: Theory, tools and applications* (pp. 213–234). Amsterdam: Springer.

Rouillard, J., Lago, M., Abhold, K., Röschel, L., Kafyeke, T., Mattheiß, V., & Klimmek, H. (2017). Protecting aquatic biodiversity in Europe: How much do EU environmental policies support ecosystem-based management? *Ambio, 47*(1), 15–24. https://doi.org/10.1007/s13280-017-0928-4.

Russ, G. R., & Alcala, A. C. (2011). Enhanced biodiversity beyond marine reserve boundaries: The cup spillith over. *Ecological Applications, 21*(1), 241–250. https://doi.org/10.1890/09-1197.1.

Schmiing, M., Diogo, H., Santos, R., & Afonso, P. (2014). Assessing hotspots within hotspots to conserve biodiversity and support fisheries management. *Marine Ecology Progress Series, 513* (October), 187–199. https://doi.org/10.3354/meps10924.

Schmiing, M., Diogo, H., Serrao Santos, R., & Afonso, P. (2015). Marine conservation of multispecies and multi-use areas with various conservation objectives and targets. *ICES Journal of Marine Science, 72*(3), 851–862. https://doi.org/10.1093/icesjms/fsu180.

Secretariat of the CBD. (2011). *Aichi targets. Decision X/2. Convention on biological diversity*. Retrieved from https://www.cbd.int/sp/targets/.

SREA. (2017). Estatisticas do Turismo - janeiro a dezembro de 2016. Retrieved from http://srea.azores.gov.pt/Conteudos/Relatorios/lista_relatorios.aspx?idc=392&idsc=6454&lang_id=2.

Statistics Portugal. (2017). Statistics Portugal website. Retrieved from www.ine.pt/.

Tanzer, John, Carol Phua, Barney Jeffries, Anissa Lawrence, Aimee Gonzales, Paul Gamblin, Tony Roxburgh, WWF (Organization), and Zoological Society of London. (2015). *Living blue planet report: Species, habitats and human well-being*. Gland: WWF International. Retrieved from http://ocean.panda.org/media/Living_Blue_Planet_Report_2015_Final_LR.pdf.

UN. (2016). *United Nations sustainable development goals*. Retrieved from https://sustainabledevelopment.un.org/sdgs.

Permissions

All chapters in this book were first published by Springer; hereby published with permission under the Creative Commons Attribution License or equivalent. Every chapter published in this book has been scrutinized by our experts. Their significance has been extensively debated. The topics covered herein carry significant findings which will fuel the growth of the discipline. They may even be implemented as practical applications or may be referred to as a beginning point for another development.

The contributors of this book come from diverse backgrounds, making this book a truly international effort. This book will bring forth new frontiers with its revolutionizing research information and detailed analysis of the nascent developments around the world.

We would like to thank all the contributing authors for lending their expertise to make the book truly unique. They have played a crucial role in the development of this book. Without their invaluable contributions this book wouldn't have been possible. They have made vital efforts to compile up to date information on the varied aspects of this subject to make this book a valuable addition to the collection of many professionals and students.

This book was conceptualized with the vision of imparting up-to-date information and advanced data in this field. To ensure the same, a matchless editorial board was set up. Every individual on the board went through rigorous rounds of assessment to prove their worth. After which they invested a large part of their time researching and compiling the most relevant data for our readers.

The editorial board has been involved in producing this book since its inception. They have spent rigorous hours researching and exploring the diverse topics which have resulted in the successful publishing of this book. They have passed on their knowledge of decades through this book. To expedite this challenging task, the publisher supported the team at every step. A small team of assistant editors was also appointed to further simplify the editing procedure and attain best results for the readers.

Apart from the editorial board, the designing team has also invested a significant amount of their time in understanding the subject and creating the most relevant covers. They scrutinized every image to scout for the most suitable representation of the subject and create an appropriate cover for the book.

The publishing team has been an ardent support to the editorial, designing and production team. Their endless efforts to recruit the best for this project, has resulted in the accomplishment of this book. They are a veteran in the field of academics and their pool of knowledge is as vast as their experience in printing. Their expertise and guidance has proved useful at every step. Their uncompromising quality standards have made this book an exceptional effort. Their encouragement from time to time has been an inspiration for everyone.

The publisher and the editorial board hope that this book will prove to be a valuable piece of knowledge for researchers, students, practitioners and scholars across the globe.

List of Contributors

Theodore H. DeWitt
US Environmental Protection Agency, Pacific Ecological Systems Division, Newport, OR, USA

Walter J. Berry
US Environmental Protection Agency, Atlantic Coastal Environmental Sciences Division, Narragansett, RI, USA

Timothy J. Canfield
US Environmental Protection Agency, Groundwater Characterization and Remediation Division, Ada, OK, USA

Richard S. Fulford, Matthew C. Harwell, Leah A. Sharpe and Susan H. Yee
US Environmental Protection Agency, Gulf Ecosystem Measurement and Modeling Division, Gulf Breeze, FL, USA

Joel C. Hoffman
US Environmental Protection Agency, Great Lakes Toxicology and Ecology Division, Duluth, MN, USA

John M. Johnston
US Environmental Protection Agency Office Research and Development, Environmental Processes Division, Athens, GA, USA

Tammy A. Newcomer-Johnson
US Environmental Protection Agency, Watershed and Ecosystem Characterization Division, Cincinnati, OH, USA

Paul L. Ringold
US Environmental Protection Agency, Pacific Ecological Systems Division, Corvallis, OR, USA

Marc J. Russell
US Environmental Protection Agency, Center for Computational Toxicology and Exposure, Gulf Breeze, FL, USA

Nathaniel S. Lewis and Darryl E. Marois
Pacific Ecological Systems Division, U.S. Environmental Protection Agency, Newport, OR, USA

Chanda J. Littles
Environmental Resources Branch, U.S. Army Corps of Engineers, Portland, OR, USA

Justin Bousquin
US Environmental Protection Agency, Gulf Ecosystem Measurement and Modeling Division, Gulf Breeze, FL, USA

Marisa Mazzotta
US Environmental Protection Agency, Atlantic Coastal Environmental Sciences Division, Narragansett, RI, USA

Mark Myer
City of New Orleans Mosquito Termite and Rodent Control Board, New Orleans, LA, USA

Gonzalo Delacámara
IMDEA Water Institute, Alcalá de Henares, Madrid, Spain

Timothy G. O'Higgins
Marine and Renewable Energy Ireland (MaREI), Environmental Research Institute, University College Cork (UCC), Cork, Ireland

Manuel Lago
Ecologic Institute, Berlin, Germany

Simone Langhans
Department of Zoology, University of Otago, Dunedin, New Zealand
Basque Centre for Climate Change (BC3), Leioa, Spain

Kathleen C. Williams and Joel C. Hoffman
Great Lakes Toxicology and Ecology Division, United States Environmental Protection Agency, Office of Research and Development, Center for Computational Toxicology and Exposure, Duluth, MN, USA

Robert B. McKane, Allen F. Brookes, Jonathan J. Halama and Paul B. Pettus
U.S. Environmental Protection Agency, Corvallis, OR, USA

Kevin S. Djang
Inoventures LLC, c/o U.S. Environmental Protection Agency, Corvallis, OR, USA

Bradley L. Barnhart
National Council for Air and Stream Improvement, Inc., Corvallis, OR, USA

Marc Russell
US Environmental Protection Agency, Center for Computational Toxicology and Exposure, Gulf Breeze, FL, USA

Kellie B. Vache and John P. Bolte
Oregon State University, Corvallis, OR, USA

Donna R. Harwell
Palm Beach State College, Palm Beach Gardens, FL, USA

Stephen Flood
MaREI Centre for Energy, Climate and Marine and Renewable Energy, Environmental Research Institute, University College Cork, Cork, Ireland
School of Geography, Environment & Earth Sciences, Victoria University of Wellington, Wellington, New Zealand

Barry O'Dwyer
MaREI, Environmental Research Institute, University College Cork, Cork, Ireland

Fiona E. Culhane and Leonie A. Robinson
School of Environmental Sciences, University of Liverpool, Liverpool, UK

Maneul Lago
Ecologic Institute, Berlin, Germany

Susan Yee, Deborah L. Santavy and Leah Sharpe
Gulf Ecosystem Measurement and Modeling Division, US Environmental Protection Agency, Gulf Breeze, FL, USA

Giancarlo Cicchetti and Kenneth Rocha
Atlantic Coastal Environmental Sciences Division, US Environmental Protection Agency, Narragansett, RI, USA

Susan K. Jackson
Health and Ecological Criteria Division, Office of Water, US Environmental Protection Agency, Washington, DC, USA

Margherita Pryor
Water Division, Region 1, US Environmental Protection Agency, Boston, MA, USA

Emily Shumchenia
E&C Enviroscape, LLC, Ashaway, RI, USA

Matthew C. Harwell and Jeannine L. Molleda
Gulf Ecosystem Measurement and Modeling Division, U.S. EPA, Gulf Breeze, FL, USA

Chloe A. Jackson
ORISE Research Participant, Pacific Ecological Systems Division, Newport, OR, USA

Hugh McDonald, Lina Röschel, Holger Gerdes and Keighley McFarland
Ecologic Institute, Berlin, Germany

Helene Hoffman
Zukunft - Umwelt - Gesellschaft (ZUG) gGmbH, Berlin, Germany

Adriana Ressurreição
CCMAR - Centre of Marine Sciences, Faro, Portugal

Ben Boteler
Institut für transformative Nachhaltigkeitsforschung /Institute for Advanced Sustainability Studies e.V. (IASS), Potsdam, Germany

Heliana Teixeira
DBIO & CESAM, Universidade de Aveiro, Aveiro, Portugal

Index

www.ingramcontent.com/pod-product-compliance
Lightning Source LLC
LaVergne TN
LVHW051048220926
848485LV00001BA/42
9781641167840